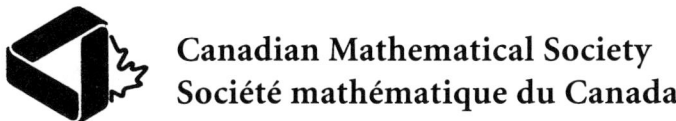

Canadian Mathematical Society
Société mathématique du Canada

Editors-in-Chief
Rédacteurs-en-chef
Jonathan Borwein
Peter Borwein

Springer
New York
Berlin
Heidelberg
Barcelona
Hong Kong
London
Milan
Paris
Singapore
Tokyo

CMS Books in Mathematics
Ouvrages de mathématiques de la SMC

Michal Křížek   Florian Luca
Lawrence Somer

17 Lectures on Fermat Numbers

From Number Theory to Geometry

With a Foreword by Alena Šolcová

With 71 Illustrations

 Springer

Michal Křížek
Mathematical Institute
Academy of Sciences
Prague 1, CZ-115 67
Czech Republic

Florian Luca
Mathematical Institute
National Autonomous University
of Mexico
Morelia, CP 58 089
Mexico

Lawrence Somer
Department of Mathematics
Catholic University of America
Washington, DC 20064
USA

Editors-in-Chief
Rédacteurs-en-chef
Jonathan Borwein
Peter Borwein
Centre for Experimental and Constructive Mathematics
Department of Mathematics and Statistics
Simon Fraser University
Burnaby, British Columbia V5A 1S6
Canada

Mathematics Subject Classification (2000): 11Axx, 11A41, 11Dxx, 11NO5, 11Rxx

Library of Congress Cataloging-in-Publication Data
Krizek, M.
    17 lectures on Fermat numbers : from number theory to geometry / Michal Krizek,
Florian Luca, Lawrence Somer.
      p.   cm. — (CMS books in mathematics ; 9)
    Includes bibliographical references and index.
    ISBN 0-387-95332-9 (alk. paper)
    1. Fermat numbers. I. Title: Seventeen lectures on Fermat numbers. II. Luca, Florian.
III. Somer, Lawrence. IV. Title. V. Series.
    QA246.K75   2001
    512'.7—dc21   2001042960

Printed on acid-free paper.

Production managed by Terry Kornak; manufacturing supervised by Jerome Basma.
Photocomposed copy prepared from the authors' PostScript files.
Printed and bound by Edwards Brothers, Inc., Ann Arbor, MI.
Printed in the United States of America.

9 8 7 6 5 4 3 2 1

ISBN 0-387-95332-9   SPIN 10844773

Springer-Verlag   New York Berlin Heidelberg
A member of BertelsmannSpringer Science+Business Media GmbH

Pierre de Fermat (1601–1665)

Foreword

The French mathematician Pierre de Fermat (1601–1665) is famous primarily because of his extensive work in number theory. Contributions such as Fermat's little theorem, Fermat's last theorem, and Fermat numbers, to list just a few, made him one of the giants among mathematicians of the seventeenth-century. Number theory was his true love. He made the first significant advances in it since the classical era. However, his name is also connected with the prehistory of differential and integral calculus. He derived a method for finding tangents to curves. His manuscript *Ad locos planos et solidos isagoge* went beyond René Descartes's work in analytic geometry and allowed Fermat to define many important curves such as the ellipse, hyperbola, parabola, cubic curve, cycloid, and the Fermat spiral (see Figure 1).

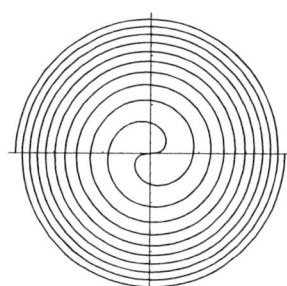

Figure 1. The Fermat spiral $r^2 = \theta$ in polar coordinates.

Later he solved simple problems in probability theory with Blaise Pascal, and in the field of optics he was the discoverer of the famous Fermat principle.

Briefly About His Life

Fermat was born into the family of a prosperous leather merchant in Beaumont de Lomagne, near Toulouse. Although there is disagreement concerning the date of his birth,[1] varying widely from 1590 to 1608, the generally accepted date of his birth is August 17, 1601 (he was baptized on August 20). Pierre had a brother and two sisters.

Fermat was initially educated at home and later at the local Franciscan monastery. Ultimately, he studied law, probably in Toulouse. According to the letter of

---

[1]See, e.g., K. Barner: *How old did Fermat become?* Univ. Gesamthochschule Kassel, Fachbereich Mathematik/Informatik, Preprint No. 17/00, 2000, 1–22.

Gilles Roberval[2] dated September 22, 1636, Fermat was influenced by the mathematical work of François Viète during his stay in Bordeaux in 1629. He could speak French, Latin, Greek, Spanish, and Italian. On May 1, 1631, Fermat became a bachelor of law in Orléans, and subsequently he obtained the post of counsellor at the local parliament in Toulouse. At that time he changed his name from Pierre Fermat to Pierre de Fermat.

Although Fermat's profession was law, mathematics was his consuming avocation, which he pursued for the pleasure of discovery and not to build a reputation. Most of his mathematical results were found as marginal notes in the books he read, and some of his ideas are preserved in letters to his friends. Indeed, he published only one important manuscript during his lifetime, and signed it using the cryptic initials M.P.E.A.S.[3] Their meaning remains inexplicably unknown. It was his eldest son, Clément-Samuel, who took care of his father's intellectual legacy and posthumously published selections of the mathematical and scientific work in 1670 and 1679.

Fermat traveled little, and may never have visited Paris in his lifetime. He died on January 9, 1665, in Castres and was buried on January 12.[4] This date is written down at the church of the Augustines in Toulouse. More extended biographies of Fermat are available in a number of sources.[5]

Fermat's Mathematics and Notation, Symbolism, and Terminology

The mathematics of the first half of the seventeenth century was not a consistent professional discipline, but was divided into different "schools." Fermat belonged to the school of François Viète, whose works constituted his main source of inspiration and directed his research.

The language of the original sources offers an insight into the essence of Fermat's mathematical thoughts. Fermat borrowed his cumbersome system of mathematical symbols from Viète. For instance, Fermat wrote the expression

$$x^3 + 13x^2 + 5x + 2$$

as

$$1C + 13Q + 5N + 2.$$

Powers of unknown quantities were sometimes also expressed in words (e.g., cubus).

Fermat used the word *geometrae* when referring to mathematicians at large, but preferred to be called an *analyst* himself. Other terms for what we today

---

[2] Gilles Personne de Roberval (1602–1675) began to study mathematics at the age of 14. Then he traveled widely, visiting many places in France. On his travels he went to Bordeaux and met Fermat. Later, in 1632, Roberval was appointed professor of philosophy in Collège Gervais in Paris, and then, in 1634, he was appointed to the Ramus chair of mathematics in the Collège Royale. He was a founding member of the Académie in Paris.

[3] See R. Huron: *L'aventure mathématique de Fermat,* in Pierre de Fermat, Toulouse et sa région, CNRS, Toulouse, 1966, 13–34.

[4] See p. 13 of the above article[3] or P. Chabbert: *La ville où mourut Fermat, Castres vers 1665,* in Pierre de Fermat, Toulouse et sa région, CNRS, Toulouse, 1966, 219–227.

[5] *Dictionary of Scientific Biography,* Ch. C. Gillispie (ed.), New York, 1972–90; *Encyclopaedia Britannica;* M. S. Mahoney, *The mathematical career of Pierre de Fermat (1601–1665),* Princeton Univ. Press, 1973, 1994, etc.

would call a mathematician were *geometer, Rechenmeister, wisconstler, cossist,* and *algebraist*. One thing these variously named professionals agreed upon was that they were not *mathematicians*. In the sixteenth and seventeenth centuries the notion *mathematician* retained the same meaning as during the Middle Ages; it meant *astrologer* or *astronomer*.

Whatever they called themselves, these individuals used mathematical methods, and they searched for relationships between equations and geometry, and more generally, between mathematical methods and nature. The main areas in which they achieved development were solving equations, describing curves and their properties, trigonometry, numerical methods, and calculus and its applications.

In 1621, Claude Gaspard Bachet de Méziriac[6] published a Greek and Latin version of *Diophantus's Arithmetics* in Paris. The *Arithmetics* became the favorite book of the young Fermat, who at that time was engaged in the study of perfect numbers, amicable (friendly) numbers, figurate numbers, magic squares, Pythagorean numbers, divisibility, primality, and other areas in number theory, and it inspired him to solve numerous problems in this field.

Figure 2. *Diophantus's Arithmetics.*

Bachet de Méziriac's edition was republished with Fermat's commentaries in 1670 in Toulouse by his son, Clément-Samuel de Fermat. The first volume was called *Diophanti Alexandrini Arithmeticorum Libri Sex, ... with comments of C. G. Bachet and observations of D. P. Fermat* (refer to Figure 2). The second volume, *Varia opera mathematica D. Petri de Fermat senatoris Tolosani...,* which contained Fermat's scientific correspondence on mathematical and physical problems, was published in 1679.

As an example of elementary geometric themes studied by Fermat, we introduce the notion of the *Fermat point:*[7]

Given a triangle, find the point at which the sum of the distances to all three vertices is a minimum.

Here is Fermat's solution to this problem (see Figure 3):

---

[6]C. G. Bachet de Méziriac (1591–1638) discovered a method of constructing magic squares. He was the earliest writer who discovered the solution of indeterminate Diophantine equations by means of continued fractions.

[7]The Fermat point is sometimes also called the Torricelli point, after E. Torricelli (1608–1647), who lived in the same time period as Fermat.

(1) Given any triangle ABC.
(2) Construct an equilateral triangle on each side.
(3) Join the vertices by straight lines as shown in Figure 3.
(4) These three lines intersect at one point, which is the required Fermat point F.

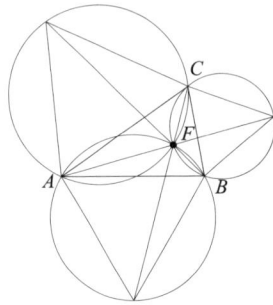

Figure 3. Construction of the Fermat point F in the triangle ABC such that the sum $|AF| + |BF| + |CF|$ attains the minimum value.

Polygonal (figurate) numbers[8] were studied as early as the Pythagoreans, who transferred geometric ideas to number theory. One of the most interesting theorems concerning figurate numbers was postulated by Bachet de Méziriac in his edition of Diophantus's works in 1621:

Every positive integer is the sum of at most four squares of integers.

This conjecture was later proved by Lagrange.[9] Fermat stated its generalization without any proof:

Every positive integer is the sum of at most n n-gonal numbers.

The proof of Fermat's conjecture by Cauchy[10] in 1812 was one of Cauchy's major discoveries. The four-squares theorem was included around 1830 in Jacobi's[11] theory of theta functions.

A Tale of the Solution of Pell's Equation

The Diophantine equation $Dy^2 + 1 = x^2$, where D is a square-free integer, is known as Pell's equation because Euler mistakenly attributed a solution of it to the seventeenth-century English mathematician John Pell (1611–1685). This type of equation had been solved earlier by Indian and Greek mathematicians (Brahmagupta (about 600 A.D.), Theon of Alexandria (end of the 4th century A.D.), Archimedes (287–212 B.C.)). Brahmagupta was probably the first mathematician who gave a solution of Pell's equation in integers. His approach was ingenious and general. The problem arose in China in the field of astronomy from the need to determine the orbits of planets.

---

[8]Recall that the polygonal numbers are of the form $(k/2 - 1)n^2 - (k/2 - 2)n$ for $k \geq 3$ and $n \geq 2$. They can be represented by regular geometric figures.

[9]Nouv. Mém. de l'Acad. de Berlin **1** (1770), 123–133, publ. 1772.

[10]Augustin Louis Cauchy (1789–1857).

[11]Carl Gustav Jacob Jacobi (1804–1851).

Fermat developed a method for solving this equation. He tried to find a general rule, and he challenged his friend Frénicle de Bessy to solve the following equation[12] for minimal values of x and y:

$$61y^2 + 1 = x^2.$$

Let us point out that Fermat chose this case for its difficulty, since the smallest values satisfying the above equation are $y = 226\,153\,980$ and $x = 1\,766\,319\,049$. This problem was later solved by Lagrange. His method required the calculation of twenty-one successive steps of the continued fraction for the square root of 61. Fermat also suggested another equation,

$$109\,y^2 + 1 = x^2,$$

the smallest solution of which is given by

$$y = 15\,140\,424\,455\,100 \quad \text{and} \quad x = 158\,070\,671\,986\,249.$$

Fermat's Last Theorem

The famous remark in the margin of *Diophantus's Arithmetics* states that for the equation $x^n + y^n = z^n$, no solutions are possible for integer values with $n > 2$ (cf. Figure 4). Fermat never published a proof of this statement for any $n \neq 4$.

Figure 4. Czech stamp issued on the occasion of the World Mathematical Year 2000.

The term "Fermat's last theorem" first appears during the first half of the nineteenth century when Gabriel Lamé wrote:[13]

Of all theorems on numbers stated by Fermat, just one remains incompletely demonstrated.[14]

Many mathematicians attempted to prove Fermat's last theorem, with the notable exception of Carl Friedrich Gauss, who refused to try. In May 1816, he wrote a letter to Heinrich Olbers in which he mentioned the theorem:

I confess that the Fermat theorem holds little interest for me as an isolated result....[15]

---

[12]Fermat's letters to Frénicle de Bessy of February 1657, *Œuvres de Fermat*, II, pp. 333–334.

[13]Lamé, G.: *Mémoire sur le dernier théorème de Fermat*, C. R. Acad. Sci. Paris **9**, 1839, pp. 45–46.

[14]*De tous les théorèmes sur les nombres, énoncés par Fermat, un seul reste incomplètement démontré.*

[15]*Ich gestehe zwar, dass das Fermatsche Theorem als isolierter Satz für mich wenig Interesse hat ...* , Singh, S., *Fermat's Last Theorem*, Fourth Estate, London, 1997, p. 105.

Fermat had written the following in the margin of his copy of *Diophantus's Arithmetics*:

For this, I have found a truly wonderful proof, but the margin is too small to contain it.

Fermat's original copy of *Diophantus's Arithmetics* with his handwritten notes has never been found, and in any case, few mathematicians believe that Fermat had in fact discovered such a proof for an arbitrary $n > 2$ and $n \neq 4$. Fermat's successors were successful only in proving special cases of Fermat's last theorem: Ch. Huygens and L. Euler (for $n = 3$, $n = 4$), A. M. Legendre (1825, $n = 5$), P. G. L. Dirichlet (1828, $n = 5$), G. Lamé (1840, $n = 7$), and E. E. Kummer[16] (1849, for $n < 100$ excluding the so-called irregular primes $n = 37, 59, 67$). In 1819, S. Germain demonstrated that if p and $2p + 1$ are both prime, then the so-called first case of Fermat's last theorem holds for the exponent p. In 1983, G. Faltings proved Mordell's conjecture, which in particular says that for each exponent $n > 2$ there exist at most finitely many triples (x, y, z) of natural numbers such that $\gcd(x, y, z) = 1$ and $x^n + y^n = z^n$. A general proof of Fermat's last theorem was not found until 1994.

About 350 years after the formulation of Fermat's last theorem, Andrew Wiles, a researcher at Princeton University, presented a proof during a seminar in Cambridge. The papers *Modular elliptic curves and Fermat's last theorem* by A. Wiles and *Ring-theoretic properties of certain Hecke algebras* by R. Taylor and A. Wiles were published in the May 1995 issue of *Annals of Mathematics*. The proof is now generally accepted.

Analytic Geometry

Fermat made new advances and created new methods in analytic geometry in both two- and three-dimensional space. He worked on analytic geometry during the period 1629–1636, but never returned to the subject thereafter. He studied Viète's treatises *Isagoge* (Tours, 1591) and *De recognitione et emendatione aequationum* (Paris, 1615) during his stay in Bordeaux. He was already familiar with the works of Archimedes, Apollonius, and Pappus. The origin of the development of Fermat's analytic methods occurs in his correspondence. Fermat wrote, *"I found many other geometrical theorems ... on planar points"* in letters to Etienne Pascal (the father of Blaise Pascal) and Gilles Roberval (August 23, 1636).

When solving Apollonius's problems, Fermat described equations of conic sections as follows:

$$\text{the hyperbola} \quad A \text{ in } E \text{ aeq. } Z \text{ pl.,}$$
$$\text{the parabola} \quad E^2 \text{ aequale } D \text{ in } A,$$
$$\text{the straight line} \quad D \text{ in } A \text{ aequetur } B \text{ in } E.$$

He followed a notation employing consonants for known quantities and vowels for unknown quantities.

---

[16]The German mathematician Ernst E. Kummer (1810–1893) evaluated Fermat's abilities: *"Diable d'homme, quelle intuition!"*

Descartes published his *Géometrie* in 1637, and it circulated widely, while Fermat's *Introduction* remained in manuscript until the publication of Samuel Fermat's edition of *Varia opera* in 1679, over 40 years later. Thus, irrespective of whether Descartes or Fermat was the first to make the critical discoveries in analytic geometry, it justifiably still bears the eponym *Cartesian*.

Fermat's Method of Infinite Descent

The term "complete induction," as distinct from Francis Bacon's "incomplete induction," was employed by Dedekind in his treatise *Was sind und was sollen die Zahlen?* for what is today called *mathematical induction*. Fermat used a similar procedure, which he called the *method of infinite descent*. In a letter to Carcavi[17] he wrote:[18]

As ordinary methods, such as are found in books, are inadequate for proving such difficult propositions, I discovered at last a most singular method ... which I called the infinite descent. At first I used it to prove only negative assertions, such as: There is no Pythagorean triangle the area of which is a square.

Fermat supposed to the contrary that there exists a Pythagorean triangle the area of which is the square of an integer. Under this assumption he proved that there is another Pythagorean triangle with the same property, the area of which is smaller. This immediately leads to a contradiction, since there must exist a Pythagorean triangle with minimal square area.

Fermat's geometrical ideas were followed by number-theoretical results. His method of infinite descent is, in fact, an application of the order relation in number theory (the well-ordering principle). In 1679, Fermat's son Samuel published his father's statement:

Quum autem numeros a binario quadratice in se ductos et unitate auctos esse semper numeros primos apud ne constet et iamdudum Analystis illius theorematis veritas fuerit significata, nempe esse primos 3, 5, 17, 257, 65537 *etc. in infinitum...,*

which states that 3, 5, 17, 257, 65537, etc., are primes.

Fermat initially claimed to have a proof based on infinite descent, but later discovered that his proof was incorrect. The above false assertion later became famous and is treated in detail in this book. In 1640 Fermat wrote to Frénicle de Bessy:

Si vous en avez la preuve assurée vous m'obligeriez de me la communiquer car après cela, rien ne m'arrêtera en ces matières.

This can be translated as follows: "If you have its undoubted proof you should communicate it to me, for after that nothing will stop me in these matters."

---

[17]Pierre de Carcavi (1600–1684) received no university education. He was a counsellor to the parliament of Toulouse from 1632 until 1636. Then he bought an office of counsellor in the Conseil in Paris. In 1648 he was forced to sell the office to pay the debts of his father (who had been a banker). He spent his last 20 years as the custodian of the Royal Library. In fact, he first met Fermat in 1632.

[18]*Œuvres de Fermat*, 1, p. 340; 3, p. 271.

Determining Extreme Values of Algebraic Polynomials

Fermat began his first substantial mathematical research in Bordeaux in 1629. While there he produced an important work on maxima and minima, but he never put the work into final polished form. His memoir describing these methods, entitled *Method for determining minima and maxima and tangents to curved lines*, began circulating in Paris by 1636.

It was Johannes Kepler (1571–1630) who remarked that the values of a "continuous function" in a small neighborhood on either side of a maximum (or minimum) value must all be approximately equal to the maximum (or minimum). Fermat applied this principle to a few examples, e.g., to find the maximum value of $x(a - x)$.

Fermat's methods for determining extreme values of algebraic polynomials, and for drawing the tangent to any point of an algebraic curve, allowed him to obtain tangents to the ellipse, cycloid, cissoid, conchoid, and quadratrix, by making use of special coordinates. He first began using this method for obtaining tangents during his studies in 1629, but it became the focus of his attention during the period 1637–1643. In 1644, Pierre Hérigone[19] included a supplement containing Fermat's method of maxima and minima in his major work, *Cursus mathematicus*.

Fermat established basic results in quadrature quite early in his career, especially around 1636, and he reached his greatest achievements in this area in the period 1643–1657. He succeeded in finding methods to calculate the areas under parabolas and hyperbolas bounded by vertical lines and the x-axis, and in determining the centers of mass of a few simple laminae and of a paraboloid of revolution.

Fermat can be considered a predecessor of Gottfried Wilhelm Leibniz (1646–1716) in that he used new methods to find the extrema of real functions in a similar way. As early as 1630 he was able to compute the derivatives of simple functions. He could find the derivative of an exponential function with a rational power and the sum of an infinite series; he was familiar with integration by parts, etc. Fermat probably was aware of a relationship between differential and integral methods. A good example of this is his method of finding the quadrature of a parabola:[20] Choose an arbitrary $q \in (0, 1)$ and approximate the parabola $y = x^2$ by rectangles $(q^k - q^{k+1}) \times q^{2k}$ as sketched in Figure 5. Then the sum of areas of these rectangles is

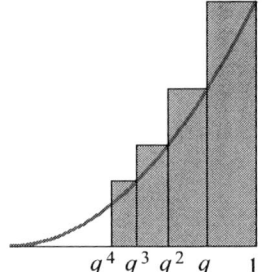

$q^4\ q^3\ q^2\ q\ \quad 1$

Figure 5. Fermat's integration method for a parabola.

$$1 \cdot (1 - q) + q^2(q - q^2) + q^4(q^2 - q^3) + \ldots = (1 - q) + q^3(1 - q) + q^6(1 - q) + \ldots$$
$$= \frac{1 - q}{1 - q^3} = \frac{1}{1 + q + q^2}.$$

---

[19]Pierre Hérigone was of Basque origin. His only important work is the six-volume *Cursus mathematicus*, which is a compendium of elementary mathematics written in French and Latin. It introduced a complete system of mathematical and logical notation, none of which is used today.

[20]A similar method was used by Diophantus: An approximation "as soon as possible" was the essence of the method.

We see that for $q \to 1$, the ratio $1/(1 + q + q^2)$ tends to $\frac{1}{3}$.

Fermat's results on finding extrema were used by Christian Huygens (1629–1695), as well as by Isaac Barrow (1630–1677), the teacher of Isaac Newton (1643–1727).

Fermat's Influence on His Contemporaries

Fermat had an enormous reputation among his contemporaries. Indeed, no less a formidable intellect and mathematician than Blaise Pascal (1623–1662) regarded Fermat as *the greatest mathematician in Europe*. John Wallis remembered him as a *satanic Frenchman*. Marin Mersenne (1588–1648) introduced him as *a scholar from Toulouse*. Correspondence with his contemporaries was important for Fermat's thinking, and it was there that he often first advanced his ideas. For example, in 1640, in one of his numerous letters to Marin Mersenne, Fermat wrote[21] that *if p is an odd prime, then 2p divides $2^p - 2$*. Shortly thereafter, he expanded this into what is now called Fermat's little theorem.

Gilles Roberval, the only professional mathematician in the circle around Fermat, greatly respected and was influenced by Fermat's ideas. Roberval developed powerful methods in the early study of integration and computed the definite integral of $\sin x$. He also solved some of the easier questions connected with the cycloid, he calculated the arc length of a spiral; and he studied many other plane curves and discussed the nature of their tangents.

Bernard Frénicle de Bessy[22] solved many of the problems stated by Fermat, introducing new ideas and posing further questions. Like Fermat, he worked on magic squares.

Two years after Fermat's death, Christian Huygens gave a lecture on Fermat's method in the Académie des Sciences in which Huygens not only used infinitesimal quantities, but attributed the method to Fermat.

Fermat's relationship with Descartes was one of controversy. There have been several occasions in the history of mathematics when an important discovery was made independently and almost simultaneously by two different persons. Fermat, in 1629, and Descartes, in 1637, independently developed methods of analytic geometry. Descartes's work *Géometrie* might, in fact, have been started in the 1620s, but then, Fermat also did not publish his ideas immediately. As noted earlier, Fermat's work in this area did not appear until 1679, fourteen years after his death, and he unfortunately used the symbolic notation of Viète, which was out of date. Both Fermat and Descartes began with an analytic solution of the same classical geometric problem: the four-line problem of Apollonius. Their main discovery was that second-degree equations correspond to conic sections.

Later, Fermat claimed that Descartes had not correctly deduced his law of refraction. Descartes, on the other hand, attacked Fermat's method of maxima, minima,

---

[21]This result was known 20 years earlier by Johannes Broscius (Jan Brożek, 1585–1652), of the Cracow Academy, *Arithmetica Integrorum*, Cracow, 1620.

[22]Bernard Frénicle de Bessy (1605–1675) was an excellent amateur mathematician who held an official position as a counsellor at the Court of the Mint. He was elected to the Académie Royale des Sciences in 1666.

and tangents. Fermat proved the correctness of his method and Descartes finally admitted this by writing:

> ... *seeing the last method that you use for finding tangents to curved lines, I can reply to it in no other way than to say that it is very good and that, if you had explained it in this manner at the outset, I would not have contradicted it at all.*[23]

However, this did not end the rancor between them, since Descartes continued to try to damage Fermat's reputation.

Fermat's relations with Pascal and Huygens were more friendly. His correspondence with Blaise Pascal began in earnest in 1654, when Pascal wrote him to ask for confirmation about his ideas on probability. Together, they made the first steps in the theory of probability, and they are among the theory's founders.

The correspondence with Christian Huygens began two years later, and grew out of Huygens's interest in probability. When he finished his treatise *De ratione in ludo aleae* (1657), Pascal and Fermat solved some of his problems.

Fermat's Legacy–Applications of His Results

Fermat had little interest in real-life applications of mathematics. There is some irony in the fact that his initial contact with the scientific community of the time was triggered by his study of free fall. Later, during his stay in Castres, he described "beautiful experiences" with a hydroscope (a baryllion) and with an aerometer.

In 1657, Fermat postulated a simple principle (see Figure 6): *Light follows a path that minimizes total travel time.*

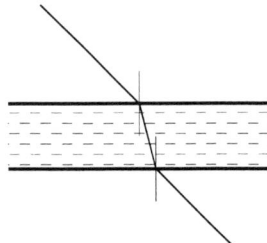

Figure 6. The Fermat principle explains the zigzag trajectory of light in glass.

Leonhard Euler (1707–1783), Joseph Louis Lagrange (1736–1813), and William Rowan Hamilton (1805–1865) later contributed to a wide extension of this principle.

Although Fermat's last theorem has had no direct practical applications to date, its investigation has triggered great development in number theory and has originated several new mathematical disciplines.

Fermat's little theorem is probably the most useful and frequently used tool in number theory. Fermat claimed that any prime number p has to satisfy

$$a^{p-1} \equiv 1 \pmod{p} \quad \text{for any positive integer } a \text{ not divisible by } p.$$

Thus, if a number does not satisfy this condition, it cannot be prime. Fermat's test is *necessary* for primality, but it is not *sufficient*. Perhaps the earliest nontrivial use

---

[23]http://www-history.mcs.st-and.ac.uk/~history/Mathematicians/Fermat.html

of the inverse of an element occurs with the operation of "multiplication modulo p", which L. Euler 1758 (and possibly Fermat before him) used to give an essentially group-theoretic proof of Fermat's little theorem.[24] The notion of congruence modulo p was invented by C. F. Gauss and introduced in the first edition of his *Disquisitiones arithmeticae*, in 1801. Fermat also developed a famous factorization method. It is efficient especially for those numbers that are not exceedingly large and are products of two primes of almost the same size.

Figure 7. Pierre de Fermat.

At the beginning of the third millennium, Fermat's name is held in great respect throughout the whole mathematical community. This book illustrates how one of Fermat's famous challenges led to developments in mathematics from number theory to geometry.

Alena Šolcová, March 17, 2001

---

[24]Stillwell, John: *Mathematics and Its History*, Springer, New York, 1989, p. 276.

Preface

Imagination is more important than knowledge.

Albert Einstein

This book was written on the occasion of the 400th anniversary of Pierre de Fermat's birth (1601). Its aim is to give the reader an overview of several of the fascinating properties of Fermat numbers and make him or her aware of many of the sometimes surprising appearances and applications of the Fermat numbers in number theory, geometry, generation of pseudorandom numbers, signal processing, etc. A special emphasis is the employment of geometric interpretations of many number-theoretic results (the Euclidean algorithm, Chinese remainder theorem, Fermat's little theorem, Pepin's test, etc.). This may be of interest to those readers for whom much of this material is otherwise familiar. We followed an old Chinese proverb that a picture is worth a thousand words.

The book is intended for a general mathematical audience. The reader is assumed to be familiar with only a few basic results from algebra. The first four chapters are a necessary basis for the rest of the book. The other chapters can be read more or less independently. The contents of the book are partly based on our own papers (see References) and also survey many results concerning Fermat numbers. Some of these results have not been published anywhere.

Although no book of this type can make the reader an expert in number theory, it introduces basic mathematical ideas and algebraic methods connected with the Fermat numbers. In order to help the reader, the bibliography has been provided with the *Mathematical Reviews* reference numbers. Most of these reviews can be found at the web site [www7]. The *Zentralblatt* reference numbers from [www8] have been added to those references where the MR reference numbers do not exist. We are aware that web sites are not subject to review and that web site addresses change quite often, so we have added them (after References) for completeness and as an aid for the reader.

This book originated when the second and third authors visited the Mathematical Institute of the Academy of Sciences of the Czech Republic during the period 1999–2000.

We would like to thank the Niedersächsische Staats- und Universitätsbibliothek in Göttingen for sending us a copy of a letter by Thomas Clausen on the factorization of the sixth Fermat number in 1855, and giving us permission to publish a part of this letter shown in Figure 1.2 of Chapter 1.

We appreciate very much Jan Brandts, Jana Grünerová, Karel Horák, Václav Kelar, and Pavel Křížek for their invaluable help in the preparation of figures for this book. All photographs were taken by Pavel Křížek. We would like to express our sincere gratitude to Gerard Alberts, Thomas Berry, Frits Beukers, Walter Carlip, Jan Chleboun, Zdeňka Crkalová, Andreas Dress, Caroline Emmet, Aleksander Grytczuk, Heiko Harborth, Eliot Jacobson, Liping Liu, Štefan Porubský, Andrzej Rotkiewicz, Bedřich Šofr, László Szalay, Kazimierz Szymiczek, Witold Więsław, and Marek Wójtowicz for fruitful discussions, and to Diego Benardete, Thomas Berry, Irving Katz, David Kramer, Andrzej Mąkowski, Alena Pravdová, and Alena Šolcová for their help in reading the manuscript. We kindly thank Ms. Helena Holovská for her assistance in the preparation of the name and subject indexes. We are indebted to Ms. Terry Kornak, Dr. Ina Lindemann, and Mr. Mark Spencer, of Springer-Verlag, for their helpful cooperation in the preparation of this book. Our great thanks go to the referees for valuable suggestions and also to our wives, Lea, Raluca, and Eva, for patience and understanding.

The work on the book was supported by grant No. 201/02/1057 of the Grant Agency of the Czech Republic and the Alexander von Humboldt Foundation. This support is gratefully acknowledged.

Michal Křížek
Florian Luca
Lawrence Somer

Glossary of Symbols

$F_m = 2^{2^m} + 1,\ m = 0, 1, \ldots$	Fermat numbers
$\mathbb{N} = \{1, 2, \ldots\}$	set of natural numbers
$\mathbb{Z} = \{\ldots, -2, -1, 0, 1, 2, \ldots\}$	set of integers
$\mathbb{Q} = \{\frac{p}{q} \mid p, q \in \mathbb{Z},\ q \neq 0\}$	set of rational numbers
$\mathbb{R}$	set of real numbers
$\mathbb{C}$	set of complex numbers
$C, C_1, C_2, \ldots$	generic constants (different at each occurrence)
(a, b)	open interval in $\mathbb{R}$
$\gcd(n_1, n_2, \ldots, n_k)$	greatest common divisor of $n_1, n_2, \ldots, n_k$
$\mathrm{lcm}(n_1, n_2, \ldots, n_k)$	least common multiple of $n_1, n_2, \ldots, n_k$
$\lfloor a \rfloor$	integer part of a real number a
$|S|$	cardinality (number of elements) of the set S
$n \equiv k \pmod{m}$	n is congruent to k modulo m
$n \not\equiv k \pmod{m}$	n is not congruent to k modulo m
$m \mid n$	m divides n
$m \nmid n$	m does not divide n
$m^j \| n$	m^j exactly divides n for $1 < m \leq n$
$k(n)$	square-free kernel of n
$P(n)$	greatest prime factor of n
$P_a(n)$	product of the primitive prime divisors of $a^n - 1$
$\mathrm{ord}(N)$	$\mathrm{ord}(N) = j$ if $2^j \| N - 1$
$\mathrm{ord}_d n$	order of n modulo d
$\mathrm{Gal}(L/K)$	group of all Galois automorphisms of L over K
$\log_b$	logarithm to the base b
$\log$	natural logarithm
e	Euler number
$\tau(n)$	number of all positive divisors of n

$\sigma(n)$	sum of all positive divisors of n
ϕ	Euler totient function
Φ_n	the nth cyclotomic polynomial
λ	Carmichael lambda function
μ	Möbius function
$\left(\frac{a}{p}\right)$	Legendre symbol for an odd prime p
$\left(\frac{a}{m}\right)$	Jacobi symbol for an odd number m
Re	real part
Im	imaginary part
i	imaginary unit
$i,\ j,\ k$	integer indices (subscripts)
$x^{\top}$	transposed vector x
$\exists$	there exist(s)
$\forall$	for all
$\mathcal{O}(\cdot)$	Landau's symbol: $f(\alpha) = \mathcal{O}(g(\alpha))$ if $|f(\alpha)| \leq C|g(\alpha)|$ as $\alpha \to 0$ or $\alpha \to \infty$
$o(\cdot)$	$f(\alpha) = o(g(\alpha))$ if $f(\alpha)/g(\alpha) \to 0$ as $\alpha \to 0$ or $\alpha \to \infty$
$\emptyset$	empty set
$\pi(x)$	number of primes less than or equal to x
$\prod$	product symbol
$\square$	Halmos symbol

Contents

1. Introduction

The French mathematician Pierre de Fermat (1601–1665) became well known not only due to Fermat's last theorem (proved in [Wiles] and [Taylor, Wiles]) and Fermat's little theorem (proved by Euler), but also due to the conjecture that all the numbers

$$(1.1) \qquad F_m = 2^{2^m} + 1 \quad \text{for } m = 0,\ 1,\ 2,\dots$$

are prime. The numbers F_m are called *Fermat numbers* after him.

If F_m is prime, we say that it is a *Fermat prime*. The first five members of sequence (1.1), i.e.,

$$(1.2) \qquad F_0 = 3, \quad F_1 = 5, \quad F_2 = 17, \quad F_3 = 257, \quad F_4 = 65537,$$

are primes.

A necessary condition for $2^n + 1$ to be prime for a positive integer n is that the exponent n be of the form 2^m for $m \in \{0,\ 1,\ 2,\dots\}$. This is because if k is a positive integer and $\ell \geq 3$ is an odd integer, then

$$(1.3) \qquad 2^{k\ell} + 1 = \left(2^k + 1\right)\left(2^{k(\ell-1)} - 2^{k(\ell-2)} + \cdots - 2^k + 1\right).$$

From this it follows that the number $2^n + 1$ is composite whenever the exponent n is divisible by an odd number $\ell \geq 3$. However, this does not happen in the sequence (1.1). Moreover, if $n = 4r + 2$, we also have the so-called Lucas formula (see [Kraïtchik, 1952], [Naur])

$$2^n + 1 = \left(2^{2r+1} - 2^{r+1} + 1\right)\left(2^{2r+1} + 2^{r+1} + 1\right).$$

In 1732 Leonhard Euler found that $F_5 = 641 \cdot 6700417$, and thus disproved the Fermat conjecture (see [Euler, 1738, p. 104] and also [Keller, 1992, p. 3]). Then a natural question arose, whether there exist infinitely many prime numbers of the form (1.1).

Until 1796 Fermat numbers were most likely a mathematical curiosity. The interest in the Fermat numbers dramatically increased when the German mathematician

Carl Friedrich Gauss (1777–1855) quite unexpectedly found through investigation of the roots of the equation $z^n = 1$ a theorem that expresses an interesting connection between the Euclidean construction of regular polygons and the Fermat primes (cf. Figure 1.1). He showed that the regular polygon can be constructed by ruler (straightedge) and compass if the number of its sides is equal to

$$n = 3, \ 4, \ 5, \ 6, \ 8, \ 10, \ 12, \ 15, \ 16, \ 17, \ldots .$$

More precisely, he proved (see [Gauss, Section VII]) that there exists a Euclidean construction of the regular polygon with n sides if

$$n = 2^i F_{m_1} F_{m_2} \cdots F_{m_j},$$

where $n \geq 3$, $i \geq 0$, $j \geq 0$, and F_{m_1}, $F_{m_2}, \ldots$, F_{m_j} are distinct Fermat primes (for $j = 0$ no Fermat primes appear in the above factorization of n). Gauss stated that the converse implication is true as well, but did not prove it. This was proved later in 1837 (see [Wantzel]). Hence, the regular polygon with n sides cannot be constructed for $n = 7, \ 9, \ 11, \ 13, \ 14, \ldots .$

Euclid (4th–3rd century B.C.): *There exists a construction of the regular polygon with n sides by ruler and compass for*
$$n = 2^i 3^j 5^k,$$
where $n \geq 3$ and $i \geq 0$ are integers and $j, k \in \{0, 1\}$.

Pierre de Fermat (1601–1665): *For $m = 0, \ 1, \ 2, \ldots$ the progression $F_m = 2^{2^m} + 1$ consists solely of primes.* (An incorrect statement.)

Leonhard Euler (1707–1783): *The Fermat number F_5 is composite.*

Carl Friedrich Gauss (1777–1855): *There exists a construction of the regular polygon with n sides by ruler and compass if and only if*

$$n = 2^i F_{m_1} F_{m_2} \cdots F_{m_j},$$

where $n \geq 3$, $i \geq 0$, $j \geq 0$, and F_{m_1}, $F_{m_2}, \ldots$, F_{m_j} are distinct Fermat primes.

Figure 1.1. Milestones in a Euclidean construction of regular polygons.

The ancient Greeks were already able to construct the regular pentagon. However, they were unsuccessful in finding an algorithm for constructing the regular 7-gon or 9-gon. A construction of the regular 17-gon was given by Gauss much later (see Chapter 17). The Euclidean constructions of the regular 257-gon and 65537-gon have also been described (see, e.g., [Gottlieb]).

Since it has not been proved whether (1.2) contains all Fermat primes (no others have been discovered), we do not know, in fact, how many regular n-gons can be theoretically constructed with ruler and compass (cf. Remark 17.17). By computer testing, it is true that (see Appendix A)

$$(1.4) \qquad F_m \text{ is composite for } 5 \leq m \leq 30,$$

but the status of F_{31} is unknown at present. In Chapter 14 we will give a heuristic argument that the number of Fermat primes is finite.

Throughout the book we shall see how Fermat numbers can be applied to prove that there exist infinitely many primes, pseudoprimes, and superpseudoprimes, and that for some k's the sequence $\{k2^n + 1\}_{n=1}^{\infty}$ contains only composite numbers (see Chapters 4, 7, 12). However, the Fermat numbers also have several practical applications. They are used in the construction of generators of pseudorandom numbers. For example, an algorithm involving the Fermat prime $F_4 = 2^{16} + 1$ was used in the popular home microcomputer ZX Spectrum, which was developed in the 1980s. An application of Fermat numbers in group theory is given in [Liu]. In Chapter 15 we will introduce further real-life applications of Fermat numbers. In particular, we present the Fermat number transform, which can be utilized in digital signal processing. We show how Fermat numbers can be employed for fast multiplication of large binary numbers with N digits requiring only $\mathcal{O}(N \log N \log \log N)$ operations. We also describe the use of Fermat numbers in hashing schemes and in an analysis of the logistic equation describing chaos.

The complete factorizations of the Fermat numbers $F_5, F_6, \ldots, F_{11}$, as given in Appendix A, were performed by the following methods. In 1732, Euler found the factors of F_5 by means of trial division by numbers of the form $k2^6 + 1$ (see [Euler, 1738, 1750]). In 1880 F. Landry, when he was eighty-two years old, factored the sixth Fermat number F_6 into two factors (see [Landry, 1880a], [Williams, 1993]). However, Kurt-R. Biermann in [Biermann, p. 185] states that in a letter of January 1, 1855, to Gauss (see Figure 1.2), which remains in the library of the University of Göttingen, Thomas Clausen (who was also known as an important astronomer) gave the complete factorization of F_6 into two prime factors, i.e., before Landry in 1880. He believed that the larger of the two factors was the largest known prime at that time, which turned out to be correct. The factorization methods used are not completely certain. For an indication of how Landry might have factored F_6, see [Williams, 1993] or [Williams, 1998, pp. 102–107] (cf. also [Landry, 1880b]). We have no knowledge of how Clausen factored F_6 (compare with Remark 4.15).

The first new complete factorizations of the Fermat numbers since Euler factored F_5 in 1732 and Clausen and Landry factored F_6 in the latter half of the nineteenth century have occurred beginning only in 1970 by the use of electronic computers. The number F_7 was factored in 1970 (see [Morrison, Brillhart, 1971, 1975]) by means of the continued fraction method. In 1980, Brent and Pollard (see [Brent,

Pollard]) factored F_8 using a modification of Pollard's rho method. The larger of the two factors of F_8, with 62 digits, was shown to be prime by Williams (see [Brent, Pollard, Section 4]) using a method given in [Williams, Judd].

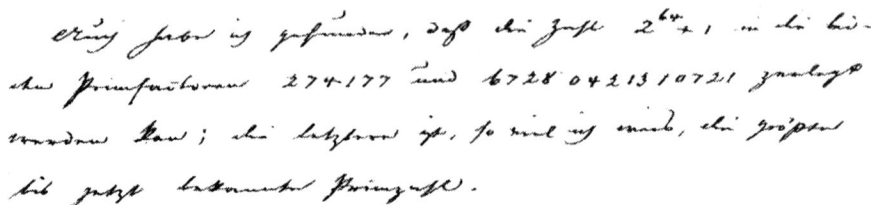

Figure 1.2. A part of the letter by Thomas Clausen to Carl Friedrich Gauss announcing the factorization of F_6 in 1855 (prior to Landry's factorization): *"Auch habe ich gefunden, daß die Zahl $2^{64}+1$ in die beiden Primfactoren 274177 und 67280421310721 zerlegt werden kann; die letztere ist, so viel ich weiß, die größte bis jetzt bekannte Primzahl."*

Surprisingly, F_{11} had already been completely factored before F_9 and F_{10}, in 1988, by Brent using the elliptic curve method (see [Brent, 1989]). It helped him very much that F_{11} has 4 relatively small prime divisors (compare with Appendix A and [Cunningham, Western]) and only one large divisor, with 564 digits. A further reason why the factorization of F_{11} was completed before the factorizations of F_9 and F_{10} is that the difficulty of completely factoring a number by means of the elliptic curve method is determined principally by the size of its second-largest prime factor. For F_{11}, the second-largest prime factor has 22 digits, whereas for F_9, the second-largest prime factor has 49 digits, and for F_{10}, the penultimate prime factor has 40 digits. The largest prime factor of F_{11}, with 564 digits, was proved to be prime in 1988 by Morain, using a refinement of the elliptic curve primality-proving algorithm developed by A. O. L. Atkin in 1986 (see [Atkin, Morain]). The test of primality required 30 days of computer time. The number F_{11} is at present the largest Fermat number to have been completely factored.

In 1903, Western found a relatively small prime factor of F_9 consisting of seven digits (see [Cunningham, Western]). The number F_9 was completely factored in 1990 by Lenstra, Lenstra, Manasse, and Pollard by the special number field sieve with the assistance of hundreds of collaborators. The factoring of F_9 was achieved by employing approximately 700 workstations throughout the world, each performing independent computation over a period of four months, with final computations using the collected results performed on a supercomputer (see [Lenstra, Lenstra, Manasse, Pollard]). The number F_9 has three prime factors. The largest two prime factors, with 49 and 99 digits, were proved to be prime by A. M. Odlyzko. Note that the number F_9 is much bigger than the number of all elementary particles in the observable part of our universe.

In 1953, Selfridge found an eight-digit prime factor of F_{10} (see [Selfridge]), and in 1962 Brillhart found another relatively small prime factor with ten digits. Finally, in 1995 the associated cofactor of F_{10} was factored into two primes using the elliptic curve method (see [Brent, 1999]). The larger prime factor, with 252 digits, was proved to be prime using the method given in [Atkin, Morain], which completed the factorization of the tenth Fermat number.

There are many computer-based results on finding further factors of Fermat numbers (see, e.g., [Atkin, Rickert], [Brent, Crandall, Dilcher, Van Halewyn], [Gostin, 1980, 1995], [Gostin, McLaughlin], [Hallyburton, Brillhart], [Riesel, 1963], [Robinson, 1957a], [Shippee], [Suyama, 1981], [Wrathall]). About 150 further composite Fermat numbers have been found for $m > 30$ as of 2001. In all, we knew more than 200 prime factors of about 185 Fermat numbers in 2001 due to high-performance (super)computing facilities; see, e.g., [Brillhart, Lehmer, Selfridge, Tuckerman, Wagstaff], [Keller, 1992], or [www1]. For instance, in 1984, W. Keller verified that the Fermat number F_{23471} is composite. He proved, by a computer analysis, that this large number is divisible by $5 \cdot 2^{23473} + 1$. The number F_{23471} has more than 10^{7000} digits. The largest composite Fermat number found before 2001 is F_{382447}. Its factor $3 \cdot 2^{382449} + 1$ was discovered by Cosgrave and Gallot in 1999.

The compositeness of Fermat numbers can be shown without knowing any explicit nontrivial factors by using Pepin's test, which is discussed in Chapter 5. In [Morehead, 1905] it was shown that F_7 is composite. This fact was verified later in [Western]. Morehead and Western also proved that F_8 is composite (see [Morehead, Western, 1909]). In 1952, the compositeness of F_{10} was confirmed with the help of a computer whose memory contained only 256 words (see [Robinson, 1954]). The compositeness of F_{13} was shown in [Paxson]. The four Fermat numbers F_{14}, F_{20}, F_{22}, and F_{24} are the only remaining composite Fermat numbers in the range $F_5, F_6, \ldots, F_{30}$ (cf. (1.4)) for which we do not yet know any nontrivial factors. The compositeness of F_{14} was announced in 1961 (see [Hurwitz, Selfridge] and also [Selfridge, Hurwitz]). In 1987, Young and Buell verified the composite character of F_{20} (see [Young, Buell]). Later, in 1993, the compositeness of F_{22} was proved (see [Crandall, Doenias, Norrie, Young]). This result was independently checked in [Trevisan, Carvalho] only nine months later. The twenty-fourth Fermat number F_{24}, which has over 5 million decimal digits, was shown to be composite in 1999 (see [Crandall, Mayer, Papadopoulos]). This was the biggest computation ever done to obtain a simple "yes-or-no" answer. It required 10^{17} computer operations. All of the results above were obtained using Pepin's test.

Testing the primality of F_m by dividing it by each prime less than $\sqrt{F_m}$ results in an algorithm that increases rapidly in completion time as m increases, and is impractical even for small m. Suppose, for instance, that we are able to perform 10^9 divisions per second. Then we would spend more time for the factorization of the innocent-looking number $F_8 = 2^{2^8} + 1$ than the age of our universe ($\approx 15 \cdot 10^9$ years). To see this, denote by $\pi(x)$ the number of primes less than or equal to x (see Figure 1.3). Then by the Gauss formula (see, e.g., [Hardy, Wright], [Schroeder]), $\pi(x)$ is approximately equal to $x/\log x$. The error is less than 15% for each $x \geq 3000$. Hadamard and de la Vallée Poussin proved the asymptotic equality, as $x \to \infty$,

$$(1.5) \qquad \pi(x) \approx \frac{x}{\log x} \qquad (prime\ number\ theorem).$$

Since the integer part of $\sqrt{F_8}$ has 39 digits, there exist at least $10^{38}/(38 \log 10) \approx 10^{36}$ primes below $\sqrt{F_8}$. Since one year has about $3.2 \cdot 10^7$ seconds, we would need at least $3 \cdot 10^{19}$ years to factor F_8. Mathematicians compare such an algorithm with an effort to break up an atom with a hammer.

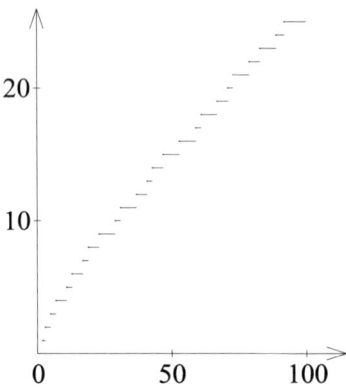

Figure 1.3. The prime-counting function $\pi(x)$.

If we have additional information about the number to be factored, we can find the factorization much more quickly. Later, we will show how to substantially reduce the set of possible divisors. Here, the key role will be played by Lucas's theorem (Theorem 6.1), which establishes a necessary condition for the form of prime numbers that divide F_m.

From Fermat's little theorem, we know that if p does not divide $a^p - a$ for some a, then p is composite (cf. Theorem 2.9). In this way we can find that a given number is composite without knowing any of its factors. If we show, for instance, that p does not divide $2^{p-1} - 1$, then p is not a prime.

Let us emphasize that factoring an integer N is harder than testing its primality. The oldest method of factoring an integer N, one that is not very efficient, is the usual trial division of N by each integer less than its square root.

A second method of factoring N is to attempt to write N as the difference of two squares, since we immediately obtain the factorization

$$(1.6) \qquad N = a^2 - b^2 = (a + b)(a - b),$$

where $a - b > 1$. For instance, if $N = F_5 = 4294\,967297$, we have $a = 3\,350529$, $b = 3\,349888$, and if $N = F_6 = 18\,446744\,073709\,551617$, then

$$a = 33\,640210\,792449,$$
$$b = 33\,640210\,518272.$$

Using the simple factorization (1.6), Fermat invented his own factorization method, which is efficient only if N is a product of two numbers of almost the same size. Assume that N is composite. If N is a square, we are finished. If it is not a square, we define $a = \lfloor \sqrt{N} \rfloor + 1$, where $\lfloor \sqrt{N} \rfloor$ denotes the integer part of $\sqrt{N}$. Set

$x = a^2 - N$. If this is a square, we are finished. Otherwise, we calculate the next candidate

$$(a+1)^2 - N = a^2 + 2a + 1 - N = x + 2a + 1$$

and check whether it is a square or not. If yes, we are again finished; otherwise, we increase a by 2, etc. After a finite number of steps we find nontrivial factors. This algorithm is called the *Fermat factorization method* (see, e.g., [Koblitz, pp. 143–144]).

Recall another simple method that can be used for factorization. Suppose that a number N can be written as a sum of two nonzero squares in two different ways, say

(1.7)
$$N = a^2 + b^2 = c^2 + d^2,$$

where $a > c \geq d > b \geq 1$ without loss of generality. Then N is composite. Indeed,

(1.8)
$$N = \frac{(a^2 - d^2)(c^2 + d^2)}{a^2 - d^2} = \frac{a^2 c^2 - d^2(c^2 - a^2 + d^2)}{a^2 - d^2}$$
$$= \frac{a^2 c^2 - d^2 b^2}{a^2 - d^2} = \frac{(ac + bd)(ac - bd)}{(a + d)(a - d)},$$

and, after canceling all factors in the last denominator, we see that this is a product of two integers, since N is an integer. We will prove this interesting fact in detail later (see the proof of Theorem 5.24).

Notice that each Fermat number F_m for $m > 0$ is of the form $\left(2^{2^{m-1}}\right)^2 + 1^2$. To show the compositeness of a Fermat number, it is enough to find an expression of the number as a sum of two squares neither of which is 1. For example, the first two composite Fermat numbers can be written in the form (1.7) as follows:

$$F_5 = 65536^2 + 1^2 = 62264^2 + 20449^2,$$
$$F_6 = 4294967296^2 + 1^2 = 4046803256^2 + 1438793759^2.$$

We shall return to these equalities in the proof of Theorem 5.24. For an algorithm implementing the above concept see [Maruyama, Kawatani].

As we noted, from the computational point of view it is much more difficult to factor numbers than to test their primality; see, e.g., [Lenstra, Pomerance], [Miller], [Pollard]. The well-known RSA method by [Diffie, Hellman] and [Rivest, Shamir, Adleman] for encrypting messages by use of large primes is based on this fact. The number of arithmetic operations for factoring a number that is a product of two large primes (of unknown character) grows exponentially with the number of its digits. If such a number has more than 200 digits, it is then practically indecomposable by modern factorization methods [Pomerance, 1996]. On the other hand, to test the primality of an arbitrary number of 200 digits takes only several minutes by supercomputing facilities.

Current efficient factoring algorithms are described or surveyed, for example, in [Brent, 1990], [Bressoud], [Crandall], [Crandall, Pomerance], [Knuth], [Koblitz], [Lenstra], [Lenstra, Lenstra, Manasse, Pollard], [Li], [Montgomery, 1987, 1994],

[Pollard], [Pomerance, 1990, 1996], [Riesel, 1994]. Among the most effective of these algorithms are the elliptic curve method, the quadratic sieve, and the number field sieve (see, e.g., [Lenstra], [Lenstra, Lenstra], [Pollard], [Pomerance, 1990], [Riesel, 1985], [Stewart, I., 1987]). All these algorithms belong to the class NP (nondeterministic polynomial). Let us point out that if a factorization of some number q into two nontrivial factors q_1 and q_2 is found, then the reverse verification that $q = q_1 q_2$ takes only a fraction of a second. This follows from the fact that for problems of the class NP the following holds: If somebody gives us the solution, then we are able to check in polynomial time (with respect to the magnitude of the input data) that it is really the true solution.

2. Fundamentals of Number Theory

The distribution of primitive roots is a deep mystery.

Carl Friedrich Gauss

Throughout this book we shall mainly work with integers unless we specify otherwise. By $\mathbb{N}$ we denote the set $\{1, 2, 3, \ldots\}$ of all *natural numbers*, i.e., all positive integers. A natural number p is said to be *prime* if $p > 1$ and p is divisible only by p and 1. Thus, the natural numbers are divided into the unit 1, prime numbers, and composite numbers. Notice that we can find arbitrarily many successive natural numbers that are all composite. Indeed, for any $n > 1$ the sequence $n! + 2$, $n! + 3, \ldots$, $n! + n$ contains $n - 1$ successive integers that are evidently composite. On the other hand we have the following theorem of Euclid:

Theorem 2.1 (Euclid). *The number of primes is infinite.*

P r o o f . Assume, to the contrary, that there exist only a finite number of primes $p_1, p_2, \ldots, p_n$. Put $N = p_1 p_2 \cdots p_n + 1$. Since the division N/p_i always gives the remainder 1, none of the p_i's divides N. Thus N is either a new prime, or N has a new prime factor different from $p_1, p_2, \ldots, p_n$, which is a contradiction. $\quad\square$

Recall that all primes in a given bounded interval starting from 1 can be found by the well-known *sieve of Eratosthenes*, which deletes the number 1 and all composite numbers in the interval, leaving the primes: $\not{1}$, 2, 3, $\not{4}$, 5, $\not{6}$, 7, $\not{8}$, $\not{9}, \ldots$.

Theorem 2.2 (Fundamental Theorem of Arithmetic). *If*

$$q_1^{\alpha_1} q_2^{\alpha_2} \cdots q_s^{\alpha_s} = r_1^{\beta_1} r_2^{\beta_2} \cdots r_t^{\beta_t},$$

where $q_1 < q_2 < \cdots < q_s$, $r_1 < r_2 < \cdots < r_t$ *are primes, and* $s, t, \alpha_i, \beta_i \in \mathbb{N}$, *then*

$$s = t, \quad q_i = r_i, \quad \alpha_i = \beta_i$$

for every $i = 1, \ldots, s$.

For a proof see, e.g., [Riesel, 1985, p. 263]. The fundamental theorem of arithmetic thus says that every natural number $n > 1$ can be written uniquely as a product of primes $q_1 < q_2 < \cdots < q_s$:

$$(2.1) \qquad n = q_1^{\alpha_1} q_2^{\alpha_2} \cdots q_s^{\alpha_s} = \prod_{i=1}^{s} q_i^{\alpha_i},$$

where $\alpha_i \geq 1$ for $i \in \{1, \ldots, s\}$.

Remark 2.3. One reason why 1 is not considered to be a prime number is that we would not have the uniqueness of exponents in (2.1). Another reason is, e.g., the famous expression of the Riemann ζ-function via the product over all primes p,

$$\zeta(z) = \sum_{n=1}^{\infty} \frac{1}{n^z} = \prod_{p} \frac{1}{1 - p^{-z}},$$

where the sum converges for an arbitrary complex number z such that $\mathrm{Re}(z) > 1$.

An important relation between two integers is their greatest common divisor and their least common multiple (cf. Figure 2.1). The *greatest common divisor* of two integers m and n, not both equal to zero, denoted by $\gcd(m, n)$, is the largest positive integer that divides both of them. The *least common multiple* of two nonzero integers m and n, denoted by $\mathrm{lcm}(m, n)$, is the smallest positive integer that is divisible by both of them.

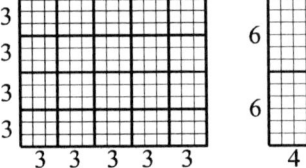

Figure 2.1. Geometric interpretation of the greatest common divisor and the least common multiple. A 12×15 rectangle can be covered by squares whose largest possible side is $\gcd(12, 15) = 3$. The smallest possible side of a square that can be decomposed into 6×4 rectangles is equal to $\mathrm{lcm}(6, 4) = 12$.

According to the fundamental theorem of arithmetic, for given $m, n \in \mathbb{N}$ we may write

$$m = \prod_{i=1}^{\infty} p_i^{k_i}, \quad n = \prod_{i=1}^{\infty} p_i^{\ell_i},$$

where p_i is the ith prime and $k_i, \ell_i \geq 0$ for all $i \in \mathbb{N}$ (in particular, if $m = 1$, then all $k_i = 0$). Then the greatest common divisor and the least common multiple for natural numbers m and n are calculated as follows:

$$\gcd(m, n) = \prod_{i=1}^{\infty} p_i^{\min(k_i, \ell_i)}, \quad \mathrm{lcm}(m, n) = \prod_{i=1}^{\infty} p_i^{\max(k_i, \ell_i)},$$

respectively. Notice that the least common multiple of two integers cannot be reasonably defined if one of them is zero. That is why we do not include zero in the set $\mathbb{N}$ of natural numbers. Another reason is that 0^0 cannot be well defined. For natural numbers m and n it is easy to see that

$$mn = \gcd(m, n)\,\mathrm{lcm}(m, n).$$

The greatest common divisor of large natural numbers $m < n$ is usually calculated by the well-known *Euclidean algorithm:* If m divides n, then $\gcd(m,n) = m$. Otherwise, we have $\gcd(m,n) = \gcd(m,r)$, where r is the remainder of n divided by m. A larger problem is thus transformed into a smaller one. Further steps are similar. The problem is successively reduced until the last step, where we are left with the remainder 0.

In Figure 2.2 we see a geometric interpretation of the following reduction:

$$\gcd(16,54) = \gcd(16,6) = \gcd(4,6) = \gcd(4,2) = 2.$$

The 16×54 square mesh is decomposed into several emphasized squares. The side of the smallest emphasized square, 2×2 in Figure 2.2, is equal to the greatest common divisor 2. Let us further note that the maximum number of steps in the Euclidean algorithm (with respect to the sizes of m and n) is attained for two successive Fibonacci numbers (compare with [Williams, 1998, p. 12] and also Figure 17.14).

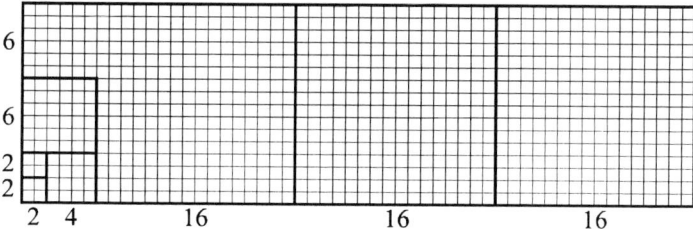

Figure 2.2. Geometric interpretation of the Euclidean algorithm.

We say that the integers m and n are *coprime* (or *relatively prime*) if

$$\gcd(m,n) = 1.$$

In this case, the associated illustration of the Euclidean algorithm finishes with the smallest square 1×1. Note that two consecutive integers are coprime.

Proposition 2.4. *Let* $d = \gcd(m,n)$ *for some integers* m *and* n *not both equal to zero. Then there exist integers* x *and* y *such that*

$$mx + ny = d.$$

P r o o f . Let S be the set of all integers of the form $ma + nb$, where a and b are integers. Since one of m or n is not zero, there are nonzero integers in S. Because $t = ma + nb$ is in S, the number $-t = (-m)a + (-n)b$ is also in S. Hence, S contains positive integers. Therefore, by the well-ordering principle, there is a smallest positive integer d_0 is S having the form $d_0 = mx + ny$. We claim that $d_0 = \gcd(m,n)$.

We first show that d_0 is a common divisor of m and n. Let $u = ma_0 + nb_0$ be any element of S. By the division algorithm, $u = qd_0 + r$, where $0 \le r < d_0$. Writing this out explicitly, we have

$$ma_0 + nb_0 = q(mx + ny) + r,$$

whence

$$r = m(a_0 - qx) + n(b_0 - qy)$$

and $r \in S$. Since $r \geq 0$ and $r < d_0$, it follows that $r = 0$ by the choice of d_0. Thus, d_0 divides u for any $u \in S$. However, $m = m \cdot 1 + n \cdot 0 \in S$ and $n = m \cdot 0 + n \cdot 1 \in S$, whence d_0 divides both m and n.

Finally, let e be any common divisor of m and n. Then e divides $mx + ny = d_0$. Consequently, $d_0 = \gcd(m, n)$. $\qquad \square$

Proposition 2.4 (see also [Burton, p. 22]) has a nice geometric interpretation, namely, the straight line $mx + ny = d$ cuts through the lattice points (x, y), that are solutions of the above linear Diophantine equation (see Figure 2.3 for $m = 2$, $n = -3$, and $d = 1$).

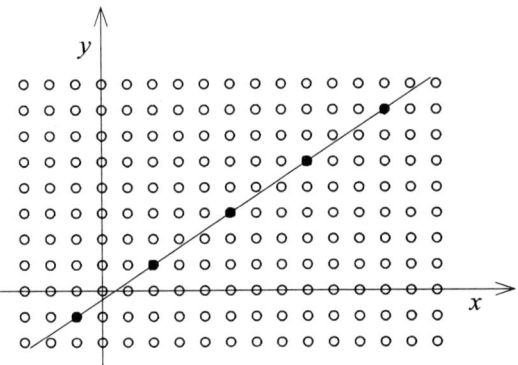

Figure 2.3. The straight line corresponding to the linear Diophantine equation $2x - 3y = 1$ cuts through the points $\ldots, (-1, -1), (2, 1), (5, 3), (8, 5), (11, 7), \ldots$.

Example 2.5. We will show how to solve the Diophantine equation

$$16x - 54y = 2$$

by using an analogue of the Euclidean algorithm (compare with Figure 2.2). Since $\gcd(16, 54) = 2$, this equation has a solution by Proposition 2.4, and we have

$$x = 3y + \frac{6y + 2}{16}.$$

To get an integer solution, $6y + 2$ has to be a multiple of 16, i.e., there exists an integer v such that

$$6y + 2 = 16v, \quad \text{and thus } y = 2v + \frac{4v - 2}{6}.$$

Similarly, we see that there must exist an integer w such that

$$4v - 2 = 6w, \quad \text{and thus } v = w + \frac{2w + 2}{4}.$$

It is obvious that w has to be odd. In particular, we can take $w = 1$. By backward substitution, we successively find that $v = 2$, $y = 5$, and $x = 17$.

Recall that the greatest common divisor and the least common multiple of more than two integers are defined in a similar manner as for two integers by induction, namely, for $k > 2$ and integers $n_1, \ldots, n_k$, we set

$$\gcd(n_1, \ldots, n_{k-1}, n_k) = \gcd\big(\gcd(n_1, \ldots, n_{k-1}), n_k\big) \quad \text{if } n_1 \neq 0,$$

$$\mathrm{lcm}(n_1, \ldots, n_{k-1}, n_k) = \mathrm{lcm}\big(\mathrm{lcm}(n_1, \ldots, n_{k-1}), n_k\big) \quad \text{if } \prod_{j=1}^{k} n_j \neq 0.$$

Throughout the book the symbol $m \mid n$ means that a natural number m divides an integer n with zero remainder (e.g., $3 \mid (-6)$, $5 \mid 5$); otherwise, we write $m \nmid n$ (e.g., $2 \nmid 3$). If $m \mid n$ and $m < n$, then m is called a *proper divisor* of n. If $1 < m < n$, then m is said to be a *nontrivial divisor* of n. We say that m^j *exactly divides* n, and write $m^j \| n$, if $m^j \mid n$, but $m^{j+1} \nmid n$.

Further, we introduce the notion of congruence, which was invented by C. F. Gauss. It has many practical applications, e.g., in cryptography (the famous RSA method; see [Rivest, Shamir, Adleman]). It is also a useful tool for generating pseudorandom numbers (see (15.7) and [Ripley, Chapter 2]) and a very efficient resource in many remarkable results (see, e.g., Remark 4.20, Theorem 7.4, and Chapter 15).

Let a, b, m be given integers and $m \geq 1$. If $m \mid (a - b)$, i.e., $a - b$ is divisible by m (cf. Figure 2.4), we write

$$a \equiv b \pmod{m}$$

and say that a *is congruent to* b *modulo* m; b is called a *residue of* a *modulo* m. It is clear that there are exactly m distinct incongruent residues modulo m.

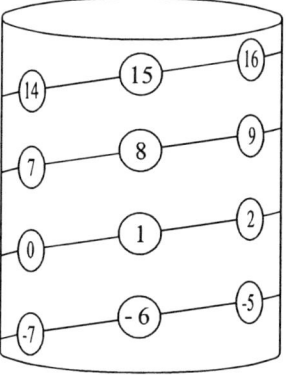

Figure 2.4. Geometric interpretation of congruence modulo 7. The real axis is uniformly rolled up as a helix on an infinite cylinder. All integers occurring in the same vertical direction are congruent modulo 7.

Let us briefly recall several basic properties of congruence. We immediately see that $a \equiv a \pmod{m}$ for any integer a. If $a \equiv b \pmod{m}$, then for any integers a, b, c, and $k \geq 0$ we immediately have

$$b \equiv a \pmod{m},$$
$$a \pm c \equiv b \pm c \pmod{m},$$
$$ac \equiv bc \pmod{m},$$
$$a^k \equiv b^k \pmod{m},$$

where the last congruence follows from the equality

$$a^k - b^k = (a - b)(a^{k-1} + a^{k-2}b + \cdots + b^{k-1}), \quad k > 1.$$

From the above we see that the relation "$\equiv$" modulo m is reflexive and symmetric. Since transitivity holds as well, it is an equivalence relation over the set of integers.

If $\gcd(c, m) = 1$, then in the congruence $ac \equiv bc \pmod{m}$ the number c can be canceled, i.e., $a \equiv b \pmod{m}$. If $a \equiv b \pmod{m}$ and $c \equiv d \pmod{m}$, then obviously

$$a + c \equiv b + d \pmod{m},$$
$$a - c \equiv b - d \pmod{m},$$

and moreover, we have

(2.2) $$ac \equiv bd \pmod{m}.$$

Indeed, if $a = b + im$ and $c = d + jm$, then $ac = bd + (jb + id + ijm)m$, and congruence (2.2) follows.

Remark 2.6. Sunzi Suanjing, in his arithmetic book, introduces the first example of the "Chinese remainder theorem" problem, one of the most important problems in number theory (see [Martzloff, p. 310]). The exact origin of this book is not known, but it was written some time in the period 280–473. This example can be stated as follows: We have an unknown number x of objects. Counting them by three, two objects will remain, counting them by five, three objects will remain, and finally, counting them by seven, two objects will remain. What is x?

Using modern Gaussian notation, we can rewrite this ancient example as a system of simultaneous congruences:

(2.3)
$$x \equiv 2 \pmod{3},$$
$$x \equiv 3 \pmod{5},$$
$$x \equiv 2 \pmod{7}.$$

Yang Hui in his book from 1275 describes five more examples similar to this (see again [Martzloff, p. 311]). All these examples can be generalized as follows:

Theorem 2.7 (Chinese Remainder Theorem). *Let $m_1, m_2, \ldots, m_k$ be pairwise coprime natural numbers. Then for the system of simultaneous congruences*

(2.4)
$$
\begin{aligned}
x &\equiv r_1 \ (\mathrm{mod}\ m_1), \\
x &\equiv r_2 \ (\mathrm{mod}\ m_2), \\
&\vdots \\
x &\equiv r_k \ (\mathrm{mod}\ m_k),
\end{aligned}
$$

where the r_i's are integers, there exists one and only one solution x modulo M, where

$$ M = m_1 m_2 \cdots m_k. $$

P r o o f . First we prove the existence of a solution x. Define M_i by the equalities

(2.5)
$$ M = m_1 M_1 = m_2 M_2 = \cdots = m_k M_k. $$

Since m_i and M_i are coprime, there exist integers y_i, $i = 1, 2, \ldots, k$, due to Proposition 2.4, such that

(2.6)
$$ M_i y_i \equiv 1 \quad (\mathrm{mod}\ m_i). $$

We claim that the general solution of (2.4) has the form

(2.7)
$$ x \equiv r_1 M_1 y_1 + r_2 M_2 y_2 + \cdots + r_k M_k y_k \quad (\mathrm{mod}\ M). $$

To verify that such a number x solves system (2.4), we choose $i \in \{1, \ldots, k\}$. From (2.5), we see that all terms, except the ith term on the right-hand side of (2.7), contain the factor m_i, and thus

$$ x \equiv r_1 M_1 y_1 + r_2 M_2 y_2 + \cdots + r_k M_k y_k \equiv r_i M_i y_i \equiv r_i \quad (\mathrm{mod}\ m_i), $$

where the last equivalence is obtained by multiplication of congruence (2.6) by r_i.

In order to prove uniqueness, suppose that x_1 and x_2 are two solutions of system (2.4). Then $x_1 \equiv x_2 \ (\mathrm{mod}\ m_i)$ for each $i = 1, \ldots, k$. Since m_i are pairwise coprime, we have $x_1 \equiv x_2 \ (\mathrm{mod}\ M)$. Therefore, the solution of (2.4) is uniquely determined modulo M.   □

Remark 2.8. Using Theorem 2.7 for system (2.3), where $\gcd(3, 5, 7) = 1$, we find the solution $x = 23$, which is the smallest in positive integers. Figure 2.5 illustrates the cube $23 \times 23 \times 23$, which is decomposed into blocks. Most of them are $3 \times 5 \times 7$ blocks associated with the moduli 3, 5, and 7. There is one $2 \times 3 \times 2$ block associated with the remainders 2, 3, and 2, and other mixed blocks associated with both moduli and remainders.

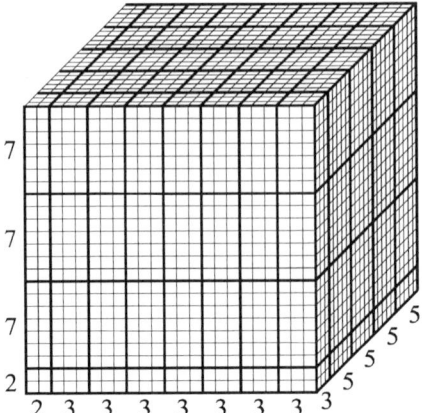

Figure 2.5. Geometric interpretation of the Chinese remainder theorem for the system of congruences (2.3).

In the thirteenth century Qin Jiushao solved congruence (2.6) using an analogue of the Euclidean algorithm (see Example 2.5 and [Martzloff, p. 317]). The problem of finding y_i satisfying congruence (2.6) can be transformed into that of finding the solution of the Diophantine equation $M_i y_i - m_i v_i = 1$ for unknowns y_i and v_i.

Another way of interpreting the Chinese remainder theorem is that every natural number not greater than $M = m_1 m_2 \cdots m_k$ can be uniquely characterized by the k remainders $(r_1, r_2, \ldots, r_k)$ after division by $(m_1, m_2, \ldots, m_k)$. This is called the Sino-representation (see [Schroeder]) and was also used as a basis of the traditional Chinese calendar.

If p is a prime and $ab \equiv 0 \pmod{p}$, then it is easy to see that $a \equiv 0 \pmod{p}$ or $b \equiv 0 \pmod{p}$. We shall use this elementary fact in proving the next theorem, which is one of the most frequently used tools in number theory, as we shall see also in this book (see the Subject Index).

Theorem 2.9 (Fermat's Little Theorem). *If a is a natural number and p a prime number, then $p \mid a^p - a$.*

P r o o f . If $p \mid a$, then p also divides $a^p - a = a(a^{p-1} - 1)$. So let

(2.8) $$\gcd(p, a) = 1.$$

We will show that then $p \mid a^{p-1} - 1$. Consider the finite sequence

(2.9) $$a, \ 2a, \ 3a, \ldots, \ (p-1)a, \ pa.$$

Dividing the two numbers ia and ja, $1 \leq j < i \leq p$, by p, we cannot obtain the same remainder, since then $p \mid a(i-j)$, which contradicts (2.8). Hence, the sequence (2.9) yields p different remainders upon division by p. Let us remove the last term in (2.9), which produces the remainder 0. We obtain the same remainders (up to order) when dividing the sequence $1, \ldots, p - 1$ by p. From this and by induction using (2.2), we arrive at $a^{p-1}(p-1)! \equiv (p-1)! \pmod{p}$. Consequently,

$$(a^{p-1} - 1)(p - 1)! \equiv 0 \pmod{p},$$

and the first term on the left-hand side is thus divisible by the prime p, since $p \nmid (p - 1)!$. $\quad\square$

Remark 2.10. There are several other proofs of Fermat's little theorem. A very interesting approach to this is given in [Gutfreund, Little]. It uses an analogy with physical particles based on symmetry properties of Ising spin configurations. For further discussion about Fermat's little theorem and also about the *Fermat quotient* $(a^{p-1} - 1)/p$ see [Lepka].

Remark 2.11. Fermat's little theorem is often formulated by means of the congruence notation as follows:

If p is a prime and a is a natural number coprime to p, then

$$(2.10) \qquad\qquad a^{p-1} \equiv 1 \pmod{p}.$$

Congruence (2.10) can be used to prove that a given number is composite without knowing any of its factors. Namely, if we find a "small" a satisfying (2.8) such that (2.10) does not hold, then p is not a prime.

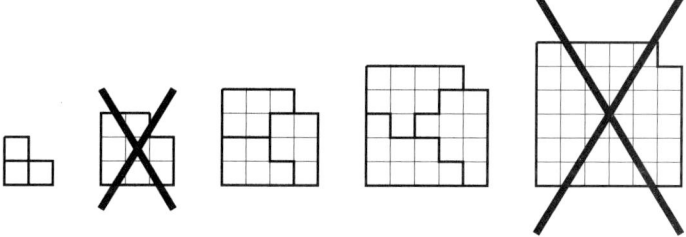

Figure 2.6. Geometric interpretation of Fermat's little theorem stated in (2.10) for $p = 3$. If $\gcd(3, a) = 1$, then $3 \mid a^2 - 1$; i.e., the square a^2 reduced by one unit square can be decomposed into 3 parts of the same integer area. If $3 \mid a$, then a similar decomposition is impossible.

Definition 2.12. Let d and a be positive integers such that $\gcd(d, a) = 1$. The smallest positive exponent e for which $d \mid a^e - 1$ (i.e., $a^e \equiv 1 \pmod{d}$) is called the *multiplicative order of the number a modulo d*, which we shall write as $e = \mathrm{ord}_d a$.

By Fermat's little theorem, $p \mid a^{p-1} - 1$, provided that $\gcd(p, a) = 1$ holds for any prime p. From this we infer the existence of the order $e \le p - 1$. For instance, we can easily verify that $\mathrm{ord}_7 2 = 3$, $\mathrm{ord}_5 3 = 4$, etc. Note, in particular, that $\mathrm{ord}_1 a = 1$ for any $a \ge 1$.

In Theorem 4.12 and Remark 4.13, we show that $\mathrm{ord}_{F_m} 2 = 2^{m+1}$.

Lemma 2.13. *If $e = \mathrm{ord}_d a$, then*

$$(2.11) \qquad\qquad d \mid a^n - 1$$

for $n = ke$, $k \in \{1, 2, \ldots\}$, and relation (2.11) holds only for these exponents.

P r o o f . If $n = ke$, then

$$a^n - 1 = a^{ke} - 1 = \left(a^e - 1\right)\left(a^{e(k-1)} + \cdots + a^e + 1\right),$$

and thus (2.11) is valid due to the previous definition.

Assume, for an instant, that $d \mid a^{ke+h} - 1$ for some $k \in \{1, 2, \ldots\}$ and $0 < h < e$. Then

$$\left(a^{ke+h} - 1\right) - \left(a^{ke} - 1\right) = a^{ke}\left(a^h - 1\right).$$

Since both the numbers in parentheses on the left are divisible by d and, by (2.11), $\gcd(d, a^{ke}) = 1$, we get that $a^h - 1$ is divisible by d. This contradicts the minimality of e. $\qquad\square$

Let p be a prime. By Fermat's little theorem, the maximum order modulo p of any integer a coprime to p is $p - 1$. We call the integer $g \not\equiv 0 \pmod{p}$ a *primitive root modulo p* if

$$\mathrm{ord}_p g = p - 1.$$

For example, 3 is a primitive root modulo 17 (see Figure 2.7), since $3^{16} \equiv 1 \pmod{17}$ and $3^k \not\equiv 1 \pmod{17}$ for all $k = 1, \ldots, 15$. In this case, the order of 3 modulo 17 is 16, i.e., $\mathrm{ord}_{17} 3 = 16$.

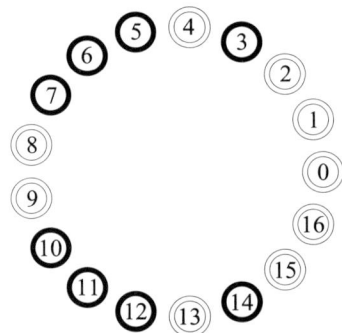

Figure 2.7. Primitive roots modulo 17 are indicated by a black circle.

We have the following theorem on the existence of primitive roots.

Theorem 2.14. *If p is a prime, then there exists a primitive root modulo p.*

For a proof see, e.g., [Gauss, article 55] or [Burton, pp. 156–157]. A generalization of this result will be given in Theorem 2.18.

Remark 2.15. We note that if g is a primitive root modulo p, where p is prime, then g is a generator of the set of all nonzero residues modulo p, that is, $\{g, g^2, g^3, \ldots, g^{p-1}\}$ consists of all the $p - 1$ nonzero residues modulo p. Hence, for any integer $a \not\equiv 0 \pmod{p}$, there exists an exponent $n \in \{1, \ldots, p - 1\}$ such that $g^n \equiv a \pmod{p}$. This idea is further developed in the well-known theory of indices (see, e.g., [Burton, pp. 165–169]).

Further, let us introduce the *Euler totient function* ϕ. For every $n \in \mathbb{N}$ the value $\phi(n)$ is defined as the number of all natural numbers not greater than n that are coprime to n, i.e.,

$$\phi(n) = |\{m \in \mathbb{N} : 1 \leq m \leq n,\ \gcd(m,n) = 1\}|,$$

where $|\cdot|$ denotes the number of elements. We can easily find that

$$\phi(1) = 1,\ \ \phi(2) = 1,\ \ \phi(3) = 2,\ \ \phi(4) = 2,\ \ \phi(5) = 4,\ \ \phi(6) = 2,\ \ \phi(7) = 6,\ \ldots$$

and that all other values of ϕ are even. If p is prime, then clearly

(2.12)
$$\phi(p) = p - 1$$

and

(2.13)
$$\phi\left(p^k\right) = (p-1)p^{k-1}$$

for every $k \in \mathbb{N}$. Another interesting property of the Euler totient function can be expressed as follows:

$$\gcd(m,n) = 1 \quad \Longrightarrow \quad \phi(mn) = \phi(m)\phi(n).$$

For a proof see [Gauss, article 38], [Burton, p. 125], or [Niven, Zuckerman, Montgomery, p. 69]. Hence, if the prime-power factorization of N is given by

$$N = \prod_{i=1}^{r} p_i^{k_i},$$

where $p_1 < p_2 < \cdots < p_r$, $k_i > 0$, then

(2.14)
$$\phi(N) = \prod_{i=1}^{r}(p_i - 1)p_i^{k_i-1} = N\left(1 - \frac{1}{p_1}\right)\left(1 - \frac{1}{p_2}\right)\cdots\left(1 - \frac{1}{p_r}\right).$$

Let us further recall Gauss's well-known formula

$$\sum_{d|N} \phi(d) = N \quad \forall N \in \mathbb{N}.$$

For a proof see, e.g., [Gauss, article 39], [Burton, p. 134], or [Narkiewicz, p. 17].

Theorem 2.16. *Let $N > 1$ be a positive integer. Then $\phi(N) < N - 1$ if and only if N is composite, and $\phi(N) = N - 1$ if and only if N is prime. In particular, the Fermat number F_m is prime if and only if*

(2.15)
$$\phi(F_m) = 2^{2^m}.$$

P r o o f . This follows easily from (1.1), (2.12), (2.13), and (2.14). (Cf. also [Vassilev-Missana].) $\square$

Theorem 2.17 (Euler). *Let $a, n \in \mathbb{N}$. Then*

$$(2.16) \qquad\qquad a^{\phi(n)} \equiv 1 \pmod{n}$$

if and only if $\gcd(a, n) = 1$.

The proof will directly follow from Carmichael's Theorem 2.22 and relation (2.17) (see also [Burton, p. 129] or [Sierpiński, 1950]). From (2.12) and (2.10) we see that Euler's Theorem 2.17 is a direct generalization of Fermat's Little Theorem 2.9 (compare with Figure 2.6). A further generalization of both Theorems 2.9 and 2.17 appears in [Laššák, Porubský].

For other connections of the Euler totient function with congruences see [Luca, 2000c], [Luca, Křížek].

Analogously to the situation for primes, we define primitive roots modulo n for any integer $n \geq 2$. By Euler's Theorem 2.17, the maximum possible order modulo n of any integer a coprime to n is equal to $\phi(n)$. If a is an integer such that $\gcd(a, n) = 1$, then a is defined to be a *primitive root modulo* n if

$$\operatorname{ord}_n a = \phi(n).$$

The next theorem determines all integers $n \geq 2$ that have primitive roots.

Theorem 2.18. *Let $n \geq 2$. There exists a primitive root modulo n if and only if $n \in \{2, 4, p^k, 2p^k\}$, where p is an odd prime and $k \geq 1$. Moreover, if n has a primitive root, then n has exactly $\phi(\phi(n))$ incongruent primitive roots.*

For a proof, see [Burton, pp. 160–164] or [Niven, Zuckerman, Montgomery, pp. 102–104]. Note that if p is prime, then p has exactly $\phi(p - 1)$ incongruent primitive roots (see Figure 2.7).

By Theorem 2.18, there exist primitive roots modulo p^k, where p is a prime and $k \geq 1$, except when $p = 2$ and $k \geq 3$. The following theorem states the maximum possible order modulo 2^k for $k \geq 3$.

Theorem 2.19. *Let a be an odd integer and $k \geq 3$. Then*

$$\operatorname{ord}_{2^k} a \mid 2^{k-2}$$

and

$$2^{k-2} < \phi(2^k) = 2^{k-1}.$$

In particular,

$$\operatorname{ord}_{2^k} 5 = 2^{k-2}.$$

For a proof see [Niven, Zuckerman, Montgomery, pp. 103–105].

By Euler's Theorem 2.17 and Lemma 2.13, if a and n are coprime positive integers, then $\operatorname{ord}_n a \mid \phi(n)$ and there exists an integer b such that $\operatorname{ord}_n b = \phi(n)$ for $n = 2$ or 4 or p^k, where p is an odd prime and $k \geq 1$. However, as we shall see below, the maximum possible order modulo n may be considerably smaller than $\phi(n)$ if n is not a prime power. To proceed, we first modify the Euler totient function $\phi(n)$ by means of the Carmichael lambda function $\lambda(n)$, which was first defined in 1912 (see [Carmichael, 1912]).

Definition 2.20. Let n be a positive integer. Then the *Carmichael lambda function* $\lambda(n)$ is defined as follows:

$$\lambda(1) = 1 = \phi(1),$$
$$\lambda(2) = 1 = \phi(2),$$
$$\lambda(4) = 2 = \phi(4),$$
$$\lambda(2^k) = 2^{k-2} = \tfrac{1}{2}\phi(2^k) \text{ for } k \geq 3,$$
$$\lambda(p^k) = (p-1)p^{k-1} = \phi(p^k) \text{ for any odd prime } p \text{ and } k \geq 1,$$
$$\lambda\big(p_1^{k_1} p_2^{k_2} \cdots p_r^{k_r}\big) = \mathrm{lcm}\big(\lambda\big(p_1^{k_1}\big),\ \lambda\big(p_2^{k_2}\big), \ldots,\ \lambda\big(p_r^{k_r}\big)\big),$$

where $p_1,\ p_2, \ldots,\ p_r$ are distinct primes and $k_i \geq 1$ for $1 \leq i \leq r$.

Remark 2.21. Note that by Theorems 2.18 and 2.19, if p is a prime and $k \geq 1$, then $\lambda\big(p^k\big)$ is equal to the maximum possible order modulo p^k. We also observe from the definition of $\lambda(n)$ that

$$(2.17) \qquad\qquad\qquad \lambda(n) \mid \phi(n)$$

for all n and that $\lambda(n) = \phi(n)$ if and only if $n \in \{1,\ 2,\ 4,\ q^k,\ 2q^k\}$, where q is an odd prime and $k \geq 1$. Notice that $\lambda(n)$ can be much smaller than $\phi(n)$ if n has many factors. For example, let $n = 2^6 \cdot 11 \cdot 17 \cdot 41 = 490688$. Then

$$\lambda(n) = \mathrm{lcm}\big(\lambda\big(2^6\big), \lambda(11), \lambda(17), \lambda(41)\big) = \mathrm{lcm}(16, 10, 16, 40) = 80,$$

whereas

$$\phi(n) = \phi\big(2^6\big)\phi(11)\phi(17)\phi(41) = 32 \cdot 10 \cdot 16 \cdot 40 = 204800.$$

The following theorem generalizes Euler's theorem. It shows that $\lambda(n)$ is a universal order modulo n.

Theorem 2.22 (Carmichael). *Let* $a, n \in \mathbb{N}$. *Then*

$$(2.18) \qquad\qquad\qquad a^{\lambda(n)} \equiv 1 \pmod{n}$$

if and only if $\gcd(a, n) = 1$. *Moreover, there exists an integer* b *such that*

$$(2.19) \qquad\qquad\qquad \mathrm{ord}_n b = \lambda(n).$$

P r o o f . If $\gcd(a, n) = d > 1$, then $d \nmid a^{\lambda(n)} - 1$, and thus (2.18) does not hold.

Conversely, let $\gcd(a, n) = 1$. Congruence (2.18) is clearly true when $n = 1$. Assume that $n \geq 2$. To show that congruence (2.18) is satisfied, it suffices by Lemma 2.13 to show that

$$(2.20) \qquad\qquad\qquad \mathrm{ord}_n a \mid \lambda(n).$$

Let the prime-power factorization of n be given by

$$(2.21) \qquad\qquad\qquad n = \prod_{i=1}^{r} p_i^{k_i},$$

where $p_1 < p_2 < \cdots < p_r$, $k_i > 0$, and let $Q_i = p_i^{k_i}$ for $1 \leq i \leq r$. Since powers of distinct primes are coprime, it follows that

(2.22) $$\mathrm{ord}_n a = \mathrm{lcm}\big(\mathrm{ord}_{Q_1} a, \ \mathrm{ord}_{Q_2} a, \ldots, \ \mathrm{ord}_{Q_r} a\big).$$

But

(2.23) $$\mathrm{ord}_{Q_i} a \mid \lambda(Q_i)$$

for $1 \leq i \leq r$ due to Remark 2.21 and Lemma 2.13. By the definition of $\lambda(n)$, (2.22), and (2.23), we see that (2.20) holds.

We now show that there exists an integer b such that (2.19) is satisfied. Consider the factorization of n given by (2.21). By the Chinese Remainder Theorem 2.7 and Theorems 2.18 and 2.19, we can find an integer b such that for $1 \leq i \leq r$,

(2.24) $$b \equiv g_i \pmod{p_i^{k_i}},$$

where g_1 is a primitive root modulo $p_1^{k_1}$ if p_1 is odd or $p_1 = 2$ and $1 \leq k_1 \leq 2$, $g_1 = 5$ if $p_1 = 2$ and $k_1 \geq 3$, and g_i is a primitive root modulo $p_i^{k_i}$ for $2 \leq i \leq r$. Since $\gcd\big(g_i, p_i^{k_i}\big) = 1$ for $1 \leq i \leq r$, it follows that $\gcd(b, n) = 1$. We claim that $\mathrm{ord}_n b = \lambda(n)$. By (2.24), Remark 2.21, and Theorem 2.19, we have

(2.25) $$\mathrm{ord}_{Q_i} b = \lambda(Q_i)$$

for $1 \leq i \leq r$. Since

$$\mathrm{ord}_n b = \mathrm{lcm}\big(\mathrm{ord}_{Q_1} b, \ \mathrm{ord}_{Q_2} b, \ldots, \ \mathrm{ord}_{Q_r} b\big),$$

we see by (2.25) and the definition of $\lambda(n)$ that $\mathrm{ord}_n b = \lambda(n)$. $\square$

An analogue of the following theorem was first proved by Dirichlet in 1837 (see [Dirichlet]). It represents the first sophisticated use of analytic techniques in number theory.

Theorem 2.23 (Dirichlet). *Let $d \geq 2$ and $a \neq 0$ be coprime integers. Then the arithmetic progression*

$$a, \ a + d, \ a + 2d, \ a + 3d, \ \ldots$$

contains infinitely many primes. Equivalently, the set

$$S = \{p \mid p \ \text{is prime and} \ p \equiv a \pmod{d}\}$$

has infinite cardinality. Moreover, the density of the set S in the set of primes is $1/\phi(d)$, where ϕ is the Euler totient function; that is,

$$\lim_{x \to \infty} \frac{|\{p \mid p \ \text{is prime}, \ p \equiv a \pmod{d}, \ \text{and} \ p \leq x\}|}{|\{p \mid p \ \text{is prime and} \ p \leq x\}|} = \frac{1}{\phi(d)}.$$

For a proof of Theorem 2.23 see [Ireland, Rosen, pp. 251–261].

We will now discuss the Legendre symbol and its properties. First we need to define quadratic residues and nonresidues modulo a prime.

Definition 2.24. Let $n \geq 2$ and a be integers such that $\gcd(a, n) = 1$. If the quadratic congruence

$$x^2 \equiv a \pmod{n}$$

has a solution x, then a is called a *quadratic residue modulo n*. Otherwise, a is called a *quadratic nonresidue modulo n*.

Definition 2.25. Let p be an odd prime. Then the *Legendre symbol* $\left(\frac{a}{p}\right)$ is defined to be 1 if a is a quadratic residue modulo p, -1 if a is a quadratic nonresidue modulo p, and 0 if $p \mid a$.

The quadratic character of a with respect to p is thus expressed via the Legendre symbol. It is clear that the sequence $\left\{\left(\frac{a}{p}\right)\right\}_{a=0}^{\infty}$ repeats periodically. For instance,

$$\left(\frac{0}{7}\right) = 0, \; \left(\frac{1}{7}\right) = 1, \; \left(\frac{2}{7}\right) = 1, \; \left(\frac{3}{7}\right) = -1, \; \left(\frac{4}{7}\right) = 1, \; \left(\frac{5}{7}\right) = -1, \; \left(\frac{6}{7}\right) = -1,$$

$$\left(\frac{7}{7}\right) = 0, \; \left(\frac{8}{7}\right) = 1, \; \left(\frac{9}{7}\right) = 1, \ldots .$$

If p is an odd prime, then the number of quadratic residues is equal to the number of quadratic nonresidues, which is equal to $(p-1)/2$ (for a proof see, e.g., [Riesel, 1985, p. 279]). In Figure 2.8 we see the values of the Legendre symbol for $p = 17$. Notice that this figure is the same as Figure 2.7. In Theorems 3.10 and 5.28 we will explain the reason why this is so only in the case of Fermat primes.

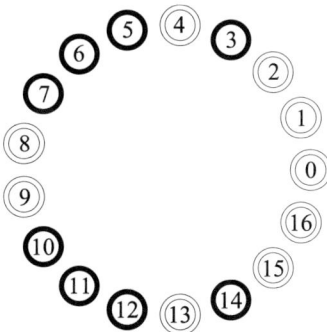

Figure 2.8. Values of the Legendre symbol $\left(\frac{a}{p}\right)$ for $p = 17$ and $a = 0, \ldots, 16$. The black circles stands for the value -1, whereas white stands for 0 or 1 (the only zero value of the Legendre symbol is at 0).

The following criterion, due to Euler, gives a method for evaluating the Legendre symbol $\left(\frac{a}{p}\right)$.

Theorem 2.26 (Euler's Criterion). *Let p be an odd prime. Then*

$$(2.26) \qquad\qquad a^{(p-1)/2} \equiv \left(\frac{a}{p}\right) \pmod{p}.$$

P r o o f . If $p \mid a$, then (2.26) is clearly satisfied. Now suppose that $p \nmid a$. By Fermat's Little Theorem 2.9,

$$(2.27) \qquad a^{p-1} - 1 = \left(a^{(p-1)/2} - 1\right)\left(a^{(p-1)/2} + 1\right) \equiv 0 \pmod{p}.$$

Since p is a prime, either $a^{(p-1)/2} \equiv 1 \pmod{p}$ or $a^{(p-1)/2} \equiv -1 \pmod{p}$. It thus suffices to prove that $a^{(p-1)/2} \equiv 1 \pmod{p}$ if and only if a is a quadratic residue modulo p.

Let a be a quadratic residue modulo p; i.e., $a \equiv b^2 \pmod{p}$ for some integer b such that $b \not\equiv 0 \pmod{p}$. Then by Fermat's little theorem,

$$a^{(p-1)/2} \equiv \left(b^2\right)^{(p-1)/2} \equiv b^{p-1} \equiv 1 \pmod{p}.$$

Conversely, assume that $a^{(p-1)/2} \equiv 1 \pmod{p}$. Let q be a primitive root modulo p. Then $a \equiv q^t \pmod{p}$ for some t such that $1 \le t \le p - 1$. Then

$$q^{t(p-1)/2} \equiv a^{(p-1)/2} \equiv 1 \pmod{p}.$$

However, $\operatorname{ord}_p q = p - 1$, and hence $p - 1 \mid t(p-1)/2$. This implies that $2 \mid t$, so let $t = 2j$. Then

$$(q^j)^2 = q^t \equiv a \pmod{p}$$

and $\left(\frac{a}{p}\right) = 1$. $\square$

Below are some basic properties of the Legendre symbol, which will be employed in further chapters.

Theorem 2.27. *Let p be an odd prime and let a and b be integers. Then*

(i) *if $a \equiv b \pmod{p}$, then $\left(\frac{a}{p}\right) = \left(\frac{b}{p}\right)$;*

(ii) $\left(\dfrac{ab}{p}\right) = \left(\dfrac{a}{p}\right)\left(\dfrac{b}{p}\right)$;

(iii) *if $\gcd(a, p) = 1$, then $\left(\dfrac{a^2}{p}\right) = 1$, $\left(\dfrac{a^2 b}{p}\right) = \left(\dfrac{b}{p}\right)$;*

(iv) $\left(\dfrac{1}{p}\right) = 1$, $\left(\dfrac{-1}{p}\right) = (-1)^{(p-1)/2}$;

(v) $\left(\dfrac{2}{p}\right) = (-1)^{(p^2-1)/8}$.

For proofs of Theorem 2.27, see [Burton, pp. 178–183] or [Niven, Zuckerman, Montgomery, pp. 132–135].

The celebrated law of quadratic reciprocity first proved by Gauss is invaluable in calculating the Legendre symbol $\left(\frac{p}{q}\right)$ when p and q are both odd primes (see [Gauss, articles 135–144]).

Theorem 2.28 (Law of Quadratic Reciprocity). *If p and q are both odd primes, then*

$$\left(\frac{p}{q}\right)\left(\frac{q}{p}\right) = (-1)^{\frac{p-1}{2}\frac{q-1}{2}}.$$

For proofs of the law of quadratic reciprocity see [Burton, pp. 188-190], [Koblitz, pp. 45–46], or [Niven, Zuckerman, Montgomery, pp. 137–138].

Theorems 2.26–2.28 enable us to substantially simplify computation of the Legendre symbol. For instance, in the case of the prime number $p = 1999$ we have

$$\left(\frac{111}{1999}\right) = \left(\frac{3 \cdot 37}{1999}\right) = \left(\frac{3}{1999}\right)\left(\frac{37}{1999}\right) = (-1)^{999}\left(\frac{1999}{3}\right) \times (-1)^{18 \cdot 999}\left(\frac{1999}{37}\right)$$
$$= (-1)\left(\frac{1}{3}\right)\left(\frac{1}{37}\right) = -1.$$

The Legendre symbol $\left(\frac{a}{p}\right)$, where p is prime, can be generalized by means of the Jacobi symbol, which is defined as follows.

Definition 2.29. Let a be an integer and let $n \geq 3$ be an odd integer. Let $n = p_1 p_2 \cdots p_r$, where the p_i's are odd primes, not necessarily distinct. Then the *Jacobi symbol* $\left(\frac{a}{n}\right)$ is defined by

$$\left(\frac{a}{n}\right) = \prod_{i=1}^{r}\left(\frac{a}{p_i}\right),$$

where $\left(\frac{a}{p_i}\right)$ is the Legendre symbol.

The properties of the Jacobi symbol are similar to those of the Legendre symbol and are given below. In particular, a version of the law of quadratic reciprocity also holds for the Jacobi symbol.

Theorem 2.30. *Let $m > 1$ and $n > 1$ be odd integers and let a and b be integers. Then*

(i) *if $a \equiv b \pmod{n}$, then $\left(\frac{a}{n}\right) = \left(\frac{b}{n}\right)$;*

(ii) $\left(\frac{ab}{n}\right) = \left(\frac{a}{n}\right)\left(\frac{b}{n}\right)$;

(iii) $\left(\frac{a}{mn}\right) = \left(\frac{a}{m}\right)\left(\frac{a}{n}\right)$;

(iv) *if $\gcd(a, n) = 1$, then $\left(\frac{a^2}{n}\right) = \left(\frac{a}{n^2}\right) = 1$;*

(v) *if $\gcd(ab, mn) = 1$, then $\left(\frac{a^2 b}{m^2 n}\right) = \left(\frac{b}{n}\right)$;*

(vi) $\left(\frac{1}{n}\right) = 1$, $\left(\frac{-1}{n}\right) = (-1)^{(n-1)/2}$;

(vii) $\left(\frac{2}{n}\right) = (-1)^{(n^2-1)/8}$;

(viii) $\left(\frac{m}{n}\right)\left(\frac{n}{m}\right) = (-1)^{\frac{m-1}{2}\frac{n-1}{2}}$.

The proof of the above theorem is based on the properties of the Legendre symbol given in Theorem 2.27. It can be found in [Niven, Zuckerman, Montgomery, pp. 143–146].

3. Basic Properties of Fermat Numbers

First we present a few recurrence formulae for the Fermat numbers. Most of these can be found in the paper [Grytczuk] (see also [Schram]).

Proposition 3.1. *The formula*

$$(3.1) \qquad F_{m+1} = F_m + 2^{2^m} F_0 F_1 \cdots F_{m-1}$$

holds for all $m \geq 1$.

P r o o f . We have

$$(3.2) \qquad F_{m+1} - F_m = 2^{2^{m+1}} + 1 - 2^{2^m} - 1 = 2^{2^m} \left(2^{2^m} - 1 \right).$$

However,

$$2^{2^m} - 1 = \left(2^{2^{m-1}} - 1 \right)\left(2^{2^{m-1}} + 1 \right) = \left(2^{2^{m-1}} - 1 \right) F_{m-1}.$$

By induction, we get easily that

$$2^{2^m} - 1 = \left(2^{2^{m-k}} - 1 \right) F_{m-k} F_{m-k+1} \cdots F_{m-1} \qquad \text{for all } k \in \{1, \dots, m\}.$$

In particular, when $k = m$, we get

$$(3.3) \qquad 2^{2^m} - 1 = F_0 F_1 \cdots F_{m-1}.$$

Now formula (3.1) is an immediate consequence of formulae (3.2) and (3.3). $\qquad \square$

Proposition 3.2. *The formula*

$$(3.4) \qquad F_m = F_0 F_1 \cdots F_{m-1} + 2$$

holds for all $m \geq 1$.

P r o o f . For $m \geq 1$, we use formula (3.3) and get

$$F_m - 2 = \left(2^{2^m} + 1\right) - 2 = 2^{2^m} - 1 = F_0 F_1 \cdots F_{m-1},$$

which is exactly formula (3.4). □

Notice that Proposition 3.2 implies that $F_m - 2$ for $m \geq 1$ is divisible by all the smaller Fermat numbers; that is,

$$F_k \mid F_m - 2 \quad \text{for all } k = 0, 1, \ldots, m - 1,$$

or equivalently,

(3.5) $$F_m \equiv 2 \pmod{F_k} \quad \text{for all } k = 0, 1, \ldots, m - 1.$$

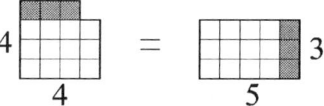

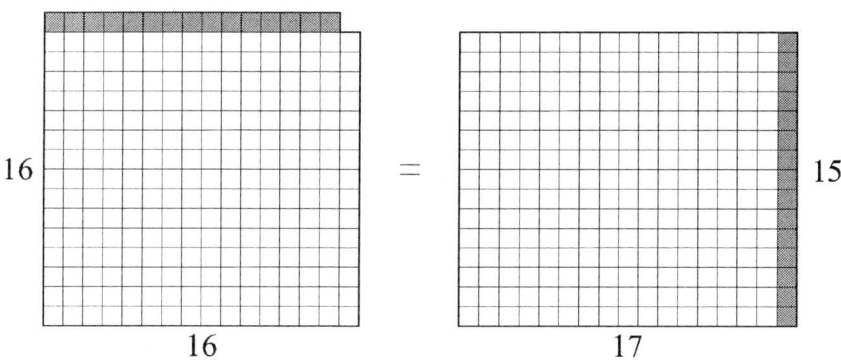

Figure 3.1. Geometric interpretation of the relation $F_m - 2 = F_0 F_1 \cdots F_{m-1}$ for $m = 2$ and $m = 3$.

We also record the following recurrence relations.

Proposition 3.3. *The formula*

(3.6) $$F_m = (F_{m-1} - 1)^2 + 1$$

holds for all $m \geq 1$. The formula

(3.7) $$F_m = F_{m-1}^2 - 2(F_{m-2} - 1)^2$$

holds for all $m \geq 2$.

P r o o f . For formula (3.6), we notice that

$$\left(F_{m-1} - 1\right)^2 + 1 = \left(2^{2^{m-1}}\right)^2 + 1 = 2^{2^m} + 1 = F_m.$$

For formula (3.7), we observe that by formula (3.6) one has

$$(F_{m-2} - 1)^2 = F_{m-1} - 1.$$

Hence,

$$
\begin{aligned}
(3.8) \qquad F_{m-1}^2 - 2(F_{m-2} - 1)^2 &= F_{m-1}^2 - 2(F_{m-1} - 1) \\
&= F_{m-1}^2 - 2F_{m-1} + 2 \\
&= (F_{m-1}^2 - 2F_{m-1} + 1) + 1 \\
&= (F_{m-1} - 1)^2 + 1 = F_m.
\end{aligned}
$$

For the last equality in (3.8) above we used formula (3.6). $\quad\square$

Using formula (3.7), we obtain the following result (see [Grytczuk]).

Proposition 3.4. *Each Fermat number F_m, where $m \geq 2$, has infinitely many representations of the form $x^2 - 2y^2$, with x and y positive integers.*

P r o o f . One such representation is given by formula (3.7), where we may take $x = F_{m-1}$ and $y = F_{m-2} - 1$. To get infinitely many such representations, notice that

$$(3x + 4y)^2 - 2(2x + 3y)^2 = x^2 - 2y^2$$

is an identity. Since $3x + 4y > x$ and $2x + 3y > y$ when both x and y are positive integers, we can construct a different representation $(3x + 4y, 2x + 3y)$ starting with the pair (x, y). The truth of the assertion of Proposition 3.4 is now clear. $\quad\square$

We can easily deduce the following result concerning the divisors of the number $H_m = 2^{2^{m+1}} + 2^{2^m} + 1$. The short proof of this result is due to A. Mąkowski.

Proposition 3.5. *The formula*

$$(3.9) \qquad H_m = \left(F_{m-1}^2 - 3F_{m-1} + 3\right) H_{m-1}$$

holds for all $m \geq 1$.

P r o o f . It is enough to use the formula

$$
\begin{aligned}
[(x + 1)^2 - 3(x + 1) + 3]\left(x^2 + x + 1\right) &= \left(x^2 - x + 1\right)\left(x^2 + x + 1\right) \\
&= \left(x^2 + 1\right)^2 - x^2 = x^4 + x^2 + 1,
\end{aligned}
$$

and substitute $x = 2^{2^{m-1}}$. $\quad\square$

Remark 3.6. From (3.5) we see that $F_m \equiv 2 \pmod 5$ for $m \geq 2$. Since Fermat numbers are odd, this immediately implies that

$$(3.10) \qquad F_m \equiv 7 \pmod{10} \quad \text{for } m \geq 2,$$

and hence the last digit of F_m is always 7 for $m \geq 2$. The last two digits of F_m can be only 17, 37, 57, or 97 for $m \geq 2$.

Remark 3.7. The number of digits of a Fermat number $F_m = 2^{2^m} + 1$ is (compare with Appendix A)

$$D(m) = \left\lfloor \log_{10}\left(2^{2^m} + 1\right) + 1 \right\rfloor \approx \left\lfloor \log_{10} 2^{2^m} + 1 \right\rfloor = \left\lfloor 2^m \log_{10} 2 + 1 \right\rfloor.$$

The following proposition can be found in [Sierpiński, 1970, Problem 213 b)].

Proposition 3.8. *No Fermat number F_m for $m > 1$ can be expressed as the sum of two primes.*

P r o o f . The numbers F_m for $m > 1$ are odd. If F_m were the sum of two primes, then one of them must be 2 and the second one $F_m - 2$. However,

$$F_m - 2 = 2^{2^m} - 1 = \left(2^{2^{m-1}} - 1\right)\left(2^{2^{m-1}} + 1\right)$$

is composite for $m > 1$, since $2^{2^{m-1}} - 1 \geq 3$.   □

The next theorem can be found in [Vasilenko].

Theorem 3.9. (i) *For $m \geq 2$ we have*

$$F_m^{(F_{m+1}-1)/2} \equiv 1 \pmod{F_{m+1}}.$$

(ii) *If $m \geq 2$, $k \geq 1$, and F_{m+k} is prime, then*

$$F_m^{(F_{m+k}-1)/2} \equiv 1 \pmod{F_{m+k}}.$$

P r o o f . (i) We first note that

$$F_m^2 = \left(2^{2^m} + 1\right)^2 = 2^{2^{m+1}} + 1 + 2 \cdot 2^{2^m} \equiv 2 \cdot 2^{2^m} \pmod{F_{m+1}}.$$

Moreover,

$$F_m^{2^2} \equiv \left(2 \cdot 2^{2^m}\right)^2 \equiv 4 \cdot 2^{2^{m+1}} \equiv 4(F_{m+1} - 1) \equiv -2^2 \pmod{F_{m+1}}.$$

Therefore, for $k \geq 3$,

$$(3.11) \qquad F_m^{2^k} = \left(F_m^{2^2}\right)^{2^{k-2}} \equiv \left(-2^2\right)^{2^{k-2}} \equiv 2^{2^{k-1}} \pmod{F_{m+1}}.$$

Put $k = 2^{m+1} - 1$ in (3.11). Then

$$(3.12) \qquad F_m^{(F_{m+1}-1)/2} \equiv 2^{2^{2^{m+1}-2}} \pmod{F_{m+1}}.$$

The number $2^{2^{m+1}-2}$ for $m \geq 2$ is divisible by 2^{m+2}. From this and (3.12) we see that there exists a positive integer M such that

$$F_m^{(F_{m+1}-1)/2} \equiv \left(2^{2^{m+1}}\right)^{2M} \equiv (-1)^{2M} \equiv 1 \pmod{F_{m+1}}.$$

(ii) By Euler's criterion (see Theorem 2.26)

$$F_m^{(F_{m+k}-1)/2} \equiv \left(\frac{F_m}{F_{m+k}}\right) \quad (\bmod\ F_{m+k}).$$

By (3.5), $F_{m+k} \equiv 2 \pmod{F_m}$. Therefore, by Theorem 2.30,

$$\left(\frac{F_m}{F_{m+k}}\right) = \left(\frac{F_{m+k}}{F_m}\right) = \left(\frac{2}{F_m}\right) = 1. \qquad \square$$

Theorem 3.10. *The set of all quadratic nonresidues of a Fermat prime is equal to the set of all its primitive roots.*

P r o o f. Let n be a quadratic nonresidue of the Fermat prime F_m. Then, by Theorem 2.26 (Euler's criterion),

(3.13) $$n^{(F_m-1)/2} = n^{2^{2^m-1}} \equiv -1 \pmod{F_m}.$$

However, according to Fermat's little theorem and Lemma 2.13,

$$\mathrm{ord}_{F_m} n \mid F_m - 1 = 2^{2^m}.$$

Suppose that

$$\mathrm{ord}_{F_m} n = 2^k,$$

where $0 \le k < 2^m$. Then

$$\mathrm{ord}_{F_m} n \mid (F_m - 1)/2.$$

Hence,

$$n^{(F_m-1)/2} \equiv 1 \pmod{F_m},$$

contradicting congruence (3.13). Therefore, $\mathrm{ord}_{F_m} n = F_m - 1$, and n is a primitive root modulo F_m.

Conversely, by Theorem 2.26 (Euler's criterion), a quadratic residue r modulo a prime p cannot be a primitive root modulo p, since $r^{(p-1)/2} \equiv 1 \pmod{p}$. $\qquad \square$

Theorem 3.10 explains why Figures 2.7 and 2.8 illustrate the same situation. In Chapter 5 we will explore this phenomenon further.

The theorem below appears in [Racliş].

Theorem 3.11 (Racliş). *Let F_m be a prime and a an integer. Then every prime Fermat number less than or equal to F_m divides $a^{F_m} - a$.*

P r o o f . Let F_k be any prime Fermat number for which $0 \le k \le m$. First suppose that $F_k \mid a$. Then clearly,

$$F_k \mid a^{F_m} - a = a(a^{F_m-1} - 1).$$

Now assume that $F_k \nmid a$. By Fermat's little theorem,

(3.14) $$F_k \mid a^{F_k-1} - 1.$$

However,

$$F_k - 1 \mid F_m - 1.$$

Therefore,

(3.15) $$a^{F_k - 1} - 1 \mid a^{F_m - 1} - 1.$$

Hence, by (3.14) and (3.15),

$$F_k \mid a^{F_m - 1} - 1 \mid a^{F_m} - a. \qquad \square$$

The following theorem is mentioned in [Kiss].

Theorem 3.12 (Bolyai). *Every Fermat number F_m for $m \geq 1$ is of the form $6n - 1$.*

P r o o f . We have to show that $6 \mid F_m + 1$. Indeed, by (3.4),

$$F_m + 1 = F_0 F_1 \cdots F_{m-1} + 3 = 3(F_1 \cdots F_{m-1} + 1),$$

where the number in parentheses is evidently even. $\square$

By Theorem 3.12, no Fermat number F_m is divisible by 3 for $m \geq 1$. In Theorem 4.1 we prove a stronger result.

Remark 3.13. Any odd number can be written as the difference of two squares, $2n + 1 = (n + 1)^2 - n^2$. Hence, any Fermat number can be written as the difference of two squares, namely

$$F_m = 2^{2^m} + 1 = \left(2^{2^m - 1} + 1\right)^2 - \left(2^{2^m - 1}\right)^2.$$

There are other possible ways of expressing F_m as a difference of two squares if it is composite (compare with (1.6)).

Theorem 3.14. *No Fermat prime can be expressed as the difference of two pth powers, where p is an odd prime.*

P r o o f . Suppose that p is an odd prime and let

(3.16) $$F_m = a^p - b^p = (a - b)\left(a^{p-1} + a^{p-2}b + \cdots + ab^{p-2} + b^{p-1}\right),$$

where $a > b$. Since F_m is prime, we see that $a - b = 1$. By Fermat's Little Theorem 2.9,

$$F_m = a^p - b^p \equiv a - b \equiv 1 \pmod{p}.$$

Thus,

$$p \mid F_m - 1 = 2^{2^m},$$

which is impossible. Consequently, F_m cannot be represented as a difference of two pth powers. $\square$

Remark 3.15. Theorem 3.14 was posed as problem 18 in [Burton, p. 222] for the case $p = 3$.

Remark 3.16. According to a well-known theorem of Sophie Germain, every number of the form $a^4 + 4$ is composite if $a > 1$. Indeed,

$$a^4 + 4 = \left(a^2 + 2\right)^2 - 4a^2 = \left(a^2 + 2 + 2a\right)\left(a^2 + 2 - 2a\right).$$

Setting $a = 2^{2^{m-2}}$ for $m \geq 2$, we find that $a^2 + 1 = F_{m-1}$, $a^4 + 1 = F_m$, and therefore,

$$F_m + 3 = \left(F_{m-1} + 2F_{m-2} - 1\right)\left(F_{m-1} - 2F_{m-2} + 3\right),$$

where both factors are even and greater than 2 for $m > 2$.

Remark 3.17. Every Fermat number F_m in the binary system has the form $1000 \ldots 0001$ with $2^m - 1$ zeros inside (cf. [Leyendekkers, Shannon]).

Remark 3.18. Basic properties of the Fermat numbers are also surveyed, for example, in [Arya, 1989, 1990], [Dilcher], [Keller, 1991, 1992], [Křížek], [Maruyama, Kawatani], [Sierpiński, 1955], and in many standard books on number theory such as [Burton], [Conway, Guy], [Dickson], [Gauss], [Guy, 1994], [Hardy, Wright], [Klein], [Narkiewicz], [Niven, Zuckerman, Montgomery], [Rademacher], [Reid], [Ribenboim, 1988, 1989], [Riesel, 1994], [Robbins], [Scharlau, Opolka], [Schroeder], [Shanks, 1962, 1978, 1985], [Sierpiński, 1950, 1964b, 1970, 1988], [Stewart, I., 1973, 1989], [Weil], [Williams, 1998].

4. The Most Beautiful Theorems on Fermat Numbers

Whenever there is a number, there is a beauty.

Morris Kline

A power of two increased by one can divide another power of two increased by one, e.g., $2 + 1 \mid 8 + 1$ or $4 + 1 \mid 64 + 1$. However, Euler's contemporary Christian Goldbach proved that the Fermat numbers are pairwise coprime.

Theorem 4.1 (Goldbach). *No two different Fermat numbers have a common divisor greater than 1.*

P r o o f . Assume that

$$(4.1) \qquad\qquad q \mid F_m \quad \text{and} \quad q \mid F_{m-k}$$

for $m \geq k \geq 1$. From this and (3.5) we obtain $q \mid F_m - 2$. Hence, according to (4.1), we see that $q \mid 2$. However, since F_m is odd, we get $q = 1$. □

Remark 4.2. For another proof based on congruences, see [Robbins, p. 92]. According to Goldbach's Theorem 4.1, every Fermat number from the sequence $F_0, F_1, \ldots, F_m$ is divisible by a prime that does not divide the other Fermat numbers. Thus, there exist at least $m + 1$ different primes not greater than F_m. From this it follows, moreover, that there are infinitely many prime numbers (Euclid's Theorem 2.1 on the infinitude of primes). This proof appears in [Hardy, Wright, p. 14].

According to Goldbach's theorem and (3.4), the number $F_m - 2$ has at least m distinct nontrivial mutually coprime divisors.

There is a beautiful connection between number theory and geometry (see also [Hilton, Pedersen]).

Theorem 4.3 (Gauss). *There exists a Euclidean construction of a regular polygon (i.e., by ruler and compass) if and only if the number of its sides is equal to $n = 2^i p_1 p_2 \cdots p_j$, where $i \geq 0$, $j \geq 0$, $n \geq 3$ are integers and $p_1, p_2, \ldots, p_j$ are distinct Fermat primes.*

The proof, which is based on Galois theory, will be given in Chapter 16.

For instance, the choice $i = 2$, $j = 2$, $p_1 = 3$, and $p_2 = 5$ in Gauss's Theorem 4.3 leads to the regular 60-gon. Gauss's construction of the regular heptadecagon corresponds to $i = 0$, $j = 1$, $p_1 = 17$ (this problem is discussed in more detail in Chapter 17).

Remark 4.4. Up to now, we know exactly five Fermat primes. Hence, by Gauss's Theorem 4.3 there exist at least 31 regular n-gons with an odd number of sides that can be constructed by ruler and compass, since

$$\binom{5}{1} + \binom{5}{2} + \binom{5}{3} + \binom{5}{4} + \binom{5}{5} = 2^5 - 1 = 31.$$

The numbers of their sides are

(4.2)   3, 5, 15, 17, 51, 85, 255, 257,... .

The only five known constructible regular polygons whose number of sides is a prime are illustrated in Figure 4.1.

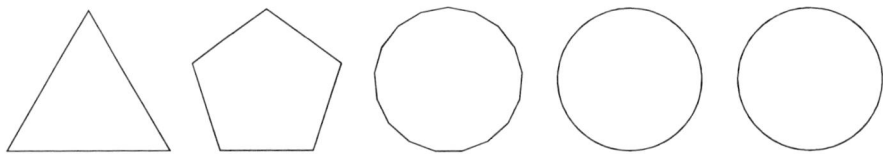

Figure 4.1. The regular triangle, pentagon, heptadecagon, 257-gon and 65537-gon.

Theorem 4.5. *The regular polygon with $n \geq 3$ sides can be constructed with ruler and compass if and only if there exists an integer $q \geq 1$ such that*

$$\phi(n) = 2^q.$$

P r o o f . Suppose that the regular polygon with $n \geq 3$ sides has a Euclidean construction. Then by Gauss's Theorem 4.3,

$$n = 2^i F_{m_1} F_{m_2} \cdots F_{m_j},$$

where $i \geq 0$, $j \geq 0$, F_{m_s} is prime for $s = 1,\ldots,j$ and $0 \leq m_1 < m_2 < \cdots < m_j$. Then by (2.13) and (2.14),

$$\phi(n) = \phi(2^i)\phi(F_{m_1})\phi(F_{m_2}) \cdots \phi(F_{m_j}) = 2^a 2^{2^{m_1}} 2^{2^{m_2}} \cdots 2^{2^{m_j}},$$

where $a = 0$ if $i = 0$ and $a = i - 1$ if $i \geq 1$. Then clearly, $\phi(n) = 2^q \geq 2$ for $q = a + 2^{m_1} + 2^{m_2} + \cdots + 2^{m_j}$.

Conversely, assume that $\phi(n) = 2^q$ for some $q \geq 1$ and consider the prime-power factorization given by

$$n = 2^i p_1^{k_1} p_2^{k_2} \cdots p_j^{k_j},$$

where $n \geq 3$, $i \geq 0$, $j \geq 0$, the p_s's are distinct odd primes for $s = 1,\ldots,j$, and $k_s \geq 1$ for $s = 1,\ldots,j$. Then by (2.13) and (2.14),

(4.3) $$\phi(n) = 2^a (p_1 - 1)p_1^{k_1-1} \cdots (p_j - 1)p_j^{k_j-1},$$

where $a = 0$ if $i = 0$ and $a = i - 1$ if $i \geq 1$. Let us suppose that $k_s \geq 2$ for some $s \in \{1, \ldots, j\}$. Then by (4.3), $p_s \mid \phi(n)$, which contradicts the assumption that $\phi(n) = 2^q$. Thus, $k_s = 1$ for $s = 1, 2, \ldots, j$. It now follows from (4.3) that

$$p_s - 1 = 2^{n_s},$$

where $n_s \geq 1$ for $s = 1, \ldots, j$. We claim that $n_s = 2^{m_s}$ for some $m_s \geq 0$.

Suppose to the contrary that there exists $s \in \{1, \ldots, j\}$ such that $n_s = k\ell$ for some odd prime ℓ and some positive integer k. Then by (1.3), $2^{k\ell} + 1$ is composite which contradicts the fact that p_s is prime. Therefore, $n_s = 2^{m_s}$, where $m_s \geq 0$, and $p_s = F_{m_s}$ is a Fermat prime for $s = 1, \ldots, j$. $\square$

Remark 4.6. We demonstrate now another remarkable connection between number theory and geometry. Write out Pascal's triangle modulo 2 (compare with Figure 8.1):

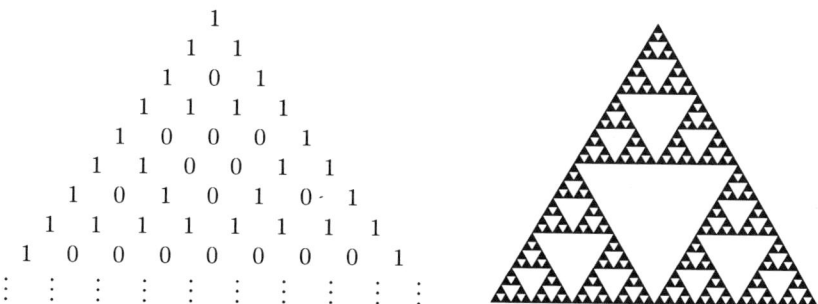

```
                1
              1   1
            1   0   1
          1   1   1   1
        1   0   0   0   1
      1   1   0   0   1   1
    1   0   1   0   1   0 · 1
  1   1   1   1   1   1   1   1
1   0   0   0   0   0   0   0   1
:   :   :   :   :   :   :   :   :   :
```

Figure 4.2. Pascal's triangle modulo 2.

Then by reading the first 32 rows as the binary expansions of numbers, we get the monotonically increasing sequence

$$1, \ 3, \ 5, \ 15, \ 17, \ 51, \ 85, \ 255, \ 257, \ldots,$$

since each row in the above triangle starts with 1 and is one bit longer than the previous row. Notice that this is exactly the same sequence as that in (4.2) (except for the first term), which gives all odd-sided constructible regular polygons in the first 31 rows. Is this not a small miracle? This interesting property of the Fermat numbers was found by William Watkins (see [Conway, Guy, p. 140] and also [Gardner, p. 207]). Its proof is due to [Hewgill]; see Theorem 8.1. Notice also that Figure 4.2 has a similar structure as the famous Sierpiński fractal set.

According to [Rosen], Gauss also discovered how to divide the lemniscate into five parts of equal length with ruler and compass (i.e., how to construct the division points). This result was later generalized by Niels Henrik Abel.

Theorem 4.7 (Abel). *The lemniscate can be divided into n equal parts with ruler and compass if $n = 2^i p_1 p_2 \cdots p_j$, where $i \geq 0$ and $j \geq 0$ are integers and $p_1, p_2, \ldots, p_j$ are distinct Fermat primes.*

For a proof see [Rosen]. Note that the lemniscate (of Bernoulli) is a curve whose points have a constant product (equal to $\alpha^2/2$) of their distances from two fixed points $(\pm(\alpha/2)\sqrt{2}, 0)$, where $\alpha > 0$ is a real parameter ($\alpha = \sqrt{2}$ in Figure 4.3).

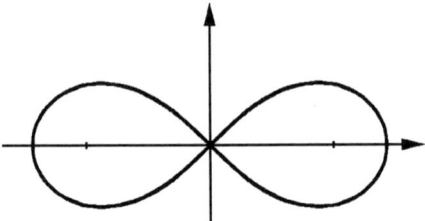

Figure 4.3. Lemniscate $\left(x^2 + y^2\right)^2 = \alpha^2\left(x^2 - y^2\right)$.

According to Fermat's Little Theorem 2.9, $p \mid a^p - a$ for any prime p. It is easy to show that the converse assertion is not true. To see this, we take for instance $a = 2$ and $n = 341$. By inspection, $2^{10} \equiv 1 \pmod{341}$, and thus $2^{340} \equiv 1 \pmod{341}$. Multiplying the last congruence by 2, we find that $341 \mid 2^{341} - 2$. Hence, the converse of Fermat's little theorem (compare with [Robinson, 1957b]) does not hold, since $n = 341 = 31 \cdot 11$ is composite. There are only two more composite numbers with three digits, $n = 561$ and $n = 645$, for which $n \mid 2^n - 2$ (more details can be found in Chapter 12).

Definition 4.8. A composite number n such that

$$n \mid a^n - a$$

is called a *pseudoprime* to the base a.

Lemma 4.9. For any $n \le 2^m - 1$ we have

$$F_n \mid 2^{F_m} - 2.$$

P r o o f . We see that

$$2^{F_m} - 2 = 2\left(2^{2^{2^m}} - 1\right) = 2\left(2^{2^{2^m}} + 1 - 2\right) = 2(F_{2^m} - 2) = 2F_0 F_1 F_2 \cdots F_{2^m - 1},$$

where the last equality follows from recurrence (3.4).   □

Theorem 4.10. *All Fermat numbers are primes or pseudoprimes to the base 2. Moreover, if $2^n + 1$ is a pseudoprime to the base 2, then n is a power of 2.*

P r o o f . Since $m \le 2^m - 1$, we obtain from Lemma 4.9 that

(4.4) $F_m \mid 2^{F_m} - 2.$

Hence, each F_m is a prime or pseudoprime to the base 2 by Definition 4.8.

Suppose now that n is not a power of 2 and that $2^n + 1$ is a pseudoprime to the base 2. Then $2^n + 1$ is composite and

(4.5) $2^{2^n} \equiv 1 \pmod{2^n + 1},$

since $\gcd(2, 2^n + 1) = 1$. Note that

$$2^n \equiv -1 \pmod{2^n + 1}$$

and that $2^n < 2^n + 1$. Let $e = \operatorname{ord}_{2^n+1} 2$. Then $e \geq n + 1$, and by Lemma 2.13, if $2^t \equiv 1 \pmod{2^n + 1}$, then $e \mid t$. Since

$$2^{2n} \equiv (-1)^2 \equiv 1 \pmod{2^n + 1},$$

it follows that $e = 2n$. Therefore, by (4.5) and Lemma 2.13, we have $2n \mid 2^n$. However, this is impossible, since $2n$ is not a power of 2. □

Remark 4.11. According to T. Banachiewicz (see [Sierpiński, 1970, Problem 8]), Fermat probably knew relation (4.4) and believed that this relation implies that all F_m are prime.

According to [Kiss, p. 72], János Bolyai (1802–1860), one of the founders of non-Euclidean geometry, was the first who showed that F_5 is a pseudoprime to the base 2. His proof was based on the observation that

$$2^{2^{32}} \equiv 1 \pmod{F_5},$$

where $F_5 = 2^{32} + 1 = 641 \cdot 6700417$.

Theorem 4.12. Let $q = p^n$ be a power of an odd prime p, where $n \geq 1$. Then the Fermat number F_m is divisible by q if and only if

$$\operatorname{ord}_q 2 = 2^{m+1}.$$

P r o o f . Suppose that $q \mid F_m$. Then q obviously also divides the number $2^{2^{m+1}} - 1 = \left(2^{2^m} + 1\right)\left(2^{2^m} - 1\right)$. From this and Lemma 2.13 for $a = 2$ it follows that $\operatorname{ord}_q 2 \leq 2^{m+1}$ and $2^{m+1} = k \operatorname{ord}_q 2$ for some $k \in \mathbb{N}$. Thus, k has to be a power of 2 and $e = \operatorname{ord}_q 2 = 2^j$. However, if j were less than $m + 1$, then by Lemma 2.13, the number

$$2^{e2^{m-j}} - 1 = 2^{2^m} - 1$$

would be divisible by the odd number $q > 1$, which contradicts the assumption $q \mid F_m$. Therefore, $\operatorname{ord}_q 2 \geq 2^{m+1}$.

Conversely, assume that $\operatorname{ord}_q 2 = 2^{m+1}$. Then

$$q \mid 2^{2^{m+1}} - 1 = \left(2^{2^m} + 1\right)\left(2^{2^m} - 1\right).$$

Since p is odd and $p > 1$, the prime p, and thus also the prime power q, divides at most one of the numbers $2^m + 1$ and $2^m - 1$. But $q \nmid 2^{2^m} - 1$, since then $\operatorname{ord}_q 2 \leq 2^m$. Thus,

$$q \mid 2^{2^m} + 1 = F_m. \qquad \square$$

Remark 4.13. We get by completely similar arguments to those in the proof of Theorem 4.12 that if $d > 1$ is any divisor of F_m (not necessarily prime; see also [Keller, 1991]), then

$$\operatorname{ord}_d 2 = 2^{m+1},$$

where the symbol $\mathrm{ord}_d 2$ is defined in the same way as in Definition 2.12. In particular,

$$\mathrm{ord}_{F_m} 2 = 2^{m+1}.$$

From Theorem 4.12 and the factorizations of Fermat numbers given in Appendix A, we know that for each $m \in \{5, 6, 7, 8\}$ there are exactly two primes p for which the number 2 has the order 2^{m+1} modulo p.

Since $2^{m+1} < 2^{2^m}$ for $m > 1$, any primitive root of any prime divisor $d > 1$ of a Fermat number cannot be a power of 2 for $m > 1$.

Theorem 4.14 (Euler). *If p is a prime and $p \mid F_m$, then p is of the form $p = k2^{m+1} + 1$, where k is a natural number.*

P r o o f . From Fermat's Little Theorem 2.9 for $a = 2$ and from Lemma 2.13 and Theorem 4.12 we have $p - 1 = k \, \mathrm{ord}_p 2 = k2^{m+1}$.   □

Remark 4.15. As mentioned in the literature (cf., e.g., [Lenstra, Lenstra, Manasse, Pollard]), Fermat probably knew the previous theorem, although its proof was given later by Euler in 1747 (see [Burton, p. 219]). However, it is then unclear why Fermat did not apply this theorem to the factorization of F_5. It was sufficient to divide F_5 only by primes of the form $64k + 1$, and for $k = 10$ Fermat could have convinced himself that his hypothesis on the primality of F_m did not hold.

Concerning the factor $2142 \cdot 2^7 + 1 = 274177$ of F_6 (see Figure 1.2), we note that there are 370 primes of the form $k2^7 + 1$ for $k \leq 2142$.

Remark 4.16. We note that Goldbach's Theorem 4.1 and Euler's Theorem 4.14 together yield an immediate proof of a special case of Dirichlet's Theorem 2.23 that for any $N \geq 2$, there exist infinitely many primes of the form $2^N k + 1$. In particular, there are infinitely many primes of the form $4k + 1$.

It is of interest that one can show algebraically that $641 \mid F_5$ without resort to explicit divisions or examination of powers of 2 modulo 641. The following demonstration is due to G. Bennett (see [Hardy, Wright, 1945, pp. 14–15] or [Burton, p. 216]) and would undoubtedly have been appreciated by Fermat.

Proposition 4.17. *F_5 is divisible by 641.*

P r o o f . Put $a = 2^7$ and $b = 5$. Hence,

$$ab + 1 = 2^7 \cdot 5 + 1 = 641.$$

Then

$$1 + ab - b^4 = 1 + (a - b^3)b = 1 + 3b = 2^4.$$

But this implies that

$$(4.6) \qquad F_5 = 2^{32} + 1 = 2^4 a^4 + 1 = (1 + ab - b^4)a^4 + 1$$
$$= (1 + ab)a^4 + (1 - a^4 b^4)$$
$$= (1 + ab)\big(a^4 + (1 - ab)(1 + a^2 b^2)\big),$$

which yields $641 \mid F_5$.   □

Another sophisticated proof of Proposition 4.17 is due to [Kraïtchik, 1926, p. 221] (see also [Chang], [Hardy, Wright, 1945, p. 22], [Hardy, Wright, 1954, 1960, pp. 14–15]): Since $641 = 2^4 + 5^4$, we have

$$2^{32} = 2^4 2^{28} = (641 - 5^4)2^{28} = 641i - (5 \cdot 2^7)^4 = 641i - (641 - 1)^4 = 641j - 1,$$

where i and j are integers.

An improvement of the above proofs combining the ideas of Bennett and Kraïtchik is given below (see [Kraïtchik, 1952], [Coxeter, p. 27], [Hardy, Wright, 1979, pp. 14–15]). Observe that

$$641 = 5^4 + 2^4 = 5 \cdot 2^7 + 1$$

divides each of $5^4 2^{28} + 2^{32}$ and $5^4 2^{28} - 1$ and so it also divides their difference, which is F_5.

In a similar elementary way (see [Dyson]), it can be explained why F_6 is divisible by

$$274177 = (2^6 - 1)(2^4 + 1)2^8 + 1.$$

In 1878, the French mathematician François Édouard Anatole Lucas proved that the number k in Euler's Theorem 4.14 is always even (see [Lucas, 1878b]).

Theorem 4.18 (Lucas). *If $m > 1$ and a prime p divides F_m, then p is of the form*

(4.7)
$$p = k2^{m+2} + 1,$$

where k is a natural number.

For a proof see the beginning of Chapter 6. It is almost sure that this theorem was not known to Euler, since in [Euler, 1750, p. 33] he writes only about numbers of the form $d = k2^{m+1} + 1$ in searching for divisors of F_m (see also [Keller, 1992, p. 3]).

Corollary 4.19. *If $m > 1$, then any divisor $d > 1$ of a Fermat number F_m is of the form $k2^{m+2} + 1$, where k is a natural number.*

P r o o f. This follows immediately from Theorem 4.18, since any prime divisor p of d is congruent to 1 (mod 2^{m+2}), and the product of numbers each congruent to 1 (mod 2^{m+2}) is also congruent to 1 (mod 2^{m+2}). □

Remark 4.20. The usefulness of Lucas's Theorem 4.18 can be illustrated by an example, which was treated by A. E. Western already in 1903. He was searching for a natural number k such that $k2^{20} + 1 \mid F_{18}$. The divisibility need only be verified for those k for which $k2^{20} + 1$ is a prime. Since the numbers $k2^{20} + 1$ are composite for all $k \leq 13$ except for $k = 7$ and $k = 13$, Western easily discovered that divisibility is attained when $k = 13$.

How can we verify that $p = 13 \cdot 2^{20} + 1 = 13631489$ divides the Fermat number F_{18}, which by Remark 3.7 has almost eighty thousand digits? This can be easily done via the following chain of congruences:

$$2^{2^5} = 65536^2 \quad \equiv 1048261 \pmod{p},$$

$$2^{2^6} \equiv 1048261^2 \equiv 3164342 \pmod{p},$$

$$2^{2^7} \equiv 3164342^2 \equiv 9153547 \pmod{p},$$

$$\vdots \qquad \qquad \vdots$$

$$2^{2^{17}} \equiv 1598622^2 \equiv 1635631 \pmod{p},$$

$$2^{2^{18}} \equiv 1635631^2 \equiv 13631488 \pmod{p}.$$

Hence, $2^{2^{18}} + 1 \equiv 0 \pmod{13631489}$.

The above procedure can be employed quite easily to find the factor

$$1071 \cdot 2^8 + 1 = 274177$$

of F_6, since there are only 188 primes of the form $k2^8 + 1$ for $k \leq 1071$.

In Chapter 6 we show that a Fermat number F_m is composite if and only if there exists a positive integer h such that $3h2^{m+2} + 1 \mid F_m$. Moreover, by Goldbach's Theorem 4.1 and Lucas's Theorem 4.18 we will derive (see [Křížek, Chleboun, 1994]) the following result:

Theorem 4.21. Let F_m be composite,

$$(4.8) \qquad\qquad F_m = (k2^n + 1)(\ell 2^j + 1),$$

where k and ℓ are odd. Then $k \geq 3$, $\ell \geq 3$, $n = j \geq m + 2$, k and ℓ are coprime, either $3 \mid k$ or $3 \mid \ell$ (but not both), and

$$(4.9) \qquad\qquad \max(k, \ell) \geq F_{m-2}.$$

The proof will be given in Chapter 6 (see Theorem 6.7). Any nontrivial factor $k2^n + 1$ (not necessarily prime) of a composite Fermat number F_m thus satisfies the relations

$$(4.10) \qquad\qquad k \geq 3 \quad \text{and} \quad n \geq m + 2.$$

Relation (4.9) says that k and ℓ cannot be simultaneously small relative to F_m.

To check immediately that the found factor of F_m is prime, we can employ Suyama's theorem (see [Keller, 1992, p. 10]). Its assumptions are satisfied for a majority (about 85%) of all known prime factors of the Fermat numbers and for all 151 known prime factors of F_m for $29 \leq m \leq 382\,447$ as of the beginning of 2000.

Theorem 4.22 (Suyama). Let $p = k2^n + 1$ divide F_m and let $k2^{n-(m+2)} < 9 \cdot 2^{m+2} + 6$. Then p is a prime.

P r o o f . Suppose, to the contrary, that p is a product of two nontrivial factors. Each of them fulfills (4.10), and thus $k2^n + 1 \geq (3 \cdot 2^{m+2} + 1)^2$. Hence, $k2^{n-(m+2)} \geq 9 \cdot 2^{m+2} + 6$. □

Example 4.23. For the divisor $45592577 = 11131 \cdot 2^{12} + 1$ of F_{10} we have $n = 12$, $m = 10$, and $n - (m + 2) = 0$. Since $k = 11131$ is less than $9 \cdot 2^{12} + 6 = 36870$, this divisor has to be a prime.

5. Primality of Fermat Numbers

I have found that numbers of the form $2^{2^m}+1$
are always prime numbers and have long since
signified to analysts the truth of this theorem ...

Pierre de Fermat in his letter to
Father Marin Mersenne
on December 25, 1640,
[Mahoney, p. 140].

Remark 5.1. Notice that the number $2^{2^3}+1 = 2^8+1$ is prime, but the numbers 2^3+1 and $2^{2^8}+1$ are composite (cf. Appendix A). This example shows that if 2^n+1 is prime, then $2^{2^n}+1$ need not be prime and vice versa (see [Sierpiński, 1970, Problem 141]).

We will now introduce two necessary and sufficient conditions for primality that will later be used to give a proof of a criterion due to Pepin. This criterion is also necessary and sufficient for testing the primality of the specific numbers F_m. The following theorem, due to Lucas (see [Lucas, 1891]), gives a practical test for the primality of a number N, provided that one can completely factor $N-1$, which is the case for the Fermat numbers $2^{2^m}+1$.

Theorem 5.2 (Lucas's Test). *The integer $N > 1$ is prime if and only if there exists an integer $a > 1$ such that*
(i) $a^{N-1} \equiv 1 \pmod{N}$,
(ii) $a^n \not\equiv 1 \pmod{N}$ *for every $n < N$ such that n divides $N-1$.*

P r o o f . Conditions (i) and (ii) are equivalent to the statement that $\operatorname{ord}_N a = N-1$. If N is prime, then by Theorem 2.14, there exists a primitive root modulo N, whose order is therefore equal to $N-1$.

Now suppose that both conditions (i) and (ii) hold for the integer a. By Euler's Theorem 2.17,

$$a^{\phi(N)} \equiv 1 \pmod{N}.$$

Thus, $\operatorname{ord}_N a \leq \phi(N)$. However, if N is composite, then $\phi(N) < N-1$ by Theorem 2.16. Since $\operatorname{ord}_N a = N-1$, we obtain that N is prime. $\qquad\square$

Selfridge (see [Brillhart, Selfridge, p. 89]) developed the following refinement of Lucas's primality test. This test also requires the complete factorization of $N-1$.

Theorem 5.3 (Selfridge's Test). *Let $N > 1$ and let the prime-power factorization of $N - 1$ be given by*

$$N - 1 = \prod_{i=1}^{r} p_i^{k_i}.$$

Then N is prime if and only if for each prime p_i, $i \in \{1, \ldots, r\}$, there exists an integer $a_i > 1$ such that

(i) $a_i^{N-1} \equiv 1 \pmod{N}$,

(ii) $a_i^{(N-1)/p_i} \not\equiv 1 \pmod{N}$.

P r o o f . If N is prime, then there exists a primitive root a that satisfies conditions (i) and (ii) for $i = 1, \ldots, r$.

Now assume that both conditions (i) and (ii) hold for $i = 1, \ldots, r$. It suffices to show that $\phi(N) = N - 1$. Since $\phi(N) \leq N - 1$, we need to show that $N - 1 \mid \phi(N)$. This will follow if we establish that $p_i^{k_i} \mid \phi(N)$ for $i = 1, \ldots, r$. Let $e_i = \mathrm{ord}_N a_i$. Then $e_i \mid N-1$ and $e_i \nmid (N-1)/p_i$. Therefore, $p_i^{k_i} \mid e_i$. Since $a^{\phi(N)} \equiv 1 \pmod{N}$ by Euler's Theorem 2.17, we get, by Lemma 2.13, $e_i \mid \phi(N)$. Consequently, $p_i^{k_i} \mid \phi(N)$ for $i = 1, \ldots, r$. $\square$

Remark 5.4. In [Lehmer, 1927], Theorem 5.3 is proved for the case in which a_i is fixed for $1 \leq i \leq r$. Thus, a_i is a primitive root modulo N in this instance.

As promised earlier, we now present Pepin's interesting and useful test for the primality of the Fermat numbers. This easily applied criterion can be found, e.g., in [Dickson], [Lenstra, Lenstra, Manasse, Pollard].

Theorem 5.5 (Pepin's Test). *For $m \geq 1$ the Fermat number F_m is prime if and only if*

$$(5.1) \qquad 3^{(F_m-1)/2} \equiv -1 \pmod{F_m}.$$

P r o o f . Clearly, (5.1) holds if $m = 1$. We will thus assume that $m \geq 2$. We claim that the Jacobi symbol satisfies

$$\left(\frac{3}{F_m}\right) = -1.$$

We first note that by (3.5), $F_m \equiv 2 \pmod{3}$. Since $F_m \equiv 1 \pmod{4}$, we see, by Theorem 2.30, that

$$(5.2) \qquad \left(\frac{3}{F_m}\right) = \left(\frac{F_m}{3}\right) = \left(\frac{2}{3}\right) = -1.$$

Let c be any integer such that $\left(\frac{c}{F_m}\right) = -1$ for all $m \geq 2$. It suffices to show that F_m is prime if and only if

$$(5.3) \qquad c^{(F_m-1)/2} \equiv -1 \pmod{F_m}.$$

We first assume that F_m is prime. Then, by Theorem 2.26 (Euler's criterion),

$$c^{(F_m-1)/2} \equiv \left(\frac{c}{F_m}\right) \equiv -1 \pmod{F_m}.$$

Now assume that congruence (5.3) holds. Then

$$c^{F_m-1} = \left(c^{(F_m-1)/2}\right)^2 \equiv (-1)^2 \equiv 1 \pmod{F_m}$$

and

$$c^{(F_m-1)/2} \equiv -1 \not\equiv 1 \pmod{F_m}.$$

Since 2 is the only prime factor of $F_m - 1$, it follows from Theorem 5.3 that F_m is prime. $\square$

Remark 5.6. The compositeness of $F_5 = 2^{32} + 1$ can be easily verified without knowing any of its divisors as follows (cf. [Knuth, p. 375]): The number $3^{2^{31}}$ (mod $2^{32} + 1$) can be computed by performing 31 operations of squaring (as in Remark 4.20) modulo F_5, and its value is 10324303. Therefore, by (5.1), $2^{32} + 1$ is not prime. In [Canals], the compositeness of $F_5, F_6, \ldots, F_{14}$ has been verified by Pepin's test given above. The computational time increases by a factor of 8 when F_m is replaced by F_{m+1}.

As an interesting addendum to Pepin's test, in [Rotkiewicz, 1964b] it is proved that there exist infinitely many composite integers n such that

$$3^{(n-1)/2} \equiv -1 \pmod{n}.$$

Remark 5.7. Pepin, in his original paper of 1877, used the base 5 rather than the base 3 (see [Pepin]). Note that by (3.5), $F_m \equiv 2 \pmod 5$ for $m \geq 2$. Then by Theorem 2.30,

$$\left(\frac{5}{F_m}\right) = \left(\frac{F_m}{5}\right) = \left(\frac{2}{5}\right) = -1.$$

We can now apply the criterion given in (5.3) with $c = 5$ to test whether F_m is prime. Namely, if $m \geq 2$, then F_m is prime if and only if

$$(5.4) \qquad 5^{(F_m-1)/2} \equiv -1 \pmod{F_m}.$$

Pepin also noted that the base 10 could be used instead of the base 5 in his test. Proth suggested that 3 could be used in place of 5, which is the present version of Pepin's test (see [Proth, 1878a, 1878b]). A proof of Pepin's test with 3 used as the base is given in [Lucas, 1879]. For a good historical account of the development of Pepin's test, see [Williams, 1998, p. 101].

Remark 5.8. Pepin's test has been employed to prove that the Fermat numbers F_{13}, F_{14}, F_{20}, F_{22}, and F_{24} are composite (see [Paxson], [Hurwitz, Selfridge], [Young, Buell], [Crandall, Doenias, Norrie, Young], [Crandall, Mayer, Papadopoulos]).

Remark 5.9. According to [Aigner], [Burton, p. 222], and [Vasilenko], the bases 3 and 5 in congruences (5.1) and (5.4) can be replaced by 7 in Pepin's test. By the proof of Theorem 5.5, it suffices to show that the Jacobi symbol $\left(\frac{7}{F_m}\right)$ is equal to -1 for all integers m greater than or equal to some small fixed integer m_0. In fact, we show that $\left(\frac{7}{F_m}\right) = -1$ for all $m \geq 1$. Since

$$F_m = 16^{2^{m-2}} + 1 \equiv F_{m-2} \pmod 7,$$

we have (compare with [van Maanen, p. 349])

$$F_m \equiv F_0 \equiv 3 \pmod 7 \text{ for } m \text{ even and } \quad F_m \equiv F_1 \equiv 5 \pmod 7 \text{ for } m \text{ odd.}$$

Noting that $F_m \equiv 1 \pmod 4$ for $m \geq 1$, we see, by the law of quadratic reciprocity for the Jacobi symbol (property (viii) of Theorem 2.30), that

$$\left(\frac{7}{F_m}\right) = \left(\frac{F_m}{7}\right) = \left(\frac{3}{7}\right) = -1 \quad \text{for any even integer } m \geq 2$$

and

$$\left(\frac{7}{F_m}\right) = \left(\frac{F_m}{7}\right) = \left(\frac{5}{7}\right) = -1 \quad \text{for any odd integer } m \geq 1.$$

Remark 5.10. In [Aigner], the *elite primes* are characterized as the primes p having the property that all Fermat numbers F_m beyond a certain point are quadratic nonresidues modulo p. Since $F_m \equiv 1 \pmod 4$ for $m \geq 1$, we see, by the law of quadratic reciprocity for the Jacobi symbol (Theorem 2.30), that if p is an elite prime, then $\left(\frac{p}{F_m}\right) = -1$ for all but finitely many Fermat numbers F_m. Thus, if $p > 7$ is an elite prime, it may be substituted for 3, 5, or 7 in Pepin's test. Aigner lists all 14 elite primes less than 35000000. The next two elite primes after 3, 5, and 7 are 41 and 15361, and the largest elite prime below 35000000 is 14172161.

It is not difficult to verify that $\left(\frac{41}{F_m}\right) = -1$ for all F_m for which $m \geq 2$. This can be done by showing that

$$\left(\frac{41}{F_{m+4}}\right) = \left(\frac{41}{F_m}\right) \quad \text{for } m \geq 2.$$

Then, by checking that $\left(\frac{41}{F_m}\right) = -1$ for $2 \leq m \leq 5$, we establish that 41 is an elite prime.

It is of interest that the Mersenne numbers (see Appendix B) can be used in Pepin's test for the restricted class of Fermat numbers of the form F_p, where p is a prime congruent to 3 modulo 4, or F_{q+1}, where q is a prime congruent to 3 or 5 modulo 8. Before presenting these results we will need Theorem 5.11, which exhibits a property of the Mersenne numbers that is analogous to that given in Theorem 4.10, which states that all Fermat numbers are primes or pseudoprimes to the base 2.

Theorem 5.11. *All Mersenne numbers are primes or pseudoprimes to the base 2.*

P r o o f . Let $M_p = 2^p - 1$, where p is a prime, be a composite Mersenne number. Then p is an odd prime. It suffices to prove that

$$2^{M_p - 1} \equiv 1 \pmod{M_p},$$

or equivalently,

(5.5) $M_p \mid 2^{M_p - 1} - 1.$

We prove a stronger result, namely,

(5.6) $$M_p \mid 2^{(M_p-1)/2} - 1,$$

which implies (5.5). By Fermat's little theorem,

$$\frac{M_p - 1}{2} = 2^{p-1} - 1 \equiv 0 \pmod{p}.$$

Hence, $(M_p - 1)/2 = kp$ for some positive integer k. Noting that

$$2^p - 1 \mid 2^{kp} - 1,$$

we see that (5.6) holds.   □

The next two theorems can be found in [Vasilenko].

Theorem 5.12. *Let p be a prime of the form $4k + 3$ and let $M_p = 2^p - 1$ be the associated Mersenne number. Then the Fermat number F_p is prime if and only if*

$$M_p^{(F_p-1)/2} \equiv -1 \pmod{F_p}.$$

P r o o f . According to the proof of Theorem 5.5, it suffices to prove that $\left(\frac{M_p}{F_p}\right) = -1$. By Theorem 5.11,

$$2^{2^p-1} \equiv 2 \pmod{M_p},$$

which implies that $F_p = 2 \cdot 2^{2^p-1} + 1 \equiv 5 \pmod{M_p}$. Hence, by Goldbach's Theorem 4.1, $\gcd(F_p, M_p) = 1$. We note that if $p \equiv 3 \pmod 4$, then $M_p = 2^p - 1 \equiv 2 \pmod 5$. Thus, by Theorem 2.30,

$$\left(\frac{M_p}{F_p}\right) = \left(\frac{F_p}{M_p}\right) = \left(\frac{5}{M_p}\right) = \left(\frac{M_p}{5}\right) = \left(\frac{2}{5}\right) \equiv -1,$$

and the theorem is proved.   □

Theorem 5.13. *Let p be a prime of the form $8k+3$ or $8k+5$ and let $M_p = 2^p - 1$ be the associated Mersenne number. Then the Fermat number F_p is prime if and only if*

$$M_p^{(F_{p+1}-1)/2} \equiv -1 \pmod{F_{p+1}}.$$

P r o o f . As in the proof of Theorem 5.12, it suffices to establish that $\left(\frac{M_p}{F_{p+1}}\right) = -1$. By the same argument used in the proof of Theorem 5.12, it can be shown that $F_p \equiv 5 \pmod{M_p}$. Then, by (3.6),

$$F_{p+1} = (F_p - 1)^2 + 1 \equiv 4^2 + 1 \equiv 17 \pmod{M_p}.$$

First assume that $p \equiv 3 \pmod 8$. Then $M_p \equiv 2^p - 1 \equiv 7 \pmod{17}$ and

$$\left(\frac{M_p}{F_{p+1}}\right) = \left(\frac{F_{p+1}}{M_p}\right) = \left(\frac{17}{M_p}\right) = \left(\frac{M_p}{17}\right) = \left(\frac{7}{17}\right) \equiv -1.$$

Now suppose that $p \equiv 5 \pmod 8$. We note that $M_p = 2^p - 1 \equiv 14 \pmod{17}$. It now follows that

$$\left(\frac{M_p}{F_{p+1}}\right) = \left(\frac{M_p}{17}\right) = \left(\frac{14}{17}\right) \equiv -1. \qquad \square$$

In the next four theorems we present primality tests for the Fermat numbers in addition to Pepin's test given in Theorem 5.5.

Theorem 5.14. *Let* $m \geq 2$, $r = 2^m$, *and* $T_1 = 3$. *If we define the sequence* $\{T_i\}$ *by* $T_{i+1} = 2T_i^2 - 1$, $i = 1, 2, 3, \ldots$, *then* F_m *is a prime if the first term of this sequence that is divisible by* F_m *is* T_{r-1}. *Also,* F_m *is composite if none of these terms up to and including* T_{r-1} *is divisible by* F_m. *Finally, if* k *(less than* r*) denotes the rank of the first term that is divisible by* F_m, *then the prime divisors of* F_m *have to be of the form* $q2^{k+1} + 1$.

For the original proof see [Lucas, 1877, p. 138].

Theorem 5.15. *Let* $m \geq 2$, $r = 2^m$, *and* $S_1 = 6$. *If we define the sequence* $\{S_i\}$ *by* $S_{i+1} = S_i^2 - 2$, *then* F_m *is a prime when* $F_m \mid S_k$ *for some* k *such that* $r/2 \leq k \leq r - 1$. *Also,* F_m *is composite if* $F_m \nmid S_k$ *for all* $k \leq r - 1$. *Finally, if* $F_m \mid S_k$ *with* $k < r/2$, *then any prime divisor of* F_m *must have the form* $q2^{k+1} + 1$.

For the original proof see [Lucas, 1878a, p. 313]. The statements of Theorems 5.14 and 5.15 are taken from [Williams, 1998, pp. 99–100], where modifications and corrections of misprints have been made.

The following theorem, which appears in [Inkeri, p. 16], is similar to Lucas's Theorems 5.14 and 5.15.

Theorem 5.16 (Inkeri). *The Fermat number* F_m, $m \geq 2$, *is prime if and only if* F_m *divides the term* R_{2^m-2} *of the sequence defined by*

$$R_0 = 8, \quad R_i = R_{i-1}^2 - 2, \quad i = 1, 2, \ldots .$$

Contrary to Theorems 5.14 and 5.15, which present only sufficient conditions, Theorem 5.16 gives both necessary and sufficient conditions for primality.

The following theorem from [McIntosh] gives another necessary and sufficient condition for the primality of F_m.

Theorem 5.17 (McIntosh). *The number* F_m *is prime if and only if* F_m *does not divide* $T(F_m - 2)$, *where* $T(n)$ *is defined by means of the power series*

$$\tan z = \sum_{n=0}^{\infty} \frac{T(n)z^n}{n!}.$$

For the proof see [McIntosh]. A similar proof was given later in [Grytczuk, Grytczuk].

If F_m is partially factored, the theorem below due to Suyama and modified slightly by H. W. Lenstra (see [Suyama, 1984b] and [Keller, 1992, pp. 11–12]) gives a sufficient condition for the cofactor C to be composite.

Theorem 5.18. *Let $m \geq 5$ and let $F_m = FC$, where $F > 1$ and $C > 1$. If*

$$3^{F-1} \not\equiv 3^{F_m-1} \pmod{C},$$

then C is composite. If

$$3^{F-1} \equiv 3^{F_m-1} \pmod{C},$$

then C is a pseudoprime to the base 3^F or a prime.

P r o o f . We first show that if C is prime, then

$$3^{F-1} \equiv 3^{F_m-1} \pmod{C}.$$

According to Goldbach's Theorem 4.1, $\gcd(3, F_m) = 1$. Now it follows, by Fermat's little theorem, that

$$3^{F-1} \equiv \left(3^{F-1}\right)^C \equiv 3^{FC-C} \equiv \frac{3^{FC}}{3^C} \equiv \frac{3^{FC}}{3} \equiv 3^{FC-1} \equiv 3^{F_m-1} \pmod{C}.$$

Now suppose that

(5.7) $$3^{F_m-1} = 3^{FC-1} \equiv 3^{F-1} \pmod{C}.$$

Noting that $\gcd(3^{F-1}, F_m) = \gcd(3^{F-1}, C) = 1$, we can cancel out 3^{F-1} in congruence (5.7) to obtain

$$3^{FC-F} = 3^{F(C-1)} = \left(3^F\right)^{C-1} \equiv 1 \pmod{C},$$

which implies that C is a pseudoprime to the base 3^F or a prime. $\qquad\square$

Remark 5.19. If the test given in Theorem 5.18 leads to the conclusion that C is a pseudoprime to the base 3^F or a prime, then we can test C for primality either by a deterministic primality test such as the ones given in [Adleman, Pomerance, Rumely] and [Cohen, Lenstra] or by the strong probabilistic primality test given in [Rabin].

We have also the following necessary and sufficient condition for the primality of F_m:

$$(F_m - 1)! \equiv -1 \pmod{F_m},$$

or equivalently,

$$F_m \mid (F_m - 1)! + 1.$$

This result follows from the well-known theorem due to Wilson.

Theorem 5.20 (Wilson). *A number p is prime if and only if*

(5.8) $$(p - 1)! \equiv -1 \pmod{p}.$$

Remark 5.21. Wilson's theorem can be proved in many ways (see, e.g., [Dickson, Chapter III], [Robbins, p. 89]). Here we briefly sketch an interesting proof from [Petr], which has a nice geometric interpretation. Consider piecewise linear closed oriented curves (paths) connecting all the vertices of a regular p-gon. Curves with opposite orientation are considered to be different. Thus, there exist just $(p-1)!$ such curves. Indeed, starting from an arbitrary fixed vertex, the number of ways in which these curves could pass through the $p-1$ remaining points equals the number of permutations of this set of points. *Regular curves* (see Figure 5.1a) are those that remain invariant after rotation about the angle $2\pi/p$. The other curves are called *irregular* (see Figure 5.1b). The number of regular curves is clearly $p-1$.

Petr's proof proceeds as follows. Let X be the set of all the closed oriented Hamiltonian paths passing through the vertices of a regular polygon with p sides centered at the origin. Let C be the cyclic group of order p. We think of C as the transformation group generated by the clockwise (or counterclockwise) rotation through the angle $2\pi/p$. Then, C acts on X. The orbit of every curve in X under the action of C is either of cardinality 1 (and this happens precisely if the curve is regular) or of cardinality $|C| = p$. Thus, by counting the number of elements of X we get

$$(p-1)! = |X| = (p-1) + pN,$$

where N is the number of orbits of irregular curves. Therefore, $(p-1)! + 1 = p(N+1)$, which yields the desired congruence (5.8).

a)   b)

Figure 5.1. Regular and irregular curves.

In proving the converse implication of Wilson's theorem, we will show that if n is composite, then $(n-1)! \equiv 2 \pmod{n}$ when $n = 4$ and $(n-1)! \equiv 0 \pmod{n}$ when $n > 4$. By inspection,

$$(4-1)! = 6 \equiv 2 \pmod{4}.$$

Now assume that n is composite, $n > 4$, and n is not a perfect square. Then we can factor n as

$$n = dk,$$

where $2 \le d < k < n-1$. Hence, $dk \mid (n-1)!$ and $(n-1)! \equiv 0 \pmod{n}$. Finally, we suppose that $n > 4$ and $n = k^2$. Then $3 \le k < 2k \le n-1$. Therefore, $k(2k) \mid (n-1)!$. Since $2k^2 \equiv 0 \pmod{n}$, we have $(n-1)! \equiv 0 \pmod{n}$ in this case also.

Remark 5.22. Since there is no efficient algorithm for computing the factorial, Wilson's theorem is highly impractical to use as a test for the primality of F_m.

Remark 5.23. By [Burton, pp. 242–243] or [Sierpiński, 1950, p. 73], every prime of the form $4k + 1$ can be written as a sum of two nonzero squares in exactly one way. This result is called *Fermat's assertion*.

Theorem 5.24. *Let $m \geq 1$. Then the Fermat number F_m is a prime if and only if it can be written as a sum of two nonzero squares in essentially only one way, namely,* $F_m = \left(2^{2^{m-1}}\right)^2 + 1^2$.

P r o o f . Let F_m be prime and let (cf. (1.7))

$$F_m = a^2 + b^2 = c^2 + d^2,$$

where a, b, c, d are positive integers. Then

$$a^2 c^2 - b^2 d^2 = F_m(a^2 - d^2) \equiv 0 \pmod{F_m},$$

which yields

$$ac \equiv bd \pmod{F_m} \quad \text{or} \quad ac \equiv -bd \pmod{F_m}.$$

Since a, b, c, d are all less than $\sqrt{F_m}$, we have

$$ac - bd = 0 \quad \text{or} \quad ac + bd = F_m.$$

If the second equality holds, then

$$F_m^2 = (a^2 + b^2)(c^2 + d^2) = (ac + bd)^2 + (bc - ad)^2 = F_m^2 + (bc - ad)^2,$$

i.e., $bc - ad = 0$. Therefore,

$$ac = bd \quad \text{or} \quad ad = bc.$$

Assume, for instance, that $ac = bd$. Then $a \mid d$, because $a \mid bd$ and $\gcd(a, b) = 1$. Therefore, there exists a positive integer k such that $d = ka$. From this we obtain $c = kb$ and thus

$$F_m = c^2 + d^2 = k^2(a^2 + b^2),$$

which yields $k = 1$, $a = d$, and $b = c$.

Assuming $ad = bc$, we get in a similar way that $a = c$ and $b = d$.

On the other hand, if F_m is composite, then $F_m = qr$ for some $q > 1$ and $r > 1$. Each prime factor of F_m is congruent (either by Lucas's Theorem 4.18, or by Euler's Theorem 2.17) to 1 modulo 4. By Remark 5.23 and the identity

$$F_m = qr = (a^2 + b^2)(c^2 + d^2) = (ac + bd)^2 + (ad - bc)^2 = (ac - bd)^2 + (ad + bc)^2,$$

it follows that every number that is a product of two primes both congruent to 1 modulo 4 can be written as a sum of two squares. The above identity is sometimes called *Viète's identity*. By induction, it follows easily that every number which is a product of primes congruent to 1 modulo 4 is a sum of two squares.

We need to show that the squares in the last two equalities above are essentially different. Clearly, $ac + bd \neq ac - bd$. So let us suppose that $ac + bd = ad + bc$ and $ad - bc = ac - bd$. Then $2ad = 2ac$, i.e., $d = c$, which contradicts the fact that r is odd. $\square$

Furthermore, we present another necessary and sufficient condition for the primality of Fermat numbers (see [Křížek, Somer]), which generalizes Theorem 3.10. In particular, we prove that an odd integer $n \geq 3$ is a Fermat prime if and only if the set of primitive roots modulo n is equal to the set of quadratic nonresidues modulo n. The necessity of this condition for the primality of F_m is well known (see, e.g., [Burton, Problem 17(b), p. 222]), whereas its sufficiency is new.

For a natural number n set

$$M(n) = \{a \in \{1, \ldots, n-1\} \mid a \text{ is a primitive root modulo } n\}$$

and

$$K(n) = \{a \in \{1, \ldots, n-1\} \mid \gcd(a, n) = 1 \text{ and } a \text{ is a quadratic nonresidue (mod } n)\}.$$

Notice that $M(1) = K(1) = \emptyset$, $M(2) = \{1\}$, and $K(2) = \emptyset$.

Lemma 5.25. *If $n \geq 3$, then*

$$(5.9) \qquad\qquad M(n) \subset K(n).$$

P r o o f . Let $n \geq 3$. Then $\phi(n)$ is even. If $\gcd(n, a) = 1$ and $a \in \{1, \ldots, n-1\}$ is a quadratic residue modulo n, then there exists an integer x such that

$$x^2 \equiv a \pmod{n}.$$

By Euler's Theorem 2.17,

$$a^{\phi(n)/2} \equiv x^{\phi(n)} \equiv 1 \pmod{n},$$

and a is not a primitive root modulo n. Thus (5.9) holds. $\square$

Proposition 5.26. *Every integer $n \geq 3$ has at least $\phi(n)/2$ quadratic non-residues. If $n = p \geq 3$ is prime, it has precisely $\phi(p)/2 = (p-1)/2$ quadratic nonresidues.*

P r o o f . Set $A = \{a \mid 1 \leq a \leq n-1, \gcd(a, n) = 1\}$. Thus $|A| = \phi(n)$. If $a \in A$, then also $-a \in A$ and $a^2 \in A$ after reduction modulo n. Modulo n, $a \not\equiv -a$ for each $a \in A$, because $2a \equiv 0$ for an $a \in A$ would imply that n divides 2. Also, $a \not\equiv b$ implies $-a \not\equiv -b$. When a runs through A, the squares a^2 reduced modulo n produce at most $\phi(n)/2$ quadratic residues because $a^2 \equiv (-a)^2$. Hence, we have at least $\phi(n)/2$ quadratic nonresidues.

In the case $n = p$, both the bounds for quadratic residues and nonresidues turn into equalities, because modulo p, $a^2 \equiv b^2$ is equivalent to $(a - b)(a + b) \equiv 0$, and since the modulus is prime, we have $a \equiv \pm b$. $\square$

Remark 5.27. Theorem 5.28 gives a necessary and sufficient condition for the primality of Fermat numbers, which states that the sets $M(n)$ and $K(n)$ for an odd $n \geq 3$ are equal if and only if n is a Fermat prime (compare with Figures 2.7 and 2.8). Later, in Theorem 5.29, we show that $M(n) = K(n)$ for an even natural number n if and only if n equals 4 or two times a Fermat prime.

Theorem 5.28. Let $n \geq 3$ be a positive odd integer. Then n is a Fermat prime if and only if $M(n) = K(n)$.

P r o o f . Let $n = F_m$ be a Fermat prime. Then, by Theorem 3.10, we have $M(F_m) = K(F_m)$. We also provide the following alternative shorter proof of this result. We note that by Theorem 2.18, (2.15), (2.13), and Proposition 5.26, we obtain

$$(5.10) \qquad |M(F_m)| = \phi(\phi(F_m)) = \phi(2^{2^m}) = 2^{2^m-1} = \frac{F_m - 1}{2} = |K(F_m)|.$$

Since $M(n)$ and $K(n)$ have the same cardinality by (5.10), we see by (5.9) that $M(n) = K(n)$.

Conversely, assume by way of contradiction that $n \geq 3$ is not a Fermat prime and that $M(n) = K(n)$. By Proposition 5.26,

$$|K(n)| \geq \frac{\phi(n)}{2} \geq 1 \quad \text{for } n \geq 3.$$

Hence, $M(n) \neq \emptyset$, since $M(n) = K(n)$. It follows from Theorem 2.18 that $n = p^s$ for some odd prime p and a positive integer s.

Assume first that $s = 1$. Then there exist $k \geq 1$ and odd $q \geq 3$ such that

$$(5.11) \qquad\qquad\qquad p - 1 = 2^k q$$

(since if $q = 1$ and if $k = r\ell$ for $r \geq 3$ odd and $\ell \geq 1$, then $p = 2^{r\ell}q + 1$ is divisible by $2^\ell + 1$ and is hence composite). Then by Theorem 2.18, (2.13), (5.11), (2.14), (5.11) again, and Proposition 5.26, we obtain

$$(5.12) \qquad |M(p)| = \phi(\phi(p)) = \phi(p-1) = \phi(2^k q) = \phi(2^k)\phi(q)$$
$$\leq 2^{k-1}(q-1) = \frac{1}{2}2^k(q-1)$$
$$< \frac{p-1}{2} = |K(p)|.$$

Hence, $M(p) \neq K(p)$.

Now assume that $s \geq 2$ and let $p - 1 = 2^k q$, where $k \geq 1$ and $q \geq 1$ is odd. By Proposition 5.26,

$$(5.13) \qquad\qquad |K(p^s)| \geq \frac{\phi(p^s)}{2} = \frac{(p-1)p^{s-1}}{2}.$$

Consequently, we obtain

$$(5.14) \qquad |M(p^s)| = \phi(\phi(p^s)) = \phi((p-1)p^{s-1}) = \phi(2^k q)\phi(p^{s-1})$$
$$= \phi(2^k)\phi(q)\phi(p^{s-1}) = 2^{k-1}\phi(q)(p-1)p^{s-2}$$
$$< 2^{k-1}qp^{s-1} = \frac{(p-1)p^{s-1}}{2} \leq |K(p^s)|.$$

From this and (5.12) we get

(5.15) $|M(p^s)| < |K(p^s)|$ for $s \geq 1$,

and the theorem is therefore proved.   □

Theorem 5.29. *Let n be a positive even integer. The number n is equal to 4 or to twice a Fermat prime if and only if $M(n) = K(n)$.*

Before proving Theorem 5.29, we will need the following lemma.

Lemma 5.30. *Suppose that $r \geq 3$ is odd. Then*

$$|M(2r)| = |M(r)| \text{and} |K(2r)| = |K(r)|.$$

P r o o f . The first equality holds if $M(r)$ is empty by Theorem 2.18. So let $M(r) \neq \emptyset$. By Theorem 2.18, $r = p^s$ for some odd prime p and integer $s \geq 1$, and $M(2r) \neq \emptyset$. Then using Theorem 2.18 again,

(5.16) $|M(2r)| = \phi(\phi(2r)) = \phi(\phi(2)\phi(r)) = \phi(\phi(r)) = |M(r)|.$

Moreover, by Proposition 5.26, $K(r) \neq \emptyset$. Note that if $a \in \{1, \ldots, r\}$ is a quadratic nonresidue modulo r such that $\gcd(a, r) = 1$, then exactly one of a and $a + r$ is odd, and hence exactly one of these two numbers is a quadratic nonresidue modulo $2r$. It now follows that

$$|K(2r)| = |K(r)|.$$

From this and (5.16) we see that the lemma holds.   □

P r o o f   o f   T h e o r e m   5 . 2 9 . Obviously,

$$M(4) = \{3\} = K(4).$$

Further, let F_m be prime. According to (5.10), $|M(F_m)| = |K(F_m)|$. Hence, by Lemma 5.30, $|M(2F_m)| = |K(2F_m)|$, and thus, by (5.9), $M(2F_m) = K(2F_m)$.

Let, to the contrary, $n \neq 4$, $n \neq 2F_m$, where F_m is prime, and $M(n) = K(n)$. First notice that $n \neq 2$, since $M(2) = \{1\}$ and $K(2) = \emptyset$.

Further, assume that $M(n) \neq \emptyset$. Then, by Theorem 2.18, $n = 2p^s$, where p is an odd prime, $s \geq 1$, and it is not the case that $s = 1$ and p is a Fermat number. According to (5.15), $|M(p^s)| < |K(p^s)|$, and thus by Lemma 5.30,

$$|M(2p^s)| < |K(2p^s)|.$$

Finally, assume that $M(n) = \emptyset$ and $n \geq 6$. By Proposition 5.26, we get $K(n) \neq \emptyset$, and hence $M(n) \neq K(n)$.   □

The next theorem determines those integers $n \geq 2$ for which the cardinality of the set $K(n) \setminus M(n)$ is equal to 1.

Theorem 5.31. *Let $n \geq 2$ be an integer. Then*

(5.17) $|M(n)| = |K(n)| - 1$

if and only if $n = 9$, $n = 18$, or either n or $n/2$ is equal to an odd prime p for which $(p-1)/2$ is also an odd prime. Moreover, if (5.17) holds, then $n - 1 \in K(n)$ but $n - 1 \notin M(n)$.

P r o o f . By Theorems 5.28 and 5.29, we may assume that $n \neq 4$, F_m, or $2F_m$, where F_m is prime. Also, clearly $n \neq 2$. Suppose first that $n = p$, where p is an odd prime that is not a Fermat number. Analogously to (5.11), let $p - 1 = 2^k q$, where $q \geq 3$ is odd and $k \geq 1$. Then, by Proposition 5.26, $|K(p)| = (p-1)/2 = 2^{k-1}q$. Moreover, by Theorem 2.18,

$$|M(p)| = \phi(\phi(p)) = \phi(p-1) = \phi(2^k q) = \phi(2^k)\phi(q) = 2^{k-1}\phi(q)$$
$$\leq 2^{k-1}(q-1) = 2^{k-1}q - 2^{k-1} = |K(p)| - 2^{k-1} \leq |K(p)| - 1.$$

Thus, $|M(p)| = |K(p)| - 1$ if and only if $\phi(q) = q - 1$ and $k = 1$. This occurs if and only if $(p-1)/2 = q$, where q is an odd prime. Since $K(p) \neq \emptyset$, it now follows by Lemma 5.30 that for $n = 2p$, where p is an odd prime, we have $|M(2p)| = |K(2p)| - 1$ if and only if $(p-1)/2$ is an odd prime.

We next assume that $n = p^s$, where p is an odd prime and $s \geq 2$. Let $p - 1 = 2^k q$, where $q \geq 1$ is odd and $k \geq 1$. Then, by (5.14),

$$|M(p^s)| = 2^{k-1}\phi(q)p^{s-2}(p-1) \leq 2^{k-1}qp^{s-1} - 2^{k-1}qp^{s-2}.$$

Moreover, by Proposition 5.26,

$$|K(p^s)| \geq \frac{(p-1)p^{s-1}}{2} = 2^{k-1}qp^{s-1}.$$

Hence, $|M(p^s)|$ can equal $|K(p^s)| - 1$ only if $\phi(q) = q$ and $2^{k-1}qp^{s-2} = 1$. This can occur if and only if $q = k = 1$ and $s = 2$. Therefore, $p - 1 = 2$, which implies that $n = 3^2 = 9$. By inspection, we find that $K(9) = \{2, 5, 8\}$, $M(9) = \{2, 5\}$, and thus $|M(9)| = |K(9)| - 1$. Since $M(9) \neq \emptyset$, it follows by Lemma 5.30 that when $n = 2p^s$, where p is an odd prime and $s \geq 2$, then $|M(2p^s)| = |K(2p^s)| - 1$ if and only if $p = 3$ and $s = 2$, i.e., $n = 18$.

The only remaining cases to consider are, by Theorem 2.18, those for which $M(n) = \emptyset$. We will show that then $|K(n)| \geq 2$, and hence $|M(n)| \neq |K(n)| - 1$. By Theorem 2.18, if $M(n) = \emptyset$, then either $n = 2^s$, where $s \geq 3$, or $n = p^s t$, where p is an odd prime, $s \geq 1$, $\gcd(p, t) = 1$, and $t \geq 3$. Assume first that $n = 2^s$, where $s \geq 3$. Then, by Proposition 5.26 and (2.13),

$$|K(n)| \geq \frac{\phi(2^s)}{2} = \frac{2^{s-1}}{2} \geq 2.$$

If $n = p^s t$, where p is an odd prime, $s \geq 1$, $\gcd(p, t) = 1$, and $t \geq 3$, then, by Proposition 5.26 and (2.14),

$$|K(n)| \geq \frac{\phi(p^s t)}{2} = \frac{\phi(p^s)\phi(t)}{2} \geq \frac{2 \cdot 2}{2} = 2.$$

Finally, to prove the last assertion, suppose that (5.17) holds. One can check that if p is an odd prime such that $(p-1)/2$ is also an odd prime, then $p \equiv 3$

(mod 4), and hence -1 is a quadratic nonresidue modulo p. Since n is divisible by a prime p for which -1 is a quadratic nonresidue modulo p, we have $n - 1 \in K(n)$. Clearly, $n - 1 \notin M(n)$, because $n \geq 7$ (thus $\phi(n) > 2$) and $(n - 1)^2 \equiv 1 \pmod{n}$. $\square$

Remark 5.32. Odd primes p for which $2p + 1$ is also a prime are called *Sophie Germain primes* (cf. [Ribenboim, 1996, pp. 329–332]). By Theorem 5.31, $|M(n)| = |K(n)| - 1$ if and only if $n \in \{9, 18\}$ or either n or $n/2$ equals p, where $(p - 1)/2$ is a Sophie Germain prime. It is not known whether the number of Sophie Germain primes is infinite.

Remark 5.33. The set $M(F_m)$ for $m > 1$ is contained in the set of those numbers that are not powers of 2 modulo F_m.

Corollary 5.34. Let $n \geq 3$ be an integer. Then n is a Fermat prime if and only if the number of primitive roots modulo n is equal to $\lfloor \frac{n}{2} \rfloor$. If $n \geq 3$ and n is not a Fermat prime, then the number of primitive roots modulo n is less than $\lfloor \frac{n}{2} \rfloor$.

P r o o f. By (5.10),

$$|M(n)| = \frac{n - 1}{2} = \left\lfloor \frac{n}{2} \right\rfloor,$$

provided that n is a Fermat prime.

Now suppose that $n \geq 3$ and n is not a Fermat prime. Then by Theorem 2.18, $|M(n)| \geq 1$ if and only if $n = 4$, p^s, or $2p^s$, where p is an odd prime and $s \geq 1$. If $n = 4$, then $M(n) = \{3\}$ and $M(n) < \lfloor \frac{4}{2} \rfloor = 2$. If $n = p$, then $|M(n)| < \lfloor \frac{n}{2} \rfloor$ by (5.12). If $n = p^s$, where $s \geq 2$, then the same sharp inequality holds due to (5.14). Finally, the case $n = 2p^s$ follows from Lemma 5.30. $\square$

Notice that $|M(2)| = 1 = \lfloor \frac{2}{2} \rfloor$. Therefore, Fermat primes and the number 2 are the only numbers with the highest possible number of primitive roots.

Further, we introduce an interesting connection between Fermat primes and certain graphs that is due to [Szalay]. Let H be a finite set and let f be a map of H into itself. Starting with an arbitrary element x_0 from H, we define a sequence of successive elements of H by

$$x_{n+1} = f(x_n), \quad n = 0, 1, \dots.$$

This iteration scheme is called a *discrete iteration*. Since H is finite, the sequence $\{x_n\}$ has to be cyclic starting from some element x_k. If $x_k, x_{k+1}, \dots, x_t$ are distinct and

$$x_{k+1} = f(x_k),$$

$$\vdots$$

$$x_t = f(x_{t-1}),$$
$$x_k = f(x_t),$$

then the elements $x_k, x_{k+1}, \dots, x_t$ constitute a *cycle*. Recall that a cycle of length 1 is called a *fixed point*.

Furthermore, assume for simplicity that

$$H = \{0, 1, \ldots, N-1\},$$

where $N > 1$. The *iteration graph* of f is a directed graph whose vertices are elements of H and such that there is a directed edge from x to $f(x)$ for all $x \in H$.

Definition 5.35. The iteration graph is called a *binary graph* if its symmetrization has exactly two components (i.e., the associated nondirected graph consists of two disjoint connected subgraphs) and the following three conditions hold:
1. The vertex 0 is an isolated fixed point.
2. The vertex 1 is a fixed point and there exists a directed edge from the vertex $N-1$ to 1.
3. For each vertex from the set $\{1, 2, \ldots, N-1\}$ there exist either two edges or no edge directed toward this vertex.

Now we will consider a special discrete iteration. For each $x \in H$ let $f(x)$ be the remainder of x^2 modulo N, i.e.,

(5.18) $$f(x) \in H \quad \text{and} \quad f(x) \equiv x^2 \pmod{N}.$$

This corresponds to the iteration scheme $x_{n+1} \equiv x_n^2 \pmod{N}$.

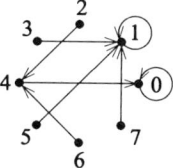

Figure 5.2. Iteration graph of f defined by (5.18) for $N = 8$.

In [Szalay], the following theorem is presented (compare with Figures 5.2 and 5.3).

Theorem 5.36 (Szalay). *The iteration graph of the map f defined by (5.18) is a binary graph if and only if the modulus N is a Fermat prime.*

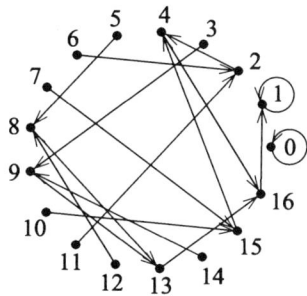

Figure 5.3. Iteration graph of f defined by (5.18) for $N = 17$.

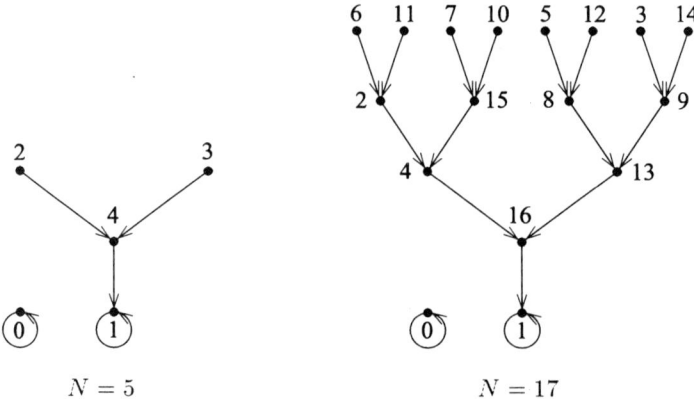

$$N = 5 \qquad\qquad\qquad N = 17$$

Figure 5.4. Binary graphs corresponding to Fermat primes.

In the right part of Figure 5.4, we see a graph topologically equivalent to the graph of Figure 5.3. This nicely illustrates why it is called a binary graph. The larger component is a binary tree if we remove the edge directed from 1 toward 1. A similar graph is also given in [Voorhees, p. 179].

Notice that the primitive roots modulo N are in the "upper part" of each graph in Figure 5.4. If F_m is a prime, then, by Pepin's test (5.1), the number 3 is always a primitive root modulo F_m for $m \geq 1$. Figure 5.4 provides therefore a graphical illustration of Pepin's test.

We will now generalize Theorem 2.16, which also states a necessary and sufficient condition for the primality of F_m.

Theorem 5.37. *The number $\phi(F_m)$ is equal to 2^n for some $n \geq 1$ if and only if F_m is prime.*

P r o o f . First suppose that F_m is prime. Then

$$\phi(F_m) = 2^{2^m}$$

by Theorem 2.16.

Now assume by way of contradiction that F_m is a composite number and that $\phi(F_m) = 2^n$ for some $n \geq 1$. Then there exists an odd prime $p < F_m$ such that $p \mid F_m$. Consequently, $p - 1 \mid \phi(F_m)$ by (2.14), and hence $p - 1 = 2^c$ for some integer $c < n$. Therefore, p is a Fermat prime, which is impossible due to Goldbach's Theorem 4.1.   □

Further, we introduce another interesting necessary and sufficient condition (see [Jones, Pearce]) for the primality of Fermat numbers, which has a beautiful geometric interpretation. To this end, we first present a graphical procedure that transforms algebraic fractions to images.

Let $b > 1$ and n be positive integers. If r_i is the remainder produced at step i of the base b long division of $1/n$, then the remainder produced at the $(i + 1)$st step obviously satisfies the congruence

$$r_{i+1} \equiv br_i \pmod{n}.$$

Starting with $r_0 = 1$, we get the sequence of remainders $r_0, r_1, r_2, \ldots$ of $1/n$ obtained through long division in base b. We may graphically analyze this fraction. This analysis begins at the point (r_0, r_0), proceeds first vertically, then horizontally, to (r_1, r_1), then moves again vertically, then horizontally to (r_2, r_2), and continues in this fashion (compare with Figure 5.5). If the remainder becomes zero at the ith step, we stop the process. In this way, the sequence of remainders entirely determines the associated graph of the fraction.

Example 5.38. Consider the fraction $\frac{1}{7}$, which has a base-10 (decimal) expansion of $0.\overline{142857}$. The corresponding sequence of remainders is periodic:

$$r_0 = 1,$$
$$r_1 = 3 \equiv 10 \pmod 7,$$
$$r_2 = 2 \equiv 30 \pmod 7,$$
$$r_3 = 6 \equiv 20 \pmod 7,$$
$$r_4 = 4 \equiv 60 \pmod 7,$$
$$r_5 = 5 \equiv 40 \pmod 7,$$
$$r_0 = r_6 = 1 \equiv 50 \pmod 7,$$

etc. In Figure 5.5, we see that the associated graph possesses rotational symmetry with respect to the point $(3.5, 3.5)$ for the base $b = 10$, but the graph is asymmetric for $b = 11$.

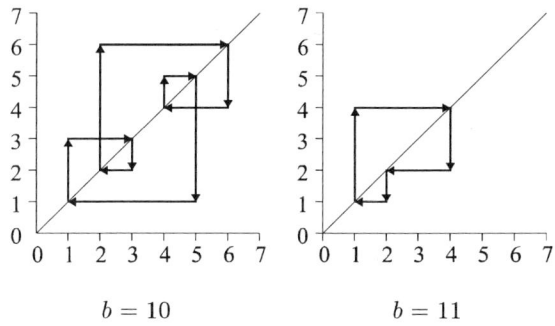

$$b = 10 \qquad\qquad\qquad b = 11$$

Figure 5.5. Graphical analysis of $\frac{1}{7}$ for two different bases.

Definition 5.39. An integer $n > 1$ is said to be *perfectly symmetric* if the associated graph of its reciprocal $1/n$ is rotationally symmetric with respect to the point $(n/2, n/2)$ in any base b, provided that $b \not\equiv 0 \pmod n$ and $b \not\equiv 1 \pmod n$.

Theorem 5.40 (Jones, Pearce). *An integer $n > 1$ is perfectly symmetric if and only if $n = 2$ or n is a Fermat prime.*

For the proof see [Jones, Pearce]. The graphical analysis of the fraction $1/F_m$ for $m = 2$ is illustrated in Figure 5.6.

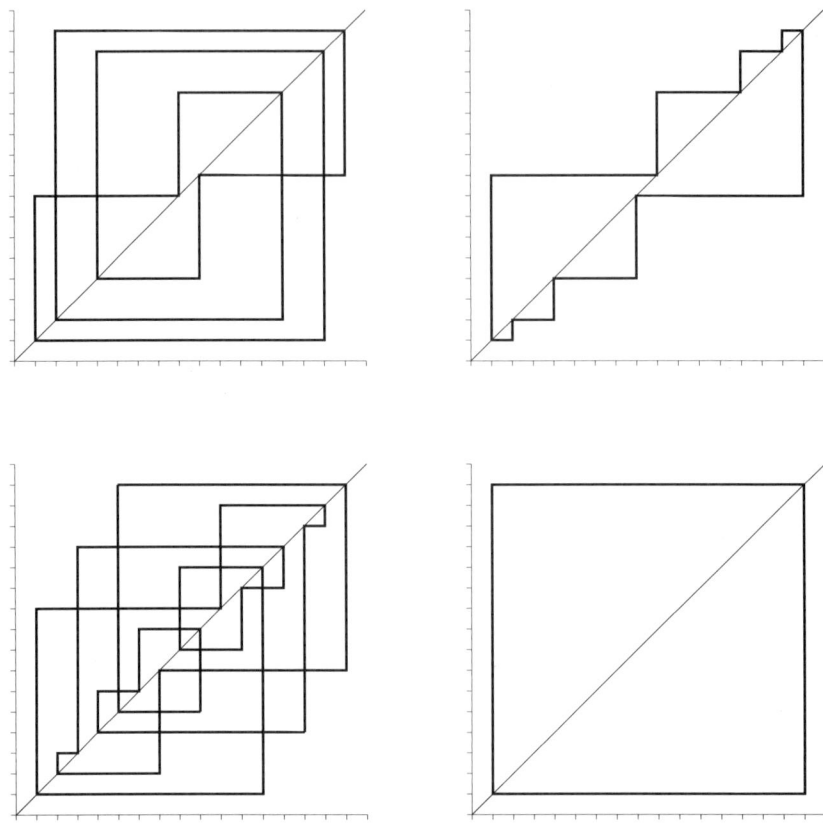

Figure 5.6. Graphical analysis of $\frac{1}{17}$ for $b = 8$, 9, 10, and 16.

Remark 5.41. Further necessary and sufficient conditions for primality are given in [Dudek] and [Vasilenko]. In [Akushskiĭ, Burtsev], an attempt to substitute modular arithmetic in place of multiprecision arithmetic into tests for primality of F_m is given. A deterministic polynomial-time algorithm to test primality of Fermat numbers is presented in [Balasubramanian].

6. Divisibility of Fermat Numbers

$$F_5 = 641 \times 6700417$$

Leonhard Euler

In 1878, Édouard A. Lucas established a criterion concerning the general form of prime divisors of the Fermat numbers, namely, that every prime divisor p of F_m, $m > 1$, satisfies the congruence (see, e.g., [Lucas, 1878b], [Dickson, p. 376])

$$p \equiv 1 \pmod{2^{m+2}}.$$

This can be reformulated as follows:

Theorem 6.1 (Lucas). *If $m > 1$ and p is a prime such that $p \mid F_m$, then p is of the form*

$$(6.1) \qquad p = k2^{m+2} + 1,$$

where k is a natural number.

P r o o f . Let $b = 2^{2^{m-2}}\left(2^{2^{m-1}} - 1\right)$. Since

$$(6.2) \qquad 2^{2^m} + 1 \equiv 0 \pmod{p},$$

we have

$$(6.3) \qquad b^2 = 2^{2^{m-1}}\left(2^{2^m} - 2 \cdot 2^{2^{m-1}} + 1\right) \equiv -2 \cdot 2^{2^m}$$
$$\equiv -2 \cdot 2^{2^m} + 2(2^{2^m} + 1) \equiv 2 \pmod{p}.$$

From this and (6.2) it follows that

$$b^{2^{m+1}} \equiv 2^{2^m} \equiv -1 \pmod{p},$$

and thus

$$b^{2^{m+2}} \equiv 1 \pmod{p}.$$

Consequently, according to Lemma 2.13, $\mathrm{ord}_p b = 2^j$ for some $j \le m + 2$. However, if $j < m + 2$ and $e = \mathrm{ord}_p(b)$, then, by Lemma 2.13 again,

$$b^{e2^{m+1-j}} - 1 = b^{2^{m+1}} - 1 \equiv 2^{2^m} - 1 \equiv 0 \pmod{p},$$

which contradicts (6.2). Hence,

(6.4) $$\mathrm{ord}_p b = 2^{m+2}.$$

The numbers p and b are coprime due to (6.3). Therefore, applying Fermat's little theorem, i.e., $b^{p-1} \equiv 1 \pmod{p}$, and using Lemma 2.13 and (6.4), we obtain

$$p - 1 = k\,\mathrm{ord}_p b = k2^{m+2}. \qquad \square$$

Remark 6.2. Other proofs of Theorem 6.1 can be found in [Ribenboim, 1996, p. 84], [Larras], and [Sierpiński, 1964b, p. 343].

Since E. A. Lucas published his theorem a great deal of attention has been devoted to primes of the form (6.1). In 1960, W. Sierpiński proved that there exist infinitely many k for which the numbers $k2^n + 1$ are composite for any natural number n (see Sierpiński's Theorem 7.4). He did not write down any such k explicitly, but today we know, e.g., that all numbers of the form $78557 \cdot 2^n + 1$ are composite. However, we do not know yet which is the smallest k with this property (see Chapter 7).

Further, we will examine divisibility of composite Fermat numbers F_m by the factor $3h2^{m+2} + 1$ (see [Křížek, Chleboun, 1994]). We prove that a Fermat number F_m is composite if and only if there exists $h \in \mathbb{N}$ such that $3h2^{m+2} + 1 \mid F_m$. In particular, we show that any factorization of a composite Fermat number $F_m = 2^{2^m} + 1$ into two nontrivial factors can be expressed in the form $F_m = (k2^n + 1)(\ell2^n + 1)$ for some odd k and ℓ, $k \geq 3$, $\ell \geq 3$, and integer $n \geq m + 2$, $3n < 2^m$. We prove that the greatest common divisor of k and ℓ is 1, $k + \ell \equiv 2^n \pmod{2^{n+1}}$, $\max(k, \ell) \geq F_{m-2}$, and either $3 \mid k$ or $3 \mid \ell$, i.e., $3h2^{m+2} + 1 \mid F_m$ for an integer $h \geq 1$. Factorizations of F_m into more than two factors are investigated as well. In particular, we prove that if the number F_m is of the form $F_m = (k2^n + 1)^2(\ell2^j + 1)$, then the exponent j is equal to $n + 1$, $3 \nmid \ell$, and $5 \nmid \ell$.

We start with several auxiliary lemmas.

Lemma 6.3. *If $2^n + 1$ divides F_m for some $n \geq 1$ and $m \geq 0$, then $F_m = 2^n + 1$.*

P r o o f . Set $Q_n = 2^n + 1$, i.e., $F_m = Q_{2^m}$. By the binomial theorem, we get

$$Q_{ij} = 2^{ij} + 1 = (Q_j - 1)^i + 1 \equiv 1 + (-1)^i \pmod{Q_j}$$

and thus

(6.5) $$\gcd(Q_{ij}, Q_j) = \begin{cases} 1 & \text{for } i \text{ even,} \\ Q_j & \text{for } i \text{ odd.} \end{cases}$$

Hence,

(6.6) $$\gcd(F_z, F_m) = 1 \quad \text{for } z \neq m;$$

i.e., no two different Fermat numbers have a common divisor greater than 1 (compare with Goldbach's Theorem 4.1).

Suppose that $Q_n \mid F_m$ for some $n < 2^m$. Then $n = i2^z$, where i is odd and $z < m$. Using (6.5) for $j = 2^z$, we see that $Q_{2^z} \mid Q_n$. However, this contradicts (6.6), since $Q_{2^z} = F_z$ and $Q_n \mid F_m$. Therefore, $n = 2^m$. $\qquad \square$

Lemma 6.4. *Let F_m be composite. If*

(6.7) $$F_m = (k2^n + 1)(\ell2^j + 1),$$

where k and ℓ are odd, then $k \geq 3$ and $\ell \geq 3$.

P r o o f . Since F_m is odd and composite, it can be written as a product of two odd numbers $k2^n + 1$ and $\ell2^j + 1$ for some natural numbers n, j and odd integers k, ℓ. However, according to Lemma 6.3 the case $k = 1$ or $\ell = 1$ is not possible. Hence, $k \geq 3$ and $\ell \geq 3$.   □

Definition 6.5. Let $q > 1$ be an odd integer. A uniquely determined exponent n from the decomposition $q = k2^n + 1$, where k is odd, is called the *order of q*.

In the next lemma we prove that the orders of two odd factors are not greater than the order of their product.

Lemma 6.6. *Let*

$$(6.8) \qquad\qquad k2^n + 1 = (k_1 2^{n_1} + 1)(k_2 2^{n_2} + 1),$$

where k, k_1, k_2 are odd. Then $n \geq \min(n_1, n_2)$, where the sharp inequality holds if and only if $n_1 = n_2$. Moreover, $k > k_1 k_2 2^{\max(n_1, n_2)}$ whenever $n_1 \neq n_2$.

P r o o f . Without loss of generality assume that $n_1 \geq n_2$. Then

$$(6.9) \qquad\qquad k2^n + 1 = (k_1 k_2 2^{n_1} + k_1 2^{n_1-n_2} + k_2)2^{n_2} + 1.$$

Since k is odd, $n \geq n_2 = \min(n_1, n_2)$. The number in the parentheses from (6.9) is even if and only if $n_1 = n_2$. If $n_1 > n_2$, then $n = n_2$, and therefore $k > k_1 k_2 2^{n_1}$ by (6.9).   □

Theorem 6.7. *Let F_m be composite and let $k2^n + 1 \neq F_m$ be an arbitrary factor (not necessarily prime), where k is odd. Then $k \geq 3$, n is an integer for which*

$$(6.10) \qquad\qquad m + 2 \leq n < \frac{2^m}{3},$$

and there exists an odd $\ell \geq 3$ such that

$$(6.11) \qquad\qquad F_m = (k2^n + 1)(\ell2^n + 1),$$

i.e., both the factors have the same order. Moreover,

$$(6.12) \qquad\qquad 2^n \| k + \ell,$$

k and ℓ are coprime, i.e.,

$$(6.13) \qquad\qquad \gcd(k, \ell) = 1,$$

$$(6.14) \qquad\qquad \max(k, \ell) \geq F_{m-2} \qquad\qquad (Chleboun's\ inequality),$$

and

$$(6.15) \qquad\qquad either \quad 3 \mid k \quad or \quad 3 \mid \ell, \quad but\ not\ both,$$

i.e., for any composite Fermat number F_m there exists an odd number h such that $3h2^n + 1 \mid F_m$.

P r o o f . Let $\ell 2^j + 1$ be the cofactor of $k2^n + 1$ with ℓ odd. By Lemma 6.4, k has to be greater than or equal to 3 and there does indeed exist an odd integer $\ell \geq 3$ such that (6.7) holds. According to (6.7), we have

$$F_m = k\ell 2^{n+j} + k2^n + \ell 2^j + 1.$$

Without loss of generality we may assume that $n \geq j$. Then

$$2^{2^m - j} = k\ell 2^n + k2^{n-j} + \ell,$$

where the terms $2^{2^m - j}$ and $k\ell 2^n$ are even because $2^m > j$ and $n \geq 1$. This implies that $n = j$, since ℓ is odd; i.e., equality (6.11) holds.

From the relation $2^{2^m - n} = k\ell 2^n + k + \ell$, we deduce that $2^m - n > n$, which implies $k + \ell \equiv 2^n \pmod{2^{n+1}}$. Hence, (6.12) holds. Moreover, if $q \mid k$ and $q \mid \ell$ for some odd q, then $q \mid 2^{2^m - n}$. Hence, $q = 1$ and we observe that (6.13) holds.

Further, we establish the bounds for n in (6.10). By (6.12), $k + \ell \geq 2^n$. Since $k \neq \ell$ due to (6.13), we have

(6.16) $\max(k, \ell) > 2^{n-1},$

and thus

$$F_m = (k2^n + 1)(\ell 2^n + 1) > (2^{n-1}2^n + 1)(2 \cdot 2^n + 1) > 2^{3n} + 1.$$

Consequently, $3n < 2^m$.

By Theorem 6.1, each prime factor of F_m is of the form $r2^{m+2} + 1$ for some integer r. Hence, if $k2^n + 1$ is a prime factor, then $m + 2 \leq n$, since k is odd. Suppose that $k2^n + 1$ is a product of two primes as given in (6.8). Then Lemma 6.6 implies $m + 2 \leq \min(n_1, n_2) \leq n$. By induction we find that $m + 2 \leq n$ for any factor of F_m; i.e., (6.10) is valid.

If $n \leq 2^{m-2}$, then by (6.11), (6.13), and (6.10)

$$\max(k, \ell) > 2^{-n}\left(\sqrt{F_m} - 1\right) > 2^{-2^{m-2}}\left(2^{2^{m-1}} - 1\right) = 2^{2^{m-2}} - 2^{-2^{m-2}},$$

and thus $\max(k, \ell) \geq F_{m-2}$, since $\max(k, \ell) \geq 2^{2^{m-2}}$ and k and ℓ are odd. If on the other hand $n \geq 2^{m-2} + 1$, then by (6.16),

$$\max(k, \ell) > 2^{n-1} \geq 2^{2^{m-2}};$$

i.e., (6.14) holds.

Finally, we prove (6.15). By (3.5),

(6.17) $F_m \equiv 2 \pmod{3}.$

We easily find that $xy \equiv 2 \pmod{3}$ if and only if $x \equiv 2 \pmod{3}$ and $y \equiv 1 \pmod{3}$ or $x \equiv 1 \pmod{3}$ and $y \equiv 2 \pmod{3}$. From this, (6.11), and (6.17) we observe that exactly one of the numbers k and ℓ is divisible by 3.   □

Corollary 6.8. *Let the assumptions of Theorem 6.7 be satisfied and let*

(6.18) $3 \mid \ell.$

Then

(6.19) $k = 3u + 1$ *for some u even* $\Longleftrightarrow$ *n is even,*
(6.20) $k = 3u + 2$ *for some u odd* $\Longleftrightarrow$ *n is odd.*

P r o o f . From (6.11), (6.18), and congruence (6.17) we see that

(6.21) $k2^n + 1 \equiv 2 \pmod{3}.$

Now the properties (6.15) and (6.18) yield that $k = 3u + y$, where $1 \le y \le 2$. If $k = 3u + 1$ and n is odd, then $k2^n + 1 \equiv 0 \pmod{3}$, which contradicts (6.21). Similarly, if $k = 3u + 2$ and n is even, then $k2^n + 1 \equiv 0 \pmod{3}$, which again contradicts (6.21). Hence, (6.19) and (6.20) both hold.   □

The following assertion can be found in the book [Sierpiński, 1970, Problem 44].

Corollary 6.9. $\gcd(m, F_m) = 1$ *for* $m = 1, 2, \ldots$.

P r o o f . By Theorem 6.7, any given nontrivial divisor of F_m is of the form $k2^{m+2} + 1$, where $k \ge 3$. Since $k2^{m+2} + 1 > 2^{m+2} + 1 > m$, the divisor is greater than m. Hence, $\gcd(m, F_m) = 1$.   □

Remark 6.10. Although the upper bound on n in (6.10) is too rough, we observe that no n satisfies (6.10) if $m \le 4$. This implies that $F_0, \ldots, F_4$ are primes without the necessity of carrying out any trial divisions. For the prime factor $641 = 5 \cdot 2^7 + 1$ of F_5, we have the equality $n = m + 2$. On the other hand, the sharp inequality $n > m + 2$ holds, e.g., for the factorization of F_8 into two primes with $n = 11$. According to (6.11) and (6.10),

$$\min(k, \ell) < (2^n \min(k, \ell) + 1)/2^n < \sqrt{F_m}/2^n < F_{m-1}/2^{m+2}.$$

Moreover, $\min(k, \ell) \ge 3$, where the equality is achieved, e.g., for prime factors of F_{38} and F_{207} (see [Brillhart, Lehmer, Selfridge, Tuckerman, Wagstaff, p. lxxxviii] or [www1]). We further note that when $m = 207$, we also have the equality $n = 209 = m + 2$. Thus, when $m = 207$, we find a factor $3 \cdot 2^{209} + 1$ of F_{207} by examining the first possible candidate.

Remark 6.11. According to (6.11) and (6.13), no Fermat number is the square of a natural number. We will generalize this result in Theorem 9.1.

Remark 6.12. By Corollary 6.8, we can completely determine the parity of n when $k2^n + 1$ is a nontrivial divisor of F_m such that k is odd and $3 \nmid k$. When k is odd and $3 \mid k$, we have only the following partial results.

By Theorem 6.7, if F_m is composite, there exists an odd number h such that $3h2^n + 1$ divides F_m. In particular, it was shown in [Morehead, 1906] (see also

[Robinson, 1958, p. 680]) that if a prime of the form $3 \cdot 2^n + 1$ divides F_m, then n is odd. The following generalization slightly extends a result from [Suyama, 1984a]:

Theorem 6.13. *Suppose that* $p = 3h^2 2^n + 1$ *divides a composite Fermat number* F_m, *where* p *is a prime,* h *is odd, and* $3 \nmid h$. *Then* $n \equiv 1 \pmod 4$ *if* $h \equiv \pm 1 \pmod 5$ *and* $n \equiv 3 \pmod 4$ *if* $h \equiv \pm 2 \pmod 5$. *In particular, if* $p = 3 \cdot 2^n + 1$, *then* $n \equiv 1 \pmod 4$.

Before proving Theorem 6.13, we will have to consider kth-power residues modulo p, where $p \equiv 1 \pmod k$ is a prime and $k \geq 2$. An integer $a \not\equiv 0 \pmod p$ is a kth-power residue $\pmod p$ if there exists an integer r such that $a \equiv r^k \pmod p$. The following lemma, due to Euler, provides a necessary and sufficient condition for determining kth-power residues modulo p and generalizes Theorem 2.26 (Euler's criterion) for quadratic residues modulo p.

Lemma 6.14. *Let* $k \geq 2$ *and* $p \equiv 1 \pmod k$ *be a prime. Then the integer* $a \not\equiv 0 \pmod p$ *is a* kth-power *residue modulo* p *if and only if*

$$(6.22) \qquad\qquad a^{(p-1)/k} \equiv 1 \pmod p.$$

P r o o f . Suppose that a is a kth-power residue $\pmod p$. Then $a \equiv r^k \pmod p$ for some integer $r \not\equiv 0 \pmod p$. Then, by Fermat's little theorem,

$$a^{(p-1)/k} \equiv \left(r^k \right)^{(p-1)/k} \equiv r^{p-1} \equiv 1 \pmod p.$$

Now suppose that $a^{(p-1)/k} \equiv 1 \pmod p$. Let g be a primitive root modulo p. Then $a \equiv g^t \pmod p$ for some integer $t \in \{1, \ldots, p-1\}$. It follows that

$$g^{t(p-1)/k} \equiv a^{(p-1)/k} \equiv 1 \pmod p.$$

However, the order of $g \pmod p$, namely $p - 1$, has to divide $t(p - 1)/k$. This implies that $k \mid t$.

Let $t = kj$. Hence,

$$(g^j)^k = g^t \equiv a \pmod p,$$

and a is a kth-power residue modulo p. □

Let us now suppose that F_m is composite and has a prime factor p of the form $k2^n + 1$, where $k \geq 3$ is odd. By Theorem 4.12,

$$2^{2^{m+1}} \equiv 1 \pmod p.$$

However, by Lucas's Theorem 6.1, $n \geq m + 2$. Therefore,

$$(6.23) \qquad\qquad 2^{2^n} = 2^{(p-1)/k} \equiv 1 \pmod p,$$

and 2 is a kth-power residue modulo p by Lemma 6.14.

We are now ready for the proof of Theorem 6.13.

P r o o f   o f   T h e o r e m   6 . 1 3 . Suppose that $p = 3h^2 2^n + 1 \mid F_m$. Then $p \equiv 1 \pmod 3$. We claim that 2 is a cubic residue modulo p. By (6.23),

$$2^{(p-1)/3} = \left(2^{2^n} \right)^{h^2} \equiv 1^{h^2} \equiv 1 \pmod p.$$

Hence, 2 is a cubic residue modulo p by Lemma 6.14. But in order for 2 to be a cubic residue modulo p, it is necessary that in the unique representation of p in the form $p = r^2 + 3s^2$, the number s be a multiple of 3 (see, for example, [Ireland, Rosen, pp. 118–119]). However, if n is even, then $r = 1$ and $s = h2^{n/2}$, contradicting the fact that $3 \mid s$. Thus n is odd.

We now consider powers of 2 modulo 5. We then obtain $\mathrm{ord}_5 2 = 4$, $2^{4i+1} \equiv 2$ (mod 5), and $2^{4i+3} \equiv 3$ (mod 5). Assuming that $h \equiv \pm 1$ (mod 5), we find that $3h^2 \equiv 3$ (mod 5).

Suppose that $n \equiv 3$ (mod 4). Then

$$p = 3h^2 2^n + 1 \equiv 3 \cdot 3 + 1 \equiv 0 \pmod{5},$$

which contradicts the fact that p is a prime and $p \geq 3 \cdot 2 + 1 = 7$. Now assume that $h \equiv \pm 2$ (mod 5) and that $n \equiv 1$ (mod 4). Then $3h^2 \equiv 3 \cdot 4 \equiv 2$ (mod 5) and consequently,

$$p = 3h^2 2^n + 1 \equiv 2 \cdot 2 + 1 \equiv 0 \pmod{5},$$

which again is a contradiction. The result now follows.   □

Remark 6.15. In [Suyama, 1984a] it was shown that if $p = 3h^2 2^n + 1$ divides a composite Fermat number F_m, where $3 \nmid h$, then n is odd.

According to [Golomb, 1976], the exponent of 2 modulo a prime $p = 3 \cdot 2^n + 1$ fails to be divisible by 3 if and only if p divides a Fermat number F_m with $m \leq n - 1$.

Theorem 6.16. Let $n_1 \leq n_2 \leq n_3$ and let

$$(6.24) \qquad F_m = \prod_{j=1}^{3} (k_j 2^{n_j} + 1),$$

where the k_j's are odd. Then $k_j \geq 3$ for $j = 1, 2, 3$,

$$(6.25) \qquad m + 2 \leq n_1 = n_2 < n_3,$$

and either no k_j is divisible by 3 or just two k_j are divisible by 3.

Moreover, if $k_1 = k_2$ (i.e., if F_m is not square-free), then $n_3 = n_1 + 1$, $3 \nmid k_3$, and $5 \nmid k_3$, i.e., $\gcd(k_3, 15) = 1$.

P r o o f . Obviously, $k_j \geq 3$ and $n_j \geq m + 2$ by Theorem 6.7. Let us rewrite (6.24) as a product of two factors:

$$(6.26) \qquad F_m = (k_1 2^{n_1} + 1)[(k_2 k_3 2^{n_3} + k_2 + k_3 2^{n_3 - n_2}) 2^{n_2} + 1].$$

The number $k_2 k_3 2^{n_3} + k_2 + k_3 2^{n_3 - n_2}$ cannot be even, since then we would have $n_3 = n_2$, and by Theorem 6.7 we would get $n_1 \geq n_2 + 1$, which contradicts the assumption $n_1 \leq n_2$. Consequently, $k_2 k_3 2^{n_3} + k_2 + k_3 2^{n_3 - n_2}$ is an odd number, and thus $k_3 2^{n_3 - n_2}$ is even. This implies that $n_3 > n_2$. By Theorem 6.7 and equality (6.26) we have $n_1 = n_2$.

From (6.26) and (6.15) we see that all three k_j cannot be divisible by 3. Suppose now that just one k_j is divisible by 3. Let, for instance, $3 \nmid k_1$, $3 \mid k_2$, and $3 \nmid k_3$. Then $k_2 k_3 2^{n_3} + k_2 + k_3 2^{n_3 - n_2}$ is not divisible by 3, which contradicts (6.15) and

(6.26). In a similar way we get a contradiction for the cases $3 \nmid k_1$, $3 \nmid k_2$, $3 \mid k_3$, and $3 \mid k_1$, $3 \nmid k_2$, $3 \nmid k_3$.

Finally, suppose that $k_1 = k_2$ in (6.24). Then obviously $3 \nmid k_3$, and from (6.11) and the relation

$$F_m = [k_1(k_1 2^{n_1-1} + 1)2^{n_1+1} + 1](k_3 2^{n_3} + 1)$$

we find that $n_3 = n_1 + 1$.

Recall that the last digit of $k_1 2^{n_1} + 1$ belongs to the set $\{1, 3, 7, 9\}$, since $5 \nmid F_m$ for $m \neq 1$ by (6.6). Hence,

$$(k_1 2^{n_1} + 1)^2 \quad (\text{mod } 10) \in \{1, 9\}.$$

From this, (6.24), and the fact that $F_m \equiv 7 \pmod{10}$ for $m > 1$ we have $k_3 2^{n_3} + 1$ (mod 10) $\in \{3, 7\}$, which yields $5 \nmid k_3$.   □

Remark 6.17. The Fermat number F_9 is a product of three prime factors $k_j 2^{n_j} + 1$, $j = 1, 2, 3$; cf. Appendix A. According to [Lenstra, Lenstra, Manasse, Pollard, p. 321], their orders are $n_1 = n_2 = 11 = m + 2$ and $n_3 = 16$, and thus by (6.11) we get

$$(6.27) \quad F_9 = (k_1 2^{11} + 1)(\ell_1 2^{11} + 1) = (k_2 2^{11} + 1)(\ell_2 2^{11} + 1) = (k_3 2^{16} + 1)(\ell_3 2^{16} + 1)$$

for some $\ell_j \geq 3$ odd. Hence, any factor $\ell 2^n + 1$ of F_m for which $n = m + 2$ need not be a prime factor yet. We also see that for a given $n \geq m + 2$ the Diophantine equation (6.11) with unknowns k and ℓ can have no solution, one solution, or more than one solution. It is also interesting that no k_j from (6.27) is divisible by 3. This can be directly verified from the explicit expressions of the prime factors of F_9 (see Appendix A for two of the explicit expressions for the prime factors of F_9 and use Theorem 6.16), and thus $3 \mid \ell_j$ for $j = 1, 2, 3$ by (6.15). According to (6.25), no Fermat number is the cube of a natural number (see also [Krishna] and Theorem 9.1).

Theorem 6.18. Let $n_1 \leq n_2 \leq \cdots \leq n_N$, $N > 1$, and let

$$(6.28) \qquad\qquad F_m = \prod_{j=1}^{N} (k_j 2^{n_j} + 1),$$

where the k_j's are odd. Then $m + 2 \leq n_j$, $k_j \geq 3$ for $j = 1, \ldots, N$, and the number of factors $k_j 2^{n_j} + 1$, whose order is n_1, is even. No two factors from (6.28) form a twin-prime pair.

P r o o f . We again have by Theorem 6.7 that $m + 2 \leq n_j$ and $k_j \geq 3$ for all $j = 1, \ldots, N$. For $N < 4$ the proof of the first part of Theorem 6.18 follows from Theorems 6.7 and 6.16. So let $N \geq 4$. Suppose, to the contrary, that $2z + 1$ (for an integer $z \geq 0$) is the number of factors of lowest order n_1, i.e., $n_{2z+1} < n_{2z+2}$ if $2z + 1 < N$. Then by Lemma 6.6 we have for $z \geq 1$ that

$$\text{ord}\big((k_{2i} 2^{n_1} + 1)(k_{2i+1} 2^{n_1} + 1)\big) > n_1 \quad \text{for } i = 1, \ldots, z,$$

where analogously to [Lenstra, Lenstra, Manasse, Pollard, p. 321] the operator ord denotes the order from Definition 6.5, i.e., $\text{ord}(k2^n + 1) = n$ for k odd. Using Lemma 6.6 again, we find by induction that

$$\text{ord}\Big(\prod_{j=2}^{2z+1} (k_j 2^{n_1} + 1) \Big) > n_1,$$

and thus also

(6.29) $$\text{ord}\Big(\prod_{j=2}^{N} (k_j 2^{n_j} + 1) \Big) > n_1$$

for $z \geq 1$. However, we easily find that (6.29) holds even if $z \geq 0$. This contradicts (6.28) and (6.11), since $\text{ord}(k_1 2^{n_1} + 1) = n_1$.

Let $n_j \leq n_i$. Then

$$\big| (k_i 2^{n_i} + 1) - (k_j 2^{n_j} + 1) \big| = \big| (k_i 2^{n_i - n_j} - k_j) 2^{n_j} \big| \geq 2^{n_j} \geq 2^{m+2}$$

whenever $n_i \neq n_j$ or $k_i \neq k_j$. From this we see that the product (6.28) cannot contain a twin-prime pair. $\square$

Remark 6.19. The factors p_8 and p_{10} of F_{10} (see Appendix A) have orders 12 and 14, respectively, where p_j denotes a prime factor with exactly j digits. Using Theorem 6.18, we predicted in [Křížek, Chleboun, 1994, p. 444] that there exists an unknown factor of F_{10} whose order is 12. This 40-digit factor was later discovered by Brent. The remaining factor p_{252} of F_{10} has order 13. The 21-digit prime factor of F_{11} (see [Brent, 1989]) is of order 14. The other four prime factors have order 13. From Theorem 6.18, we observe that there exist at least four factors of F_{12} of order $m + 2 = 14$, since three of them are already known (see Appendix A).

Finally note that the k_j's in the equality (6.28) need not be coprime (compare with (6.13)). For instance, we have $3 \mid k_j$ for two prime factors of F_{11} and $7 \mid k_j$ for two of its other three prime factors, and $7 \mid k_j$ for three of the known prime factors of F_{12}, etc.

Remark 6.20. By Lucas's Theorem 6.1, the complete prime factorization of F_m, $m > 2$, satisfies the inequality

$$2^{2^m} + 1 = (k_1 2^{n_1} + 1) \cdots (k_j 2^{n_j} + 1) > (2 \cdot 2^{m+2} + 1)^j,$$

where j is the number of all prime factors. Hence, $2^{2^m} > 2^{j(m+3)}$, which implies that $2^m > j(m + 3)$. Therefore, the number of prime factors of F_m is at most $2^m/(m + 3)$.

If $p^2 \mid F_m$, where p is a prime, considerably more can be said about the form of p than in Theorem 6.1. The following theorem was proved in [Ribenboim, 1979a, p. 88].

Theorem 6.21. Let $m \geq 2$ and suppose that $p^2 \mid F_m$, where p is a prime. Then

$$p = k2^r + 1,$$

where $r \geq m + 2$, k is odd, and

$$\frac{k^{p-1} - 1}{p} \equiv 1 \pmod{p}.$$

No Fermat numbers are known that are divisible by the square of a prime number (for more on such numbers see Chapters 13 and 14). The following theorem is usually applied to test whether F_m is divisible by the square of a prime (cf. [Ribenboim, 1979a], [Rotkiewicz, 1965], [Warren, Bray]):

Theorem 6.22. *If a prime number p divides F_m, then*

$$(6.30) \qquad p^2 \mid F_m \quad \Longleftrightarrow \quad 2^{p-1} \equiv 1 \pmod{p^2} \qquad \text{(Wieferich's congruence)}.$$

P r o o f . If $p^2 \mid F_m$ then

$$2^{2^m} \equiv -1 \pmod{p^2}.$$

From this we see that $2^{2^{m+2}} \equiv 1 \pmod{p^2}$, and thus $2^{k2^{m+2}} \equiv 1 \pmod{p^2}$ for any $k \in N$. Since $p = k2^{m+2} + 1$ for some k by Lucas's Theorem 6.1, we get $2^{p-1} \equiv 1 \pmod{p^2}$.

Conversely, suppose that $2^{p-1} \equiv 1 \pmod{p^2}$. It is well known and easily proven by use of the binomial theorem (see [LeVeque]) that if $a \geq 2$ is an integer and t is the largest integer such that $\text{ord}_{p^t}(a) = \text{ord}_p(a)$, then $\text{ord}_{p^r}(a) = \text{ord}_p(a)$ for $r \in \{1, \ldots, t\}$ and $\text{ord}_{p^t}(a) = p^{r-t}\text{ord}_p(a)$ for $r > t$. Since $p \nmid p - 1$, it follows that $\text{ord}_{p^2}(2) = \text{ord}_p(2)$. Thus, $2^k \equiv 1 \pmod{p^2}$ if and only if $2^k \equiv 1 \pmod{p}$. Since $p \mid F_m$, we have $2^{2^m} \equiv -1 \pmod{p}$ and hence $2^{2^{m+1}} \equiv 1 \pmod{p}$, which implies that $2^{2^{m+1}} \equiv 1 \pmod{p^2}$. Thus

$$p^2 \mid 2^{2^{m+1}} - 1 = \left(2^{2^m} + 1\right)\left(2^{2^m} - 1\right).$$

Since $p \mid F_m = 2^{2^m} + 1$ and $\left(2^{2^m} + 1\right) - \left(2^{2^m} - 1\right) = 2$, we see that $p \nmid 2^{2^m} - 1$. Hence, $\gcd\left(p^2, 2^{2^m} - 1\right) = 1$ and thus $p^2 \mid 2^{2^m} + 1 = F_m$. $\square$

Remark 6.23. For example, a test that the square of the prime number $p = 85 \cdot 2^{2458} + 1$ does not divide F_{2456} took 165 hours of computer time [Gostin, McLaughlin]. Although extensive computer searches have been performed (see [Crandall, Dilcher, Pomerance], in which the authors searched up to $4 \cdot 10^{12}$), we know only two primes that satisfy Wieferich's congruence in (6.30), $p = 1093$ and $p = 3511$. (For an elementary proof that 1093 and 3511 satisfy Wieferich's congruence, see [Guy, 1967]. See also [Montgomery, 1993].) The primes 1093 and 3511 are called *Wieferich primes;* they were discovered by Meissner in 1913 and Beeger in 1922, long before the advent of computers (see [Ribenboim, 1996, p. 334]). However, these two primes do not divide any Fermat number F_m, since neither has the form $k2^{m+2} + 1$ for $m \geq 5$.

Before the solution of Fermat's last theorem (see [Wiles] and [Taylor, Wiles]), investigators considered equations of the form

$$(6.31) \qquad\qquad x^p + y^p = z^p,$$

where p is an odd prime, $xyz \neq 0$, and $\gcd(x, y, z) = 1$. Traditionally, Fermat's last theorem was split into two cases. In the first case, it was assumed that $p \nmid xyz$. In the second, and much harder, case, it was assumed that $p \mid xyz$. Using equivalence (6.30), we show below a connection between the first case of Fermat's last theorem and Fermat numbers.

Theorem 6.24 (Wieferich). *If the first case of Fermat's last theorem is false for the odd prime exponent p, then p satisfies Wieferich's congruence*

$$(6.32) \qquad\qquad\qquad 2^{p-1} \equiv 1 \pmod{p^2}.$$

This theorem is given in [Wieferich] (cf. also [Ribenboim, 1979b]). The composite integer number n is said to be *powerful* if whenever $p \mid n$, then $p^2 \mid n$, where p is a prime. Theorem 6.25 below gives a connection between Wieferich primes and powerful Fermat numbers (see [Ribenboim, 1996, pp. 343–345]).

Theorem 6.25. *If there exist only finitely many Wieferich primes, then there exist infinitely many Fermat numbers that are not powerful.*

P r o o f . Assume that there exist only finitely many Fermat numbers that are not powerful. Then infinitely many Fermat numbers are powerful. By Theorem 6.22 each prime divisor of a powerful Fermat number is a Wieferich prime. Since any two distinct Fermat numbers are, by Goldbach's Theorem 4.1, coprime, there exist infinitely many Wieferich primes, contrary to the hypothesis. The result now follows.   □

We also have the following theorem, which gives a relationship between powerful Fermat numbers and primitive prime divisors of $2^n - 1$ (see [Ribenboim, 1996, pp. 341, 343–344]). The prime p is said to be a *primitive prime divisor* of $2^n - 1$ ($n \geq 1$) if $p \mid 2^n - 1$ but $p \nmid 2^m - 1$ for $1 \leq m < n$ (see also Chapter 12 for a discussion of primitive prime divisors).

Theorem 6.26. *If there exist infinitely many Fermat numbers that are not powerful, then there exist infinitely many numbers $2^n - 1$ having a primitive prime divisor p_n such that p_n^2 does not divide $2^n - 1$.*

P r o o f . Since the Fermat numbers are pairwise relatively prime, by Goldbach's Theorem 4.1 it suffices to show that if p is a prime such that $p \mid F_m$ but $p^2 \nmid F_m$, then p is a primitive prime divisor of $2^{2^{m+1}} - 1$ and $p^2 \nmid 2^{2^{m+1}} - 1$.

Note that $2^{2^{m+1}} - 1 = \left(2^{2^m} + 1\right)\left(2^{2^m} - 1\right)$. However, $p \nmid 2^{2^m} - 1$, since p is odd and p divides $F_m = 2^{2^m} + 1$. Hence, $p \mid 2^{2^{m+1}} - 1$ but $p^2 \nmid 2^{2^{m+1}} - 1$. Let $t \geq 1$ be the smallest integer such that p divides $2^{2^t} - 1$. If $t < m + 1$, then $p \nmid F_{t-1} = 2^{2^{t-1}} + 1$ by Goldbach's theorem. Therefore, $p \mid 2^{2^{t-1}} - 1$, which is a contradiction. Consequently, p is a primitive prime divisor of $2^{2^{m+1}} - 1$.   □

Remark 6.27. Almost surely, there are infinitely many Fermat numbers that are not powerful. However, there is virtually no convincing evidence one way or the other regarding the existence of infinitely many Wieferich primes. In [Ribenboim, 1996, pp. 333–346] further interesting results concerning connections among Fermat numbers, Wieferich primes, and powerful numbers are presented.

7. Factors of Fermat Numbers

Numbers constitute the only universal language.

Nathanael West

The factors $k2^n + 1$, $k, n \in \mathbb{N}$, of Fermat numbers have been intensively studied by many authors, e.g., [Artjuhov], [Baillie], [Bosma], [Brent, 1982], [Brillhart, Lehmer, Selfridge], [Cormack, Williams], [Golomb, 1976], [Keller 1983, 1992], [Křížek, Chleboun, 1994, 1997], [Papademetrios], [Shorey, Stewart], [Williams, 1988]. In 1878, F. Proth stated the following theorem (see [Proth, 1878b, 1978c] and see [Robinson, 1957b], [Sierpiński, 1964a] for proofs of this theorem), which can be applied to verify easily the primality of divisors of Fermat numbers for $k < 2^n$ (see [Robinson, 1957a] and [Robinson, 1958]; also compare with Suyama's Theorem 4.22).

Theorem 7.1 (Proth). *Let $N = k2^n + 1$ with odd $k < 2^n$, and suppose that $\left(\frac{a}{N}\right) = -1$. Then N is prime if and only if*

$$(7.1) \qquad\qquad a^{(N-1)/2} \equiv -1 \pmod{N}.$$

First note that Pepin's test stated in Theorem 5.5 is a special case of Theorem 7.1 for $N = F_m$, $m \geq 1$, and $a = 3$.

In order to prove Theorem 7.1 we will make use of the following primality test, which is proved in [Pocklington].

Theorem 7.2 (Pocklington). *Let $N = kp^n + 1$, where p is prime and $p \nmid k$. If for some integer a we have*

(i) $a^{N-1} \equiv 1 \pmod{N}$,
(ii) $\gcd(a^{(N-1)/p} - 1, N) = 1$,

then each prime factor of N is of the form $mp^n + 1$.

P r o o f . Let q be a prime factor of N and let $e = \mathrm{ord}_q a$. Then $e \mid N - 1$ by (i) and $e \mid q - 1$ by Fermat's Little Theorem 2.9. By assumption (ii), $e \nmid (N-1)/p$, because $q \mid N$. Thus, $p \nmid (N-1)/e$. Since $p^n \| N - 1$, we have $p^n \| e$, which implies that $p^n \mid q - 1$. □

We are now ready for the proof of Proth's theorem.

P r o o f o f T h e o r e m 7 . 1 . Suppose that N is prime. Since

$$\left(\frac{a}{N}\right) = -1,$$

then, by Theorem 2.26 (Euler's criterion),

$$a^{(N-1)/2} \equiv -1 \pmod{N}.$$

Now assume that congruence (7.1) holds. Then $N - 1 = k2^n$, $\gcd(k, 2^n) = 1$, and

$$a^{N-1} = \left(a^{(N-1)/2}\right)^2 \equiv (-1)^2 \equiv 1 \pmod{N}.$$

Since N is odd and $N \mid a^{(N-1)/2} + 1$, it follows that $\gcd\left(a^{(N-1)/2} - 1, N\right) = 1$. By Pocklington's Theorem 7.2, each prime factor q of N is of the form

$$q = m2^n + 1 > 2^n.$$

But

$$N = k2^n + 1 < 2^{2n}.$$

Therefore,

$$\sqrt{N} < 2^n < q,$$

which implies that N is prime.   □

Remark 7.3. Primality testing of the factors $k2^n + 1$ of F_m by Proth's Theorem 7.1 should be employed only when $n \geq m + 4$ and when the assumption

$$(7.2) \qquad\qquad k2^{n-(m+2)} < 9 \cdot 2^{m+2} + 6$$

of Suyama's Theorem 4.22 is not satisfied. The reason is that if $n \leq m + 3$, then the assumption

$$(7.3) \qquad\qquad\qquad k < 2^n$$

appearing in Proth's theorem implies (7.2). Indeed, for $n - (m + 2) \leq 1$ we have

$$k2^{n-(m+2)} < 2^{n+1} \leq 2^{m+4} < 9 \cdot 2^{m+2} + 6.$$

Note that the factors $1071 \cdot 2^8 + 1$ and $11131 \cdot 2^{12} + 1$ of F_6 and F_{10} (see Appendix A) fulfill (7.2), but they do not fulfill (7.3).

The following theorem can be found in [Sierpiński, 1960] or [Sierpiński, 1970, Problem 128]. Its proof contains an interesting application of the Fermat numbers.

Theorem 7.4 (Sierpiński). *There exist infinitely many natural numbers k such that all numbers $k2^n + 1$, $n = 1, 2, \ldots$, are composite.*

P r o o f . We know that $F_0, \ldots, F_4$ are primes and that $F_5 = 641\,p$, where p is a prime such that $p > 2^{16} + 1 = F_4$. Moreover, since $p \mid F_5$, we see that $\gcd(p, F_5 - 2) = \gcd(p, 2^{32} - 1) = 1$. Hence, $\gcd\left(p, 641\left(2^{32} - 1\right)\right) = 1$. Now, according to the Chinese remainder theorem, there exist infinitely many natural numbers k such that

$$(7.4) \qquad\qquad k \equiv 1 \pmod{641(2^{32} - 1)},$$
$$(7.5) \qquad\qquad k \equiv -1 \pmod{p}.$$

We prove that if k is a natural number greater than p that satisfies the system of congruences (7.4)–(7.5), then all numbers $k2^n + 1$ $(n = 1, 2, \ldots)$ are composite. Obviously, we may express the number n in the form $n = 2^m(2t + 1)$, where m and t are nonnegative integers. First let $m \in \{0, 1, 2, 3, 4\}$. Then from (7.4) we have

$$(7.6) \qquad k2^n + 1 \equiv 2^{2^m(2t+1)} + 1 \pmod{2^{32} - 1}.$$

Since $F_m \mid 2^{32} - 1$ for $m = 0, 1, 2, 3, 4$ and $F_m \mid 2^{2^m(2t+1)} + 1$, we find from (7.6) that $F_m \mid k2^n + 1$, and thus the inequalities $k2^n + 1 > k > p > F_4$ imply that the number $k2^n + 1$ is composite.

We now let $m = 5$. Then, by (7.4),

$$k2^n + 1 \equiv 2^{2^5(2t+1)} + 1 \pmod{641}$$

for a nonnegative integer t. Since $k > p > 641$ and since $641 \mid 2^{2^5} + 1 \mid 2^{2^5(2t+1)} + 1$, by (1.3), we see that the number $k2^n + 1$ is composite.

It remains to examine the case $m \geq 6$. Then $2^6 \mid n$, which implies that $n = 2^6 h$ for a natural number h. According to (7.5), we have

$$k2^n + 1 \equiv -2^{2^6 h} + 1 \pmod{p},$$

and because $p \mid 2^{2^5} + 1 \mid 2^{2^6} - 1 \mid 2^{2^6 h} - 1$, we obtain $p \mid k2^n + 1$. Since

$$k2^n + 1 > k > p,$$

the number $k2^n + 1$ is composite. $\square$

Definition 7.5. A natural number k is said to be a *Sierpiński number* if the sequence $\{k2^n + 1\}_{n=1}^\infty$ contains only composite numbers.

Remark 7.6. All known Sierpiński numbers k have been found by using *covering sets*. In this method, one finds a finite set of primes $p_1, p_2, \ldots, p_r$ such that every integer of the sequence $k2^n + 1$, $n = 1, 2, \ldots$, is divisible by at least one of these primes. For instance, for any positive k satisfying the congruences (7.4)–(7.5) in the proof of Sierpiński's theorem, the covering set is

$$(7.7) \qquad \{3, 5, 17, 257, 641, 65537, 6700417\}.$$

By (2.7), the smallest positive integer k that fulfills (7.4)–(7.5) satisfies the congruence

$$k \equiv M_1 y_1 - M_2 y_2 \pmod{M},$$

where $M_1 = 6\,700417$, $M_2 = 2\,753074\,036095$, $M = 18\,446744\,073709\,551615$, and $y_1 = 1\,157493\,686323$ and $y_2 = 3\,883315$ satisfy the congruences (cf. (2.6))

$$M_i y_i \equiv 1 \pmod{(M/M_i)}, \quad i = 1, 2,$$

which can be solved by use of the Euclidean algorithm (see Example 2.5). In this way we find the smallest positive solution of the system (7.4)–(7.5),

$$k = 15\,511380\,746462\,593381.$$

The smallest Sierpiński number with the same covering set as given by (7.7) is $k = 201446\,503145\,165177$ (see [Baillie, Cormack, Williams]). The smallest known Sierpiński number is 78557. It was discovered by Selfridge in 1962 and has the covering set $\{3, 5, 7, 13, 19, 37, 73\}$. Up to now we know 11 Sierpiński numbers less than 10^6:

$$78557, \ 271129, \ 271577, \ 327739, \ 482719, \ 575041,$$
$$603713, \ 808247, \ 903983, \ 934909, \ 965431.$$

In [Jaeschke, 1983, p. 382] these numbers are listed along with their covering sets. It is interesting that the second and third numbers in the above list, which are close to each other, have the same covering set $\{3, 5, 7, 13, 17, 241\}$.

It is believed that there are no Sierpiński numbers less than 78557. As of April 25, 2001, there were only 18 numbers less than 78557 whose Sierpiński-number status was unknown. These are the following numbers k (see [www5]):

$$4847, \ \ 5359, \ 10223, \ 19249, \ 21181, \ 22699, \ 24737, \ 27653, \ 28433,$$
$$33661, \ 44131, \ 46157, \ 54767, \ 55459, \ 59569, \ 65567, \ 67607, \ 69109.$$

For each of the above numbers k, all of the integers $k2^n + 1$ have been tested for primality for values of n up to at least 300000.

The investigators who have been most successful in eliminating the most candidates k less than 78557 from the list of "possible" Sierpiński numbers include Baillie, Cormack, Williams, Jacschke, Keller, Buell, and Young (see [Baillie, Cormack, Williams], [Jaeschke, 1983], [Keller, 1983, 1992], [Buell, Young], and [www5]). For more information on Sierpiński numbers see the web sites [www4] and [www5].

Remark 7.7. By Theorem 6.7, any prime factor p of F_m is of the form $k2^{m+2}+1$, where

$$(7.8) \qquad\qquad\qquad 3 \le k \le 2^{2^m - m - 2}.$$

Let $P(F_m)$ denote the largest prime factor of F_m. The following theorem (see [Le]), whose proof is highly nonelementary, depends on results from analytic number theory and uses a linear form in p-adic logarithms due to P. P. Dong. This theorem, involving Baker's method [Baker, 1966, 1977], sharply improves the bound for k given in (7.8) for very large m when $p = P(F_m)$.

Theorem 7.8 (Le). If $m \ge 2^{18}$, then

$$P(F_m) \ge m2^{m-4} + 1.$$

Le's result is substantially improved in [Grytczuk, Luca, Wójtowicz], whose proof uses elementary arguments and is true for all $m \ge 4$, not only for $m \ge 2^{18}$. We first give an auxiliary lemma.

Lemma 7.9. For $y > x > 1$ we have

$$(7.9) \qquad\qquad\qquad \frac{\log(y+1)}{\log(x+1)} < \frac{\log y}{\log x}.$$

P r o o f . Notice that inequality (7.9) is equivalent to

$$\frac{\log(y+1)}{\log y} < \frac{\log(x+1)}{\log x}.$$

Thus, it suffices to prove that the function $\log(x+1)/\log x$ is decreasing for $x > 1$. We observe that for $x > 1$,

$$\frac{d}{dx}\left(\frac{\log(x+1)}{\log x}\right) = \frac{\frac{\log x}{x+1} - \frac{\log(x+1)}{x}}{(\log x)^2} = \frac{x\log x - (x+1)\log(x+1)}{x(x+1)(\log x)^2} < 0. \qquad \square$$

Theorem 7.10. If $m \geq 4$, then $P(F_m) \geq (4m+9)2^{m+2} + 1$.

P r o o f . Assume $m \geq 4$. By Lucas's Theorem 6.1, every prime number $p \mid F_m$ satisfies $p \equiv 1 \pmod{2^{m+2}}$. Now write

$$(7.10) \qquad\qquad F_m = \prod_{i=1}^{n} p_i^{b_i},$$

where $p_1, \ldots, p_n$ are distinct primes and $b_1, \ldots, b_n$ are positive integers. For $i = 1, \ldots, n$ write

$$(7.11) \qquad\qquad p_i = k_i 2^{m+2} + 1,$$

for some positive integer k_i. On the one hand, equation (7.10) together with the fact that $k_i \geq 1$ for all $i = 1, \ldots, n$ implies

$$2^{2^m} + 1 > \left(2^{m+2} + 1\right)^{\sum_{i=1}^{n} b_i},$$

or

$$(7.12) \qquad\qquad \sum_{i=1}^{n} b_i < \frac{\log(2^{2^m}+1)}{\log(2^{m+2}+1)} < \frac{2^m}{m+2}.$$

For the last inequality in (7.12) we used (7.9).

On the other hand, by using the binomial theorem, it follows that

$$(7.13) \qquad p_i^{b_i} = \left(k_i 2^{m+2} + 1\right)^{b_i} \equiv 2^{m+2} k_i b_i + 1 \pmod{2^{2m+4}}.$$

According to formula (7.13) and the fact that $2^m > 2m+4$ for all $m \geq 4$, it follows that one may reduce equation (7.10) modulo 2^{2m+4} and get

$$1 \equiv \prod_{i=1}^{n}\left(2^{m+2} k_i b_i + 1\right) \pmod{2^{2m+4}} \equiv 1 + 2^{m+2}\sum_{i=1}^{n} k_i b_i \pmod{2^{2m+4}},$$

or

$$2^{m+2}\sum_{i=1}^{n} k_i b_i \equiv 0 \pmod{2^{2m+4}},$$

or

(7.14)
$$\sum_{i=1}^{n} k_i b_i \equiv 0 \pmod{2^{m+2}}.$$

Formula (7.14) implies that

(7.15)
$$\sum_{i=1}^{n} k_i b_i \geq 2^{m+2}.$$

We now combine inequalities (7.12) and (7.15) to get

$$2^{m+2} \leq \sum_{i=1}^{n} k_i b_i \leq \max(k_i) \sum_{i=1}^{n} b_i < \max(k_i) \frac{2^m}{m+2},$$

or

$$\max(k_i) > 4(m+2) = 4m + 8.$$

Hence,

$$\max(k_i) \geq 4m + 9,$$

and

$$P(F_m) = 2^{m+2} \max(k_i) + 1 \geq (4m+9)2^{m+2} + 1. \qquad \square$$

A lower bound of the same order for the largest prime factor of the Fermat numbers is also derived in [Stewart, C. L., 1977, p. 430], namely

$$P(F_m) > Cm2^m \quad \text{for } m > 0,$$

where C is an effectively computable positive constant. Let us point out that in [Stewart, C. L., 1983] there is an asymptotic formula for the size of the largest square-free factor of F_m.

In what follows we denote by $\mathbb{D}$ the set of all positive integers d for which there exists a Fermat number F_m such that $d \mid F_m$, i.e.,

$$\mathbb{D} = \{d \in \mathbb{N} \mid \exists m \geq 0 : d \mid F_m\}.$$

Let $\mathbb{P}$ stand for the set of all primes belonging to $\mathbb{D}$. For a positive number x we denote $\mathbb{D} \cap [1, x]$ by $\mathbb{D}(x)$ and $\mathbb{P} \cap [1, x]$ by $\mathbb{P}(x)$. Further, we investigate the density of the sets $\mathbb{D}$ (in the set of positive integers) and $\mathbb{P}$ (in the set of prime numbers).

Theorem 7.11. *The density of the set $\mathbb{D}$ is zero, that is,*

(7.16)
$$\lim_{x \to \infty} \frac{|\mathbb{D}(x)|}{x} = 0.$$

P r o o f . It is enough to show that

(7.17)
$$\limsup_{x \to \infty} \frac{|\mathbb{D}(x)|}{x} < \varepsilon \quad \text{for all } \varepsilon > 0.$$

Clearly, inequality (7.17) is equivalent to equality (7.16).

Fix $\varepsilon > 0$. Let N_ε be the smallest positive integer n such that $2^n > \varepsilon^{-1}$. Notice that one can choose N_ε to be the smallest positive integer greater than $(-\log \varepsilon / \log 2)$.

For a positive integer j let a_j be the jth element of $\mathbb{D}$. Write $a_j = k_j 2^{n_j} + 1$ for some positive integers n_j and k_j with k_j odd. Notice that there are only finitely many j's such that $n_j < N_\varepsilon$. Indeed, assume that $n_j < N_\varepsilon$ and assume that $a_j \mid F_m$ for some $m > 1$. By Lucas's Theorem 6.1, $a_j = k2^{m+2} + 1$ for some positive integer k. In particular, we have $m + 2 \leq n_j < N_\varepsilon$. It now follows that a_j is a divisor of the integer $F_0 F_1 F_2 \cdots F_{N_\varepsilon}$, which has only finitely many divisors. Thus, we can choose a positive integer M_ε such that $n_j > N_\varepsilon$ for $j > M_\varepsilon$.

We now show that

$$(7.18) \qquad\qquad |\mathbb{D}(x)| < M_\varepsilon + \varepsilon x.$$

Indeed, let us count how many of the numbers a_j with $j > M_\varepsilon$ are in the interval $[1, x]$. Notice that for $j > M_\varepsilon$, we obtain

$$a_{j+1} - a_j = k_{j+1} 2^{n_{j+1}} - k_j 2^{n_j} \geq 2^{\min(n_{j+1}, n_j)}.$$

Since $2^{\min(n_{j+1}, n_j)} > \varepsilon^{-1}$ for $j > M_\varepsilon$, it follows that

$$a_{j+1} - a_j > \frac{1}{\varepsilon} \qquad \text{for } j > M_\varepsilon.$$

Since the difference between any two consecutive terms a_j and a_{j+1} for $j > M_\varepsilon$ is at least $1/\varepsilon$, it follows that the interval $[1, x]$ contains at most εx such terms. This clearly implies inequality (7.18). Dividing both sides of inequality (7.18) by x and letting x tend to infinity, we get inequality (7.17). $\square$

In what follows, for a positive number x we denote by $\pi(x)$ the number of all primes less than or equal to x. For two coprime positive integers k and ℓ we denote by $\pi(x; k, \ell)$ the number of primes p less than or equal to x such that $p \equiv \ell \pmod{k}$. We now investigate the density of the set $\mathbb{P}$ as a set of prime numbers.

Theorem 7.12. *The density of the set $\mathbb{P}$ in the set of primes is zero; that is,*

$$(7.19) \qquad\qquad \lim_{x \to \infty} \frac{|\mathbb{P}(x)|}{\pi(x)} = 0.$$

P r o o f . We keep the notation from the proof of Theorem 7.11. Clearly, equality (7.19) is equivalent to proving that

$$(7.20) \qquad\qquad \limsup_{x \to \infty} \frac{|\mathbb{P}(x)|}{\pi(x)} < \varepsilon \qquad \text{for all } \varepsilon > 0.$$

Fix $\varepsilon > 0$. We find an upper bound for $|\mathbb{P}(x)|/\pi(x)$ in terms of ε for $x > 2$. Assume that $p \in \mathbb{P}(x)$. Since $\mathbb{P}(x) \subseteq \mathbb{D}(x)$, it follows that we may write $p = a_j$ for some $j \geq 1$. From the arguments employed in the proof of Theorem 7.11, we know that if $j > M_\varepsilon$, then $p \equiv 1 \pmod{2^{N_\varepsilon}}$. Hence,

$$(7.21) \qquad |\mathbb{P}(x)| \leq M_\varepsilon + \pi(x; 2^{N_\varepsilon}, 1).$$

Dividing both sides of formula (7.21) by $\pi(x)$, we get

$$(7.22) \qquad \frac{|\mathbb{P}(x)|}{\pi(x)} \leq \frac{M_\varepsilon}{\pi(x)} + \frac{\pi(x; 2^{N_\varepsilon}, 1)}{\pi(x)}.$$

At this stage, we recall Dirichlet's Theorem 2.23 on the density of primes in an arithmetic progression, namely, that

$$(7.23) \qquad \lim_{x \to \infty} \frac{\pi(x; k, \ell)}{\pi(x)} = \frac{1}{\phi(k)}$$

(where ϕ is the Euler totient function) whenever k and ℓ are positive coprime integers. We now let x tend to infinity in formula (7.22), and use formula (7.23), equation (2.13), and the fact that the set of prime numbers is infinite to conclude that

$$(7.24) \qquad \limsup_{x \to \infty} \frac{|\mathbb{P}(x)|}{\pi(x)} \leq \limsup_{x \to \infty} \frac{\pi(x; 2^{N_\varepsilon}, 1)}{\pi(x)} = \frac{1}{\phi(2^{N_\varepsilon})} = \frac{1}{2^{N_\varepsilon - 1}} < 2\varepsilon.$$

The last inequality in (7.24) follows from the fact that

$$2^{N_\varepsilon} > 1/\varepsilon.$$

Notice that formula (7.24) is equivalent to (7.20) if we replace ε in formula (7.24) by $\varepsilon/2$. $\square$

For related results on the density of prime divisors of Fermat numbers see [Golomb, 1955]. We note that it has also been demonstrated in [Křížek, Luca, Somer] that

$$|\mathbb{D}(x)| = \mathcal{O}(\sqrt{x}) \quad \text{and} \quad |\mathbb{P}(x)| = \mathcal{O}\left(\frac{\sqrt{x}}{\log x}\right) \quad \text{as } x \to \infty.$$

Recall that by Lucas's Theorem 6.1 and by Theorem 6.7 any proper prime factor of a Fermat number F_m must be of the form $k2^{m+2} + 1$, where $k \geq 3$ is not a power of 2. We will see below that certain potential prime factors $k2^{m+2} + 1$ of F_m can be excluded by congruence conditions modulo 12, which are obtained by considering the cases for which $k2^{m+2} + 1$ is divisible by one of the odd primes p for which $p - 1 \mid 12$, namely, the primes 3, 5, 7, and 13.

We will call a prime factor $k2^{m+2} + 1$ of a Fermat number F_m a lucky Fermat factor if k is the smallest positive integer such that $k2^{m+2} + 1$ could possibly be a prime factor subject to some fairly simple conditions. More precisely, the prime $k2^{m+2} + 1$ is a *lucky Fermat factor* if it divides F_m and $k \in \{3, 5, 6, 7, 9\}$ is the smallest value we can choose that is not excluded by congruence constraints modulo 12, which lead to divisibility of $k2^{m+2} + 1$ by 3, 5, 7, or 13.

For example, consider the factor $641 = 5 \cdot 2^7 + 1$ of F_5, which was discovered by Euler. Then 641 will be a lucky Fermat factor of F_5 if we can exclude $3 \cdot 2^7 + 1$ from being a factor of F_5 by using congruence conditions modulo 12. It is easy to see that $5 \mid 3 \cdot 2^{m+2} + 1$ if and only if $m + 2 \equiv 3$, 7, or 11 (mod 12). It can also be shown that $7 \mid 3 \cdot 2^{m+2} + 1$ if and only if $m + 2 \equiv 1$, 4, 7, or 10 (mod 12). Thus, when $m = 5$, and consequently $m + 2 = 7$, we see from the above discussion that $3 \cdot 2^7 + 1 = 385$ is divisible by both 5 and 7, since $m + 2 = 7 \equiv 7$ (mod 12). Therefore, 641 is a lucky Fermat factor.

Note also that by Suyama's Theorem 4.22, any factor $3 \cdot 2^{m+2} + 1$ of F_m is prime and thus is automatically a lucky Fermat factor.

In Theorem 7.13 below we will present sufficient conditions for a prime factor of F_m of the form $k2^{m+2} + 1$ to be a lucky Fermat factor when $k = 3, 5, 6, 7$, or 9. In the proof of this theorem the congruence conditions modulo 12 mentioned above will play a key role.

Theorem 7.13. *Let $k2^{m+2} + 1$ be a proper prime divisor of F_m, where $k \geq 3$ and k is not a power of 2.*

(1) *If $k \in \{3, 5, 6\}$, then $k2^{m+2} + 1$ is a lucky Fermat factor of F_m (with a possible multiplicity greater than one). In particular, at most one of the following numbers $3 \cdot 2^{m+2} + 1$, $5 \cdot 2^{m+2} + 1$, and $6 \cdot 2^{m+2} + 1$ can be a prime factor of F_m.*

(2) *If $k = 7$ and $m + 2 \equiv 0, 2, 6$, or 10 (mod 12), then $7 \cdot 2^{m+2} + 1$ is a lucky Fermat factor of F_m.*

(3) *If $k = 9$ and $m + 2 \equiv 11$ (mod 12), then $9 \cdot 2^{m+2} + 1$ is a lucky Fermat factor of F_m.*

P r o o f . We will show that if $k \in \{3, 5, 6, 7\}$ and $k2^{m+2} + 1$ is a prime factor of F_m, then $m + 2$ satisfies the congruence conditions modulo 12 given below. Parts (1) and (2) will then follow from these congruence constraints. Furthermore, upon examination of these congruence conditions, it will be observed that if $k \in \{3, 5, 6, 7\}$ and $m + 2 \equiv 11$ (mod 12), then $k2^{m+2} + 1$ cannot be a prime divisor of F_m, implying that when $9 \cdot 2^{m+2} + 1$ is a prime factor of F_m, it is a lucky Fermat factor.

We now consider each of the cases $k = 3$, 5, 6, 7 separately.

$k = 3$: By Theorem 6.13, $m + 2 \equiv 1$ (mod 4). Now, $5 \mid 3 \cdot 2^{m+2} + 1$ if and only if $2^{m+2} \equiv 3$ (mod 5). This congruence occurs if and only if $m + 2 \equiv 3$ (mod 4), since $\mathrm{ord}_5 2 = 4$ and $2^3 \equiv 3$ (mod 5). Note also that $\mathrm{ord}_7 2 = 3$, and thus $7 \mid 3 \cdot 2^{m+2} + 1$ if and only if $m + 2 \equiv 1$ (mod 3). Moreover, $\mathrm{ord}_{13} 2 = 12$ and $13 \mid 3 \cdot 2^{m+2} + 1$ if and only if $m + 2 \equiv 2$ (mod 12). Combining the congruences given above, we see that $m + 2 \equiv 1$ (mod 4), $5 \nmid 3 \cdot 2^{m+2} + 1$, $7 \nmid 3 \cdot 2^{m+2} + 1$, and $13 \nmid 3 \cdot 2^{m+2} + 1$ if and only if $m + 2 \equiv 5$ or 9 (mod 12).

$k = 5$: From the proof of Corollary 6.8 we find that $3 \mid 5 \cdot 2^{m+2} + 1$ if and only if $m + 2 \equiv 0$ (mod 2). By arguments similar to those given for the case $k = 3$, $7 \mid 5 \cdot 2^{m+2} + 1$ if and only if $m + 2 \equiv 2$ (mod 3) and $13 \mid 5 \cdot 2^{m+2} + 1$ if and only if $m + 2 \equiv 9$ (mod 12). Thus $3 \nmid 5 \cdot 2^{m+2} + 1$, $7 \nmid 5 \cdot 2^{m+2} + 1$, and $13 \nmid 5 \cdot 2^{m+2} + 1$ if and only if $m + 2 \equiv 1, 3$, or 7 (mod 12).

$k = 6$: This is equivalent to the case $3 \cdot 2^{m+3} + 1 \mid F_m$. By the analysis done for the case $k = 3$ above, we see that $m + 3 \equiv 5$ or $9 \pmod{12}$, or equivalently, $m + 2 \equiv 4$ or $8 \pmod{12}$.

$k = 7$: By the proof of Corollary 6.8, $3 \mid 7 \cdot 2^{m+2} + 1$ if and only if $m + 2 \equiv 1 \pmod{2}$. Moreover, $5 \mid 7 \cdot 2^{m+2} + 1$ if and only if $m + 2 \equiv 1 \pmod{4}$ and $13 \mid 7 \cdot 2^{m+2} + 1$ if and only if $m + 2 \equiv 7 \pmod{12}$. Therefore, $3 \nmid 7 \cdot 2^{m+2} + 1$, $5 \nmid 7 \cdot 2^{m+2} + 1$, and $13 \nmid 7 \cdot 2^{m+2} + 1$ if and only if $m = 0, 2, 4, 6, 8$, or $10 \pmod{12}$.

By examining the congruence constraints given above, parts (1), (2), and (3) follow immediately.   □

Remark 7.14. It follows by Theorem 7.13 and by Suyama's Theorem 4.22 that any factor of F_m of the form $3 \cdot 2^{m+2} + 1$, $5 \cdot 2^{m+2} + 1$, or $6 \cdot 2^{m+2} + 1$ is prime and thus necessarily a lucky Fermat factor of F_m. The fact that $9 \cdot 2^{9431} + 1$ is a prime factor of F_{9428} and $9431 \equiv 11 \pmod{12}$ shows that it is indeed possible for $9 \cdot 2^n + 1$ to be a prime factor of F_m when $n \equiv 11 \pmod{12}$ and n is not necessarily equal to $m + 2$. Note that $9 \cdot 2^{9431} + 1$ is not a lucky Fermat factor.

Remark 7.15. By using conditions (1), (2), and (3) from Theorem 7.13 for $k = 3, 5, 6$, and 7 and examining the web site [www1], we obtain the following 13 lucky prime factors $k2^{m+2} + 1$ of F_m:

$k2^{m+2} + 1$	F_m
$5 \cdot 2^7 + 1$	F_5
$7 \cdot 2^{14} + 1$	F_{12}
$5 \cdot 2^{25} + 1$	F_{23}
$6 \cdot 2^{40} + 1$	F_{38}
$5 \cdot 2^{75} + 1$	F_{73}
$5 \cdot 2^{127} + 1$	F_{125}
$3 \cdot 2^{209} + 1$	F_{207}
$5 \cdot 2^{1947} + 1$	F_{1945}
$5 \cdot 2^{23473} + 1$	F_{23471}
$7 \cdot 2^{95330} + 1$	F_{95328}
$3 \cdot 2^{157169} + 1$	F_{157167}
$3 \cdot 2^{213321} + 1$	F_{213319}
$3 \cdot 2^{382449} + 1$	F_{382447}

Note that $5 \cdot 2^7 + 1$ is a lucky Fermat factor of the smallest composite Fermat number. Thus, Fermat was very unlucky in not finding it.

8. Connections of Fermat Numbers with Pascal's Triangle

Conjectures are treacherous in number theory.

Albert H. Beiler

Recall that Pascal's triangle is the infinite triangle $(C(n,j))_{\substack{0 \le n \\ 0 \le j \le n}}$ having the rows indexed by $n = 0, 1, \ldots$, the columns indexed by $j = 0, 1, \ldots, n$, and entries that are simply the binomial coefficients

$$(8.1) \qquad C(n,j) = \binom{n}{j} \qquad \text{for all } 0 \le j \le n.$$

It was originally invented around the twelfth century by Chinese mathematicians (see [Martzloff]). There are many relations known among the entries of Pascal's triangle. An interesting connection between Pascal's triangle and the Fermat numbers was pointed out in [Hewgill] and [Gardner]. To describe it, let $c(n,j) \in \{0,1\}$ be such that (see Figure 4.2)

$$(8.2) \qquad \binom{n}{j} \equiv c(n,j) \ (\mathrm{mod}\ 2).$$

That is, $c(n,j)$ is simply the residue of $\binom{n}{j}$ modulo 2. If we reduce every single entry from the nth row of Pascal's triangle modulo 2 and read the corresponding row as a whole number written in binary arithmetic, we get a number that we denote by $a(n)$. Accordingly,

$$(8.3) \qquad a(n) = \sum_{j=0}^{n} c(n,j) 2^j, \qquad \text{for } n = 0, 1, \ldots \ .$$

Let us also write the positive integer n in base 2 as

$$(8.4) \qquad n = \alpha_0 + 2\alpha_1 + \cdots + 2^t \alpha_t, \qquad \text{where } \alpha_i \in \{0,1\} \text{ and } \alpha_t \ne 0.$$

With the above notation Hewgill proved the following result (see [Hewgill] and Remark 4.6). We give a shorter proof suggested by F. Beukers.

Theorem 8.1. *Under the above notation, the formula*

$$(8.5) \qquad a(n) = F_0^{\alpha_0} F_1^{\alpha_1} \cdots F_t^{\alpha_t}$$

holds for all $n \geq 0$.

P r o o f . We first observe that

$$(8.6) \qquad (X+1)^n = \sum_{i=0}^{n} C(n,i) X^i.$$

Then

$$(X+1)^n \equiv \sum_{i=0}^{n} c(n,i) X^i \pmod{2},$$

where $c(n,i) \in \{0,1\}$. Note also that

$$(X+1)^n = (X+1)^{2^0 \alpha_0 + 2^1 \alpha_1 + \cdots + 2^t \alpha_t} \equiv \prod_{j=0}^{t} \left(X^{2^j} + 1\right)^{\alpha_j} \pmod{2}.$$

Moreover,

$$(8.7) \qquad \prod_{j=0}^{t} \left(X^{2^j} + 1\right)^{\alpha_j} = \sum_{k=0}^{n} \beta_k X^k,$$

where $\beta_k \in \{0,1\}$. From the uniqueness of the representation of n in base 2, we see that

$$(8.8) \qquad \sum_{i=0}^{n} c(n,i) X^i = \prod_{j=0}^{t} \left(X^{2^j} + 1\right)^{\alpha_j}.$$

Setting $X = 2$ in (8.8), we obtain, by (8.3),

$$a(n) = F_0^{\alpha_0} F_1^{\alpha_1} \cdots F_t^{\alpha_t}. \qquad \square$$

Using a well-known result from [Lucas, 1877–78]) given below, one can easily treat the more general case in which Pascal's triangle is computed modulo a prime number p. The resulting sequence can then be compared to the generalized Fermat numbers $p^{2^m} + 1$ (see Chapter 13).

Lemma 8.2 (Lucas). *Let p be a prime number. For any positive integers n and k with $k \leq n$ let*

$$(8.9) \qquad n = n_0 + n_1 p + \cdots + n_t p^t, \qquad \text{where } n_i \in \{0, 1, \ldots, p-1\} \text{ and } n_t \neq 0,$$

and

$$(8.10) \qquad k = k_0 + k_1 p + \cdots + k_t p^t, \qquad \text{where } k_i \in \{0, 1, \ldots, p-1\}.$$

Then the congruence

$$(8.11) \qquad \binom{n}{k} \equiv \binom{n_0}{k_0} \binom{n_1}{k_1} \cdots \binom{n_t}{k_t} \pmod{p}$$

holds for all positive integers n and k.

P r o o f . We begin by noticing that if $j \in \{1, 2, \ldots, p-1\}$, then

$$(8.12) \qquad \binom{p}{j} = \frac{p!}{j!(p-j)!} = \frac{p(p-1)\cdots(p-j+1)}{j!} \equiv 0 \pmod{p},$$

because $j < p$ and p is prime. Hence, one has

$$(X+1)^p = X^p + \left(\sum_{j=1}^{p-1} \binom{p}{j} X^{p-j}\right) + 1 \equiv X^p + 1 \pmod{p}.$$

Iterating the above congruence, we get that

$$(8.13) \qquad (X+1)^{p^i} \equiv X^{p^i} + 1 \pmod{p} \qquad \text{for all } i \geq 1.$$

Using formula (8.13), we infer that

$$(8.14) \qquad \begin{aligned} (X+1)^n &= (X+1)^{n_0 + n_1 p + \cdots + n_t p^t} \\ &= (X+1)^{n_0}(X+1)^{n_1 p} \cdots (X+1)^{n_t p^t} \\ &\equiv (X+1)^{n_0}(X^p+1)^{n_1} \cdots (X^{p^t}+1)^{n_t} \pmod{p}. \end{aligned}$$

The congruence asserted by Lemma 8.2 follows by identifying the coefficients of X^k from both sides of formula (8.14) and from the uniqueness of the representation of the positive integer k in the base p given in (8.10). □

In what follows we investigate the occurrence of Fermat numbers in Pascal's triangle. First, recall that a *triangular number* is a positive integer of the form $\binom{n}{2}$ for some positive integer n (cf. Figure 9.1). In [Krishna] it is shown that the only triangular Fermat number is $F_0 = 3 = \binom{3}{2}$. This result appears also in [Radovici-Mărculescu]. Both proofs are immediate and are based on modular arguments.

In this section we extend the above result. The following theorem was originally proved in [Luca, 2000e] (see also [Flammenkamp, Luca], where a similar result is proved for the Mersenne numbers, too).

Theorem 8.3. *If*

$$(8.15) \qquad F_m = \binom{n}{k} \qquad \text{for some } n \geq 2k \geq 2,$$

then $k = 1$.

Notice first that the condition $n \geq 2k$ is not really restrictive because of the symmetry of the binomial coefficients

$$(8.16) \qquad \binom{n}{k} = \binom{n}{n-k}.$$

The above result can be interpreted by saying that the Fermat numbers sit in Pascal's triangle only in the trivial way (see Figure 8.1).

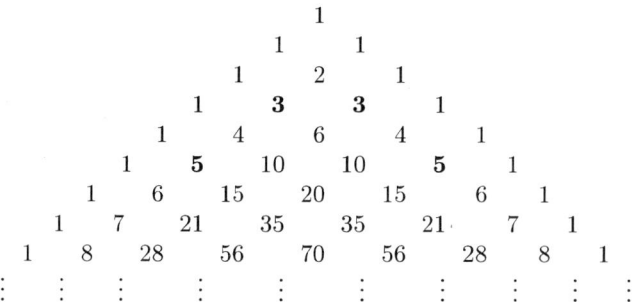

Figure 8.1. Positions of the Fermat numbers in Pascal's triangle.

P r o o f   o f   T h e o r e m   8 . 3 . Assume, to the contrary, that equation (8.15) has a solution with $k > 1$. Notice that in this case, $m > 4$, because F_m is prime for $m = 0, 1, \ldots, 4$. We first show that $k < 2^m$. Indeed, assume that $k \geq 2^m$. Since $m \geq 5$, it follows that $k \geq 2^5 = 32$. One can easily check that

$$(8.17) \qquad k! < \left(\frac{k}{2.2} \right)^k \qquad \text{for all } k \geq 10,$$

which directly follows from Stirling's formula. Equation (8.15) and inequality (8.17) now imply that

$$2^{2^m} + 1 = F_m = \binom{n}{k} = \frac{n(n-1)\cdots(n-k+1)}{k!} > \frac{(n-k)^k}{k!}$$
$$> \left(\frac{2.2(n-k)}{k} \right)^k \geq (2.2)^k \geq (2.2)^{2^m},$$

or

$$1 + \frac{1}{2^{2^m}} > \left(\frac{2.2}{2} \right)^{2^m} = \left(1 + \frac{1}{10} \right)^{2^m} > 1 + \frac{2^m}{10},$$

or

$$10 > 2^{m+2^m},$$

which is certainly impossible for $m \geq 5$. Thus, $k < 2^m$.

Assume that

$$(8.18) \qquad n = \prod_{p \mid n} p^{\alpha_p}$$

is the decomposition of n into its distinct prime factors p. Let

$$(8.19) \qquad A = \{p \mid p \equiv 1 \pmod{2^{m+1}} \text{ and } p \mid n\}.$$

Finally, let $n = n_1 d$, where

$$(8.20) \qquad n_1 = \prod_{p \in A} p^{\alpha_p}.$$

We now show that $d \mid k$. This is clear if $d = 1$. Assume that $d > 1$ and choose a prime number $q \mid d$. Since all the prime divisors of F_m are congruent to 1 modulo 2^{m+1} (see Chapter 6), it follows that $q \nmid F_m$. But since $q \mid d \mid n$, it follows that if we write n in base q according to formula (8.9), then we get that $n_0 = 0$. If $q \nmid k$, then $k_0 > 0$, and now formula (8.11) would imply that

$$F_m \equiv \binom{n}{k} \equiv \binom{0}{k_0}\binom{n_1}{k_1}\cdots\binom{n_t}{k_t} \equiv 0 \pmod{q},$$

which is impossible because q does not divide F_m. Hence, every prime divisor of d divides k as well. To show that $d \mid k$ we need to show that if $q^\alpha \| d$ for some $\alpha \geq 1$, then $q^\alpha \mid k$. Assuming that this were not so, it would follow that $q^\beta \| k$ for some $\beta < \alpha$. But in this case, $n_\beta = 0$ and $k_\beta \neq 0$, which, via formula (8.11), would imply again that q divides F_m, which is impossible. Hence, $d \mid k$. In particular, since $k < 2^m$, it follows that $d < 2^m$ as well. We now notice that since n_1 is a product of primes from A, it follows that $n_1 \equiv 1 \pmod{2^{m+1}}$. This implies that $n \equiv d \pmod{2^{m+1}}$. However, since $d \leq k < 2^m < 2^{m+1}$, Lucas's Lemma 8.2 for the prime $p = 2$ implies that

$$(8.21) \qquad\qquad F_m = \binom{n}{k} \equiv \binom{d}{k} \pmod{2}.$$

Since F_m is odd and $d \leq k$, formula (8.21) implies that $d = k$. Thus, $k \mid n$. We may now write equation (8.15) as

$$(8.22) \qquad\qquad F_m = \frac{n}{k}\binom{n-1}{k-1},$$

where n/k is an integer. At this point, one should notice that the relevant feature of the preceding argument was the shape of the prime divisors of F_m. Hence, one can iterate the above argument to get that $(k-i) \mid (n-i)$ for all $i = 0, 1, \ldots, k-1$. This is equivalent to

$$(8.23) \qquad n \equiv i \pmod{k-i} \equiv k \pmod{k-i}, \qquad \text{for all } i = 0, 1, \ldots, k-1.$$

Let

$$(8.24) \qquad\qquad N = \mathrm{lcm}(1, 2, \ldots, k).$$

From formula (8.23), we get that $n \equiv k \pmod{N}$. Write $n = k + aN$ for some positive integer a. Now equation (8.15) implies that

$$(8.25) \qquad F_m = \binom{n}{k} = \frac{n}{k} \cdot \frac{n-1}{k-1} \cdots \frac{n-k+1}{1} = \prod_{i=0}^{k-1}\left(1 + a\frac{N}{k-i}\right).$$

Let $N_i = N/(k-i)$ for $i = 0, 1, \ldots, k-1$. Notice that exactly one of the numbers N_i is odd, and all the other numbers are even. Indeed, the only odd number N_i

corresponds to $i = k - 2^\mu$, where 2^μ is the largest power of 2 less than or equal to k. We now look again at equation (8.25) and write it as

$$(8.26) \qquad 2^{2^m} + 1 = \prod_{i=0}^{k-1} (1 + aN_i) = 1 + aS_1 + a^2 S_2 + \cdots + a^k S_k,$$

where S_j is the jth fundamental symmetric polynomial in the N_i's. Since exactly one of the numbers N_i is odd, it follows that S_1 is odd and that S_j is even for all $j \geq 2$. Equation (8.26) can now be rewritten as

$$(8.27) \qquad 2^{2^m} = a\big(S_1 + aS_2 + \cdots + a^{k-1} S_k\big).$$

From formula (8.27) one can see right away that the factor $S_1 + aS_2 + \cdots + a^{k-1} S^k$ is odd and larger than 1 (here is where the condition $k > 1$ is really used), so it cannot divide the power of 2 from the left-hand side of equation (8.27).

Theorem 8.3 is therefore proved. □

Remark 8.4. One can mimic the above arguments to show that if $a > 1$ is any positive integer and $F_{a,m} = a^{2^m} + 1$ is the mth generalized Fermat number (see Chapter 13), then the equation

$$(8.28) \qquad F_{a,m} = \binom{n}{k}, \qquad \text{for some } n \geq 2k \text{ and } k \geq 2,$$

has only finitely many solutions, all of which are computable. This means that there exists a constant $C(a)$ depending only on a such that all solutions of equation (8.28) satisfy $m < C(a)$. The fact that there are sometimes nontrivial solutions of (8.28) is illustrated by the example

$$F_{3,1} = \binom{5}{2}.$$

For any positive integer k let $\phi(k)$ be the Euler totient function (see Chapter 2). Let

$$(8.29) \qquad \mathcal{C} = \{k \geq 1 \mid \phi(k) \text{ is a power of 2}\}.$$

Clearly, $1, 2 \in \mathcal{C}$, and by Theorem 4.5 we know that a positive integer $n \geq 3$ belongs to $\mathcal{C}$ if and only if the regular polygon with n sides can be constructed using only ruler and compass. Moreover, by the proof of Theorem 4.5, if n belongs to $\mathcal{C}$, then either n is a power of 2 or

$$n = 2^\alpha p_1 p_2 \cdots p_t$$

for some $\alpha \geq 0$, $t \geq 1$, and $p_1 < p_2 < \cdots < p_t$, where p_i is a Fermat prime for $i = 1, \ldots, t$.

In the remainder of this section we find all members of Pascal's triangle that belong to $\mathcal{C}$. More precisely, we find all solutions of the equation

$$(8.30) \qquad \phi\left(\binom{n}{k}\right) = 2^\alpha,$$

where k, n, and α are positive integers. By the symmetry of Pascal's triangle, shown in (8.16), it follows that it suffices to solve equation (8.30) only for $n \geq 2k$.

The following result is from [Luca, 2000d].

Theorem 8.5. *The only positive solutions of equation (8.30) for $n \geq 2k$ are the following:*

(1) $k = 1$, $n > 1$, and $n \in C$;

(2) $k = 2$ and n is either a Fermat prime, or $n \in \{2^2,\ 2 \cdot 3,\ 2^{2^2},\ 2^{2^3},\ 2^{2^4},\ 2^{2^5}\}$;

(3) $k = 3$ and $n \in \{6, 10, 17, 18, 257, 65537\}$.

P r o o f . Part (1) follows from our discussion preceding the statement of this theorem. We now split the proof into two parts.

The Case $k \geq 3$. Let n and k be positive integers such that $n \geq 2k \geq 6$ and

$$(8.31) \qquad\qquad \binom{n}{k} \in C.$$

We show that $k = 3$ and $n \in \{6, 10, 17, 18, 257, 65537\}$.

Write

$$(8.32) \qquad\qquad \binom{n}{k} = 2^\alpha F_{\beta_1} F_{\beta_2} \cdots F_{\beta_t},$$

where $\beta_1 < \beta_2 < \cdots < \beta_t$ and F_{β_i} is prime for $i = 1, \ldots, t$. Recall that a well-known result appearing in [Kummer] (see also, e.g., [Huard, Spearman, Williams]) says that the power of any prime p dividing the binomial coefficient appearing in (8.32) is given by the number of "carries" occurring in adding k and $n - k$ in base p. In particular, $2^\alpha \leq n$ and $F_{\beta_i} \leq n$ for all $i = 1, \ldots, t$. Let 2^a be the largest power of 2 that is less than or equal to n and let $m = \beta_t$. By Proposition 3.2, it now follows that $\alpha \leq a$ and

$$(8.33) \quad \binom{n}{k} \leq 2^a F_0 F_1 \cdots F_m = 2^a (F_{m+1} - 2) = 2^a F_m (F_m - 2) \leq n(n-1)(n-2).$$

Hence, if $k \geq 4$ and $n \geq 2k$, we get

$$\binom{n}{4} \leq \binom{n}{k} \leq n(n-1)(n-2),$$

which forces $n \leq 27$. One can check computationally that containment (8.31) does not occur in the range $8 \leq 2k \leq n \leq 27$.

We now suppose that $k = 3$ and $n \geq F_5$. We claim that $\beta_t \geq 5$. Suppose, to the contrary, that $\beta_t \leq 4$. Then $2^\alpha \leq n$, and $F_0 F_1 F_2 F_3 F_4 = F_5 - 2 \leq n - 1$. Since $(n-2)/6 > 1$, it follows that

$$2^\alpha F_{\beta_1} F_{\beta_2} \cdots F_{\beta_t} < \binom{n}{3} = \frac{n(n-1)(n-2)}{6}.$$

Therefore, we must have that $\beta_t \geq 5$. However, $\beta_i \neq 5$ for all $i = 1, \ldots, t$ because F_5 is not prime. Hence,

$$\frac{n(n-1)(n-2)}{6} \leq \frac{2^a}{F_5} F_0 F_1 \cdots F_m = \frac{2^a}{F_5} F_m(F_m - 2) \leq \frac{n(n-1)(n-2)}{F_5},$$

which is impossible. Hence, $n < F_5$. One can check that the only values of $n < F_5$ for which containment (8.31) occurs when $k = 3$ are indeed the claimed ones.

The Case $k = 2$. For any positive integer s let $\eta(s)$ be the number of 1's appearing in the binary representation of s. At this stage we need the following result.

Lemma 8.6. (1) *Let $n \in \mathcal{C}$. Assume that n is divisible by exactly t odd primes, where $t = 0$ if n is a power of 2. Then,*

$$\eta(n) = 2^t.$$

(2) *If $n \in \mathcal{C}$ is odd, then n is uniquely determined by $\lfloor \log_2 n \rfloor$.*

P r o o f . (1) Let $n = 2^\alpha m$, where m is odd. If $m = 1$, then $t = 0$ and $\eta(n) = 1 = 2^0$. From now on, we assume that $m > 1$. Let

$$m = F_{\beta_1} \cdots F_{\beta_t},$$

where $0 \leq \beta_1 < \cdots < \beta_t$ and F_{β_i} is prime for all $i = 1, \ldots, t$. Let $I = \{1, 2, \ldots, t\}$. Notice that

$$m = \sum_{J \subseteq I} 2^{\sum_{i \in J} 2^{\beta_i}}.$$

Since all β_i's are distinct, it follows from the uniqueness of the binary expansion of an integer that the exponents

$$\sum_{i \in J} 2^{\beta_i}$$

are all distinct for $J \subseteq I$. Hence, $\eta(m) = 2^{|I|} = 2^t$. It now follows that $\eta(n)$ is also 2^t, because the binary digits of n are just the binary digits of m shifted by α.

(2) With the previous notations we have $\alpha = 0$ and $n = m$. Hence,

$$\lfloor \log_2 n \rfloor = \sum_{i=1}^{t} 2^{\beta_i}.$$

The above formula shows that n is uniquely determined by $\lfloor \log_2 n \rfloor$. More precisely, let

$$\lfloor \log_2 n \rfloor = \sum_{i=1}^{l} 2^{\gamma_i}$$

be the binary expansion of $\lfloor \log_2 n \rfloor$. Then,

$$n = \prod_{i=1}^{l} \left(2^{2^{\gamma_i}} + 1 \right).$$

P r o o f   o f   T h e o r e m   8 . 5   (C o n t i n u a t i o n) . We are now ready to conclude the proof of Theorem 8.5. Notice that since $n(n-1)/2 \in C$, it follows that both n and $n-1$ are also in C. The proof of Theorem 8.5 ends once we establish the following result:

Lemma 8.7. *Assume that n is a positive integer such that both n and $n-1$ are in C. Then either n is a Fermat prime or*

$$(8.34) \qquad\qquad n \in \{2,\ 2^2,\ 2\cdot 3,\ 2^{2^2},\ 2^{2^3},\ 2^{2^4},\ 2^{2^5}\}.$$

P r o o f . Assume first that $n > 1$ is odd. Let t be the number of prime divisors of n and let l be the number of odd prime divisors of $n-1$. By (1) of Lemma 8.6, we get $\eta(n) = 2^t$ and $\eta(n-1) = 2^l$. However, since $n-1$ is even, it follows that $\eta(n) = 1 + \eta(n-1)$. Hence, $2^t = 1 + 2^l$. The only solutions of the above equation are $t = 1$ and $l = 0$. Hence, n is a Fermat prime.

Assume now that n is even. Since $n = 2$ or 4 certainly satisfies the hypothesis of Lemma 8.7, we may assume that $n \geq 6$. In particular, $n-1$ has at least 3 digits when written in base 2. We distinguish two cases:

Case 1. *The last two binary digits of $n-1$ are 01.* In this case,

$$n - 1 = F_{\beta_1} F_{\beta_2} \cdots F_{\beta_t},$$

where $t \geq 1$ and $\beta_t > \cdots > \beta_1 \geq 1$. We now show that $t = 1$. Indeed, assume that $t \geq 2$. We have

$$n - 1 = 2^{\sum_{i=1}^{t} 2^{\beta_i}} + \cdots + 2^{2^{\beta_1}} + 1.$$

Hence,

$$n = 2^{\sum_{i=1}^{t} 2^{\beta_i}} + \cdots + 2^{2^{\beta_1}} + 2 = 2\left(2^{2^{\beta_t}+\cdots+2^{\beta_1}-1} + \cdots + 2^{2^{\beta_1}-1} + 1\right).$$

Hence, $n/2 \in C$ is odd and

$$\lfloor \log_2(n/2) \rfloor = 2^{\beta_t} + \cdots + 2^{\beta_1} - 1 = 2^{\beta_t} + \cdots + 2^{\beta_1 - 1} + \cdots + 1.$$

From (2) of Lemma 8.6 it follows that $2^{2^{\beta_t}} + 1$ divides $n/2$; hence, it also divides n. On the other hand, $2^{2^{\beta_t}} + 1$ divides $n - 1$. This is a contradiction because n and $n-1$ are coprime. Thus, $t = 1$. It follows that

$$n - 1 = 2^{2^{\beta_1}} + 1,$$

which yields

$$n = 2\left(2^{2^{\beta_1}-1} + 1\right).$$

Since $\eta\left(2^{2^{\beta_1}-1} + 1\right) = 2 = 2^1$, it follows, by (1) of Lemma 8.6, that $2^{2^{\beta_1}-1} + 1$ is a Fermat prime. In particular,

$$2^{2^{\beta_1}-1} + 1 = 2^{2^7} + 1$$

for some $\gamma \geq 0$. Hence, $2^{\beta_1} - 1 = 2^{\gamma}$. The only solution of the above equation is $\beta_1 = 1$ and $\gamma = 0$. This gives $n = 6$.

Case 2. *The last two binary digits of $n - 1$ are* 11. Again write

$$n - 1 = F_{\beta_1} F_{\beta_2} \cdots F_{\beta_t},$$

where $t \geq 1$ and $\beta_t > \cdots > \beta_1$. Notice that $\beta_1 = 0$. We first show that all binary digits of $n - 1$ are 1. Indeed, assume that this is not the case. Let j be such that $\beta_i = i - 1$ for $i = 1, \ldots, j$, but $\beta_{j+1} \geq j + 1$. Since not all binary digits of $n - 1$ are 1, it follows that $t \geq j + 1$. Then,

$$n - 1 = \left(2^{2^{\beta_t}} + 1\right) \cdots \left(2^{2^{\beta_{j+1}}} + 1\right)\left(2^{2^{j-1}} + 1\right) \cdots \left(2^{2^0} + 1\right)$$

$$= 2^{2^{\beta_t} + \cdots + 2^{\beta_{j+1}} + 2^{j-1} + \cdots + 1} + \cdots + 2^{2^{\beta_{j+1}}} + 2^{2^{j-1} + \cdots + 1} + \cdots + 1$$

$$= 2^{2^{\beta_t} + \cdots + 2^{\beta_{j+1}} + 2^j - 1} + \cdots + 2^{2^{\beta_{j+1}}} + 2^{2^j} - 1.$$

Hence,

$$n = 2^{2^j} \left(2^{2^{\beta_t} + \cdots + 2^{\beta_{j+1}} - 1} + \cdots + 2^{2^{\beta_{j+1}} - 2^j} + 1\right).$$

Let m denote the odd part of n. Since $m \in \mathcal{C}$ and

$$\lfloor \log_2 m \rfloor = 2^{\beta_t} + \cdots + 2^{\beta_{j+1}} - 1 = 2^{\beta_t} + \cdots + 2^{\beta_{j+1}-1} + \cdots + 1,$$

it follows, by (2) of Lemma 8.6, that $2^{2^{\beta_t}} + 1$ divides m, and hence n as well. On the other hand, $2^{2^{\beta_t}} + 1$ also divides $n - 1$. This is a contradiction because n and $n - 1$ are coprime. Hence, $t = j$. In this case,

$$n - 1 = \prod_{i=0}^{j-1} \left(2^{2^i} + 1\right) = 2^{2^j} - 1.$$

If $j \geq 6$, then by Proposition 3.2, $2^{2^j} - 1$ is divisible by $2^{2^5} + 1 = F_5$, which is not prime. Hence, $j \leq 5$. On the other hand, all values $n = 2^{2^j}$ for $j \leq 5$ are acceptable because

$$n - 1 = 2^{2^j} - 1 = \prod_{i=0}^{j-1} \left(2^{2^i} + 1\right),$$

and all numbers $2^{2^i} + 1 = F_i$ are primes for $0 \leq i \leq 4$.

Hence, both Lemma 8.7 and Theorem 8.5 are proved. $\square$

In the remaining part of this chapter we point out a characterization of the Fermat primes in terms of divisibility of all binomial coefficients situated between two parallel straight lines in Pascal's triangle by their row number. This story begins with the following result, which goes back to 1972 (see [Mann, Shanks]).

Proposition 8.8. *An integer $k > 1$ is prime if and only if*

$$n \,\Big|\, \binom{n}{k - 2n} \qquad \text{for all } n \text{ such that } \frac{k}{3} \leq n \leq \frac{k}{2}.$$

Five years later, Heiko Harborth generalized the above result as follows:

Proposition 8.9. *Let $c \geq 2$ be any fixed integer. Suppose that $k > 1$ is an integer that is not divisible by any prime number less than or equal to $c^2 - c - 1$. Then k is a prime number if and only if*

$$(8.35) \qquad n \left| \binom{n}{k - cn} \right. \qquad \text{for all } n \text{ such that } \frac{k}{c+1} \leq n \leq \frac{k}{c}.$$

Notice that Proposition 8.8 is obtained from Proposition 8.9 when $c = 2$. We now supply the proof of Proposition 8.9 as it appears in [Harborth, 1977].

P r o o f . We start by proving the necessity. Let p be a prime and suppose that $p/(c+1) \leq n \leq p/c$. We first treat the case for which $n = p/c$. Since $c > 1$, it follows that $p = c$, and hence, $n = 1$. Clearly, condition (8.35) holds in this case. We next see that if $p/(c+1) \leq n < p/c$, then

$$\binom{n}{p - cn} = \frac{n}{p - cn} \binom{n-1}{p - cn - 1} \equiv 0 \pmod{n},$$

where the last congruence above follows from the facts that p is prime and $0 < n < p$, which imply that $\gcd(n, p - cn) = \gcd(n, p) = 1$.

For the sufficiency, assume that k is not prime and write it in the form $k = (cg + d)p$, where p is some prime number and $cg + d > 1$ is such that $0 \leq d < c$. Notice that $g \geq d$, because otherwise, it would follow that $cg + d \leq c(d-1) + d \leq c(c-2) + (c-1) = c^2 - c - 1$, contradicting the fact that k is not divisible by any prime smaller than $c^2 - c - 1$.

Let $n = gp$ and notice that $k/(c+1) \leq n \leq k/c$. Then

$$\binom{n}{k - cn} = \binom{gp}{0} = 1 \not\equiv 0 \pmod{gp} \qquad \text{when } d = 0,$$

$$\binom{n}{k - cn} = \binom{gp}{p} = \frac{gp(gp - 1) \cdots (gp - p + 1)}{p!} \not\equiv 0 \pmod{gp} \qquad \text{when } d = 1.$$

For $d \geq 2$ we have

$$\binom{n}{k - cn} = \binom{gp}{dp} = \frac{gp \cdots (gp - p) \cdots (gp - 2p) \cdots (gp - dp + 1)}{(dp)!} \not\equiv 0 \pmod{gp},$$

where the last incongruence holds when

$$(g - 1)(g - 2) \cdots (g - d + 1) \not\equiv 0 \pmod{p}.$$

Thus, the only case to consider is $d \geq 2$ and $g = g_1 p + r$ for some $1 \leq r < d$. Notice that $g_1 \geq 1$, because otherwise, it would follow that $g = r < d$, which is impossible. We now choose $n = g_1 p^2$ and notice that $k/(c+1) \leq n \leq k/c$. With this choice of n, we get

$$\binom{n}{k - cn} = \binom{g_1 p^2}{(cr + d)p}$$

$$= \frac{g_1 p^2 \cdots (g_1 p^2 - p) \cdots (g_1 p^2 - (cr + d)p + 1)}{((cr + d)p)!} \not\equiv 0 \pmod{n},$$

where the last incongruence above follows easily from the fact that $cr + d \leq c^2 - c - 1 < p$. Thus, for every composite number k that is not a multiple of a prime number less than or equal to $c^2 - c - 1$ we have found at least one n that does not fulfill condition (8.35). Proposition 8.9 is therefore proved. $\square$

In light of Proposition 8.9, for a given positive integer $c \geq 2$ we may denote by K_c the set of all positive integers k such that condition (8.35) is fulfilled. Notice that Proposition 8.9 tells us that there are no composite integers k in K_c for which the smallest prime divisor of k is at least $c^2 - c$, but Proposition 8.9 gives us no information about the members of K_c that are divisible by primes smaller than $c^2 - c - 1$. Of course, when $c = 2$, the number $c^2 - c - 1$ is 1, and K_2 consists of all primes. The connection with the Fermat primes appears when $c = 3$. Indeed, according to [Harborth, 1976]), we have the following description of K_3.

Theorem 8.10 (Harborth). *The set K_3 consists precisely of the numbers 1, 4, 25, all the prime numbers, and all the numbers of the form $2q$, where $q > 3$ is a Fermat prime.*

P r o o f . We have to look for composite numbers k in K_3. If $k = 3g$ with $g > 1$, then we can take $n = g = k/3$ and get

$$\binom{n}{k - 3n} = \binom{g}{0} = 1 \not\equiv 0 \pmod{g},$$

so $k \notin K_3$. In particular, if $k = 2q$, where q is a Fermat prime and k is an element of K_3, then $q > 3$, since $3 \nmid k$.

Assume now that $k = (3g + 1)p$, where $p \neq 3$ is a prime number and $g \geq 1$. Then we can take $n = gp$ and

$$\binom{n}{k - 3n} = \binom{gp}{p} = \frac{gp(gp - 1) \cdots (gp - p + 1)}{p!} \not\equiv 0 \pmod{gp},$$

where the last incongruence follows easily by comparing the orders at which p appears in gp and in $\binom{gp}{p}$, respectively.

Thus, it remains to investigate the numbers k of the form

$$k = (3g + 1)pq, \quad g \geq 0, \quad p = 3a + 2, \quad q = 3b + 2, \quad a \geq 0, \quad b \geq 0,$$

where p and q are prime numbers. With $n = (gp + a)q$ and $n = (gp + a - 1)q$, we get

$$(8.36) \qquad \binom{n}{k - 3n} = \binom{(gp + a)q}{2q} \not\equiv 0 \pmod{(gp + a)q},$$

whenever $gp + a - 1 \not\equiv 0 \pmod{q}$, and

$$(8.37) \qquad \binom{n}{k - 3n} = \binom{(gp + a - 1)q}{5q} \not\equiv 0 \pmod{(gp + a - 1)q},$$

whenever $(gp + a - 2)(gp + a - 3)(gp + a - 4)(gp + a - 5) \not\equiv 0 \pmod{q}$.

There is a slight problem in the sense that equation (8.36) makes sense only for $gp + a - 1 > 0$, while equation (8.37) makes sense only for $gp + a - 5 > 0$. For now, we investigate the case $gp + a > 5$, and we shall return to the case $gp + a \leq 5$ later. In this case, since $k \in K_3$, we conclude that none of incongruences (8.36) and (8.37) hold. Therefore,

$$(8.38) \qquad\qquad q \mid (gp + a - 1)$$

and

$$(8.39) \qquad q \mid (gp + a - 2)(gp + a - 3)(gp + a - 4)(gp + a - 5).$$

From (8.38) and (8.39) it follows that $q \leq 4$, and hence $b = 0$ and $q = 2$. Since the argument is symmetric in p and q, we conclude that if $gq + b > 5$ holds as well, then $p = q = 2$. Since $g > 0$ in this case, we can choose $n = gp^2$ and notice that

$$\binom{n}{k - 3n} = \binom{gp^2}{p^2} \not\equiv 0 \pmod{gp^2},$$

contradicting the fact that $k \in K_3$. This case is therefore impossible.

In what follows we assume that $b \geq a$ and that $gp + a \leq 5$. Notice that $g \neq 2$, because when $g = 2$ we get $n = 7pq$, where pq is congruent to 1 modulo 3, and 7 is prime, which is a case already treated. Since $gp + a \leq 5$ and $p \geq 2$, we get that $g \leq 1$. When $g = 1$, we get $5 \geq p + a = 4a + 2$, which leads to $a = 0$ and $p = 2$. In this case, $gp + a = 2$ and incongruence (8.36) holds for any q. Hence, this case is impossible. We conclude that $g = 0$ and $a \leq 5$. If $a \geq 2$, then (8.36) implies that $q \mid a - 1$. Therefore, $q < a < p$, which is impossible because we are assuming that $a \leq b$. Hence, $a \in \{0, 1\}$.

We now show that either $a = 0$, or $a = 1$ and $q = 5$. Indeed, assume that $a = 1$. If $gq + b > 5$, then the preceding arguments show that $p \leq 4$, contradicting the fact that $p = 5$. If $gq + b \leq 5$, then the preceding arguments show that $b \in \{0, 1\}$, which together with the fact that $b \geq a$ leads to $a = b = 1$ and $k = pq = 25$.

Thus, we have reduced the problem to the numbers $k = 2q$, where q is a prime. We assume that $q > 5$ and therefore $b \geq 2$. Taking $n = 2b$, we get

$$\binom{n}{k - 3n} = \binom{2b}{4} = \frac{(2b - 1)(2b - 3)}{3} \cdot \frac{b(b - 1)}{2} \not\equiv 0 \pmod{2b}$$

whenever $b \not\equiv 1 \pmod 4$. Hence, $b \equiv 1 \pmod 4$. We now construct recursively the sequence $\{b_i\}$, where $b_1 = b$ and for $i = 2, 3, \ldots$, we set $b_{i-1} = 4b_i + 1$. Thus, $k = 2q$, where

$$(8.40) \qquad\qquad q = 3 \cdot 2^{2(i-1)} b_i + 2^{2(i-1)} + 1, \quad b_i \geq 0.$$

Assume that (8.40) holds for some $i \geq 2$ and let $n = 2^{2i-1} b_i$ for some $b_i \geq 2$. Then

$$(8.41) \quad \binom{n}{k - 3n} = \binom{2^{2i-1} b_i}{2^{2i-1} + 2} \not\equiv 0 \pmod{2^{2i-1} b_i} \quad \text{for} \quad b_i \not\equiv 1 \pmod 2.$$

Incongruence (8.41) follows easily by comparing the powers of 2 appearing in $\binom{n}{k-3n}$ and n. We also let

$$n = 3 \cdot 2^{2i-3} b_i + \frac{5 \cdot 2^{2i-3} + 2}{3} \qquad \text{whenever} \quad b_i \geq 0.$$

With this choice for n, it follows that

(8.42)
$$\binom{n}{k - 3n} = \binom{n}{4n - k} = \binom{n}{\frac{2^{2i}+2}{3}}.$$

We can compute the exponent of 2 that appears in the prime factorization of the binomial coefficient on the right-hand side of formula (8.42) and conclude that it is at least 1 if and only if $b_i(3b_i + 1) \equiv 0 \pmod 4$. Combining this information with incongruence (8.41), we get that $b_i \equiv 0$ or $b_i \equiv 1 \pmod 4$. Continuing in this manner, we finally arrive at $b_i = 0$, and now formula (8.40) gives $q = 2^{2(i-1)} + 1$. Hence, q is a prime of the form $2^t + 1$, and therefore q is a Fermat prime. This completes the proof of the fact that all members of K_3 that are not primes are among the numbers 1, 4, 25 and $2q$, where $q > 3$ is a Fermat prime.

In order to complete the proof, it suffices to show that these composite numbers do indeed belong to K_3. It is an easy calculation to show that 1, 4, 25, belong to K_3. Now let $k = 2q$ and let n be in the interval $\left[\frac{q}{2}, \frac{2q}{3}\right]$. Then,

(8.43)
$$\binom{n}{k - 3n} = \binom{n}{4n - k} = \binom{n}{4n - 2q} = \frac{n}{4n - 2q}\binom{n-1}{4n - 2q - 1}.$$

Note that $4n - 2q \neq 0$, since q is an odd prime. We also notice that $0 < k - 3n < n$ if q is a Fermat prime greater than 3. Since $q > n$ is prime, it follows that $\gcd(n, 4n - 2q) = 1$ or 2 according to whether n is even or odd. If n is odd, we get that the binomial coefficient appearing on the left-hand side of formula (8.43) is a multiple of n. If n is even, it suffices to show that the binomial coefficient appearing on the right-hand side of formula (8.43) is even. But this is a straightforward computation using Lemma 8.2. Theorem 8.10 is therefore proved. $\qquad \square$

The exact structure of the sets K_c for $c \in \{4, 5, 6\}$ is also known (see [Harborth, 1977] for $c = 4$ and [Koch] for $c = 5$ and $c = 6$). In particular, it was shown that $2q \notin K_4 \cup K_5$ for any Fermat prime $q > 3$, but $2q \in K_6$ for all Fermat primes q.

Nothing seems to be known about the structure of K_c when $c \geq 7$. Since numbers that are twice a Fermat prime appear in both K_3 and K_6, it makes sense to ask the following question.

Is it true that the numbers of the form $2q$, where $q > 3$ is a Fermat prime, belong to K_c for infinitely many values of c? Do they appear in K_c for all values of c that are multiples of 3?

9. Miscellaneous Results

Number theory is a game of inspiration.
Michael Sean Mahoney

In this section we collect a few miscellaneous facts about the Fermat numbers. We start with a simple proof of the following result, which has been known since 1850. This result generalizes Remark 6.11 and Remark 6.17, which stated that no Fermat number can be a perfect square or a perfect cube.

Theorem 9.1. *A Fermat number cannot be a perfect power.*

P r o o f . Since $F_0 = 3$ and 3 is not a perfect power, we may assume that $m > 0$. Recall formula (3.6):

$$(9.1) \qquad F_m = (F_{m-1} - 1)^2 + 1.$$

From formula (9.1) it follows easily that F_m cannot be a square, because the only consecutive squares are 0 and 1. So, assume that $F_m = x^n$ for some odd number n. Then

$$(9.2) \qquad 2^{2^m} = F_m - 1 = x^n - 1 = (x - 1)(x^{n-1} + x^{n-2} + \cdots + 1),$$

where $x^{n-1} + x^{n-2} + \cdots + 1 = (x^n - 1)/(x - 1)$ is an odd integer. Since the left-hand side of (9.2) is a power of 2, this forces $(x^n - 1)/(x - 1)$ to be equal to 1. Hence, $n = 1$, and Theorem 9.1 is therefore proved. $\square$

Remark 9.2. Without too much effort one can prove that F_m cannot be a perfect square or a perfect cube by using simple congruence conditions. Noting that the last digit (which is the least nonnegative residue modulo 10) of a perfect square can only be 0, 1, 4, 5, 6, or 9, we see immediately that F_m cannot be a perfect square, since $F_0 = 3$ and $F_1 = 5$ are not squares and the last digit of F_m is 7 for $m \geq 2$ by Remark 3.6. By considering the sequence of Fermat numbers modulo 3 and modulo 7, it was proved in [Vaidya] that Fermat numbers cannot be perfect squares or perfect cubes, respectively.

Remark 9.3. The equation $F_m = x^n$, or $2^{2^m} + 1 = x^n$, is a particular case of the well-known *Catalan equation*. The Catalan equation is the Diophantine equation

$$(9.3) \qquad x^p - y^q = 1,$$

where all four unknowns x, y, p, q are positive integers larger than 1. It is conjectured that equation (9.3) has only one solution, namely, $(x, y, p, q) = (3, 2, 2, 3)$. Letting $y = 2^{2^{m-1}}$, we see that the equation

$$F_m = x^n$$

is equivalent to the Catalan equation

$$(9.4) \qquad x^n - y^2 = 1.$$

It is obvious that equation (9.4) has no solution in positive integers when $n = 2$. In 1738, L. Euler proved that equation (9.4) is not solvable in positive integers when $n = 3$, and in 1850, it was shown (see [Lebesgue]) that (9.4) is not solvable in positive integers for $n > 3$.

For a survey of what is known about the Catalan equation the reader should consult [Ribenboim, 1994].

Now recall the definition of a *polygonal number*.

Definition 9.4. Let $k \geq 3$ and $n \geq 2$ be positive integers. The *polygonal number* $p(n, k)$ is the positive integer given by

$$p(n, k) = \frac{(k - 2)n^2 - (k - 4)n}{2}.$$

When $k = 3$, the polygonal number $p(n, 3) = n(n + 1)/2$ is called *triangular*; when $k = 4$ the resulting number $p(n, 4) = n^2$ is simply a *square* number; when $k = 5$ the resulting polygonal number $p(n, 5) = n(3n - 1)/2$ is called *pentagonal* (see Figure 9.1); etc. Notice also that if $k \geq 3$ is arbitrary, then $p(2, k) = k$. Thus, k is a kth polygonal number in a natural way.

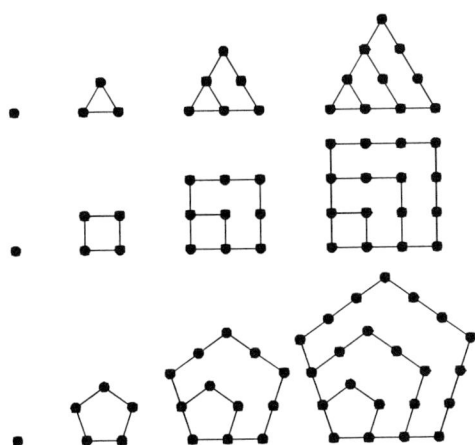

Figure 9.1. Geometric interpretation of triangular, square, and pentagonal numbers.

The purpose of the next result is to find the Fermat numbers having polygonal shape. Since by the preceding observations,

$$(9.5) \qquad\qquad p(2, F_m) = F_m \qquad \text{for all } m \geq 0,$$

it suffices to investigate the equation

$$(9.6) \qquad\qquad F_m = p(n, k) \qquad \text{for } n \geq 3 \text{ and } k \geq 3.$$

We have the following result.

Theorem 9.5. *Equation* (9.6) *has no solutions.*

P r o o f . Assume, to the contrary, that

$$(9.7) \qquad\qquad 2^{2^m} + 1 = F_m = \frac{(k-2)n^2 - (k-4)n}{2}$$

has a solution (k, m, n) with both n and k larger than 2. Rewrite equation (9.7) as

$$(9.8) \qquad 2^{2^m + 1} = (k-2)n^2 - (k-4)n - 2$$
$$= \big((k-2)n^2 - (k-2)n\big) + 2n - 2$$
$$= (k-2)n(n-1) + 2(n-1) = (n-1)\big((k-2)n + 2\big).$$

From equation (9.8) it follows that there exist nonnegative integers α and β such that

$$(9.9) \qquad\qquad \begin{aligned} n - 1 &= 2^\alpha, \\ (k-2)n + 2 &= 2^\beta, \end{aligned}$$

where $\alpha + \beta = 2^m + 1$. Since both k and n are larger than 2, it follows that $\alpha > 0$ and $\beta > 1$. If we write $n = 2^\alpha + 1$, the second formula (9.9) becomes

$$(9.10) \qquad\qquad (k-2)(2^\alpha + 1) = 2^\beta - 2 = 2(2^{\beta-1} - 1).$$

Equation (9.10) forces

$$(9.11) \qquad\qquad 2^\alpha + 1 \mid 2^{\beta-1} - 1.$$

However, it is well known (compare with Lemma 6.3 or see, for example, [Mc-Daniel]), that the divisibility relation (9.11) holds if and only if $\alpha \mid (\beta - 1)$ and $(\beta - 1)/\alpha$ is an even integer. Thus, assume that $\beta - 1 = \lambda\alpha$, where λ is even and nonzero. Since $\alpha + \beta = 2^m + 1$, we get that

$$2^m + 1 = \beta + \alpha = 1 + \lambda\alpha + \alpha = 1 + \alpha(\lambda + 1),$$

or

$$(9.12) \qquad\qquad 2^m = \alpha(\lambda + 1).$$

Equation (9.12) shows that $\lambda + 1$ is a power of 2, which is impossible because $\lambda + 1$ is an odd number larger than 1 (because λ is even and positive). This contradiction finishes the proof of Theorem 9.5. □

Remark 9.6. In [Satyanarayana], [Radovici-Mărculescu], and [Vaidya], it was proved that the only triangular Fermat number is $F_0 = 3 = p(2, 3)$. In [Krishna], the author also proved this result as well as showing that the only pentagonal Fermat number is $F_1 = 5 = p(2, 5)$. Thus, our Theorem 9.5 and (9.5) give an immediate generalization of these results. A relationship of the Fermat numbers and triangular numbers is also given in [Asadulla].

By Gauss's celebrated Theorem 4.3, the prime Fermat numbers are related to the number of sides of regular polygons that can be constructed with ruler and compass. However, Fermat primes appear not only in connection with constructing regular polygons but also in connection with Heron triangles.

Recall that a *Heron triangle* is a triangle such that the lengths of its three sides as well as its area are integers.

In the following theorem (see [Luca, 2000f]) we point out an interesting relationship between the prime Fermat numbers and the Heron triangles whose sides are prime powers (see Figure 9.2).

Theorem 9.7. *Let the lengths of all three sides of a Heron triangle be prime powers. Then the lengths of the sides are either 3, 4, 5, or F_m, F_m, $4(F_{m-1} - 1)$ for some $m \geq 1$ such that F_m is prime.*

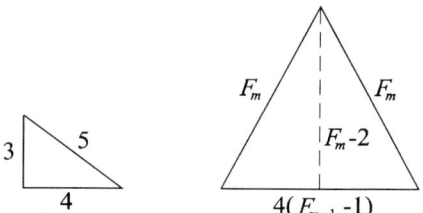

Figure 9.2. The only possible Heron triangles with prime-power sides.

P r o o f . We start our investigation with some remarks on Heron triangles. Assume that a, b, c denote the lengths of three sides of a Heron triangle and let S denote its area. For the moment, we do not assume any ordering among the integers a, b, and c. We denote by

$$p = \frac{a + b + c}{2}$$

its semiperimeter. We also use A, B, C to denote the angles opposing the sides a, b, c, respectively, and h_a, h_b, h_c to denote the lengths of the altitudes perpendicular to a, b, c, respectively. We will freely identify the angles A, B, C with the vertices of the triangle.

By Heron's formula, we have

$$(9.13) \qquad\qquad S = \sqrt{p(p-a)(p-b)(p-c)}.$$

In particular, formula (9.13) implies that p is an algebraic integer (i.e., a root of some monic polynomial with integer coefficients). Since certainly p is also rational, it follows that p is an integer. In particular, it follows that either all three numbers a, b, c are even or exactly one of them is even and the other two are odd.

Assume that a, b, c are the sides of an isosceles triangle. Let $a = b$, for example. Since

$$p = \frac{2a+c}{2} = a + \frac{c}{2}$$

is an integer, it follows that c is even. Notice that the altitude h_c coincides with the median drawn from the vertex C. In particular, $a, h_c, c/2$ are the sides of a right triangle whose hypotenuse is of length a. Hence,

$$(9.14) \qquad\qquad h_c = \sqrt{a^2 - \left(\frac{c}{2}\right)^2}.$$

Formula (9.14) implies that h_c is an algebraic integer. Since $h_c = 2S/c$ is a rational number, it follows that h_c is an integer. Hence, the triple $(a, h_c, c/2)$ is a Pythagorean triple.

Finally, we also notice that $\min(a, b, c) \geq 3$, where we are not necessarily assuming that a, b, c are the sides of an isosceles triangle. Indeed, assume that $\min(a, b, c) = c$ and that $c \leq 2$. Suppose, for example, that $c = 1$. In this case, the triangular inequality $|a - b| < c = 1$ implies that $a = b$. This is impossible, because we saw that if $a = b$, then c is even. Assume now that $c = 2$. The triangular inequality $|a - b| < c = 2$ implies that either $a = b$ or $|a - b| = 1$. In the first case, we get that $(a, h_c, c/2) = (a, h_c, 1)$ is a Pythagorean triple, which is impossible because the difference of two consecutive nonzero squares is strictly larger than 1. Finally, if $|a - b| = 1$, we may assume that $a > b$, and in this case we get that $b = a - 1$. Now

$$p = \frac{a + (a-1) + 2}{2} = \frac{2a+1}{2}$$

is not an integer, which is again impossible.

We are now ready to prove the theorem. Assume that all of a, b, c are prime powers. It follows that either all of them are even, in which case all of them are powers of 2, or only one of them is even.

We first show that there is no Heron triangle whose sides are all powers of 2. Assume that this is not so, and let $a = 2^\alpha$, $b = 2^\beta$, $c = 2^\gamma$. Since $\min(a, b, c) > 2$, we may assume that $\alpha \geq \beta \geq \gamma \geq 2$. The triangular inequality

$$2^\beta \leq 2^\alpha = a < b + c = 2^\beta + 2^\gamma$$

forces $\alpha = \beta$. Hence, $a = b$. Therefore, the triple $(a, h_c, c/2) = \left(2^\alpha, h_c, 2^{\gamma-1}\right)$ is a Pythagorean triple. We now get that

$$(9.15) \qquad\qquad 2^{2\alpha} = h_c^2 + 2^{2(\gamma-1)}.$$

Equation (9.15) implies that $2^{\gamma-1} \mid h_c$. Let $h_c = 2^{\gamma-1}k$ for some positive integer k. Dividing both sides of equation (9.15) by $2^{2(\gamma-1)}$, we get

$$2^{2(\alpha-\gamma+1)} = k^2 + 1,$$

which is impossible because the difference of two consecutive nonzero squares is strictly larger than 1.

In conclusion, exactly one of the numbers a, b, c is a power of 2 and the other two are odd prime powers. Let $a = p^\alpha$, $b = q^\beta$, and $c = 2^\gamma$ for some odd prime numbers p and q and some positive integers α, β, γ. Notice that $\gamma \geq 2$.

We first assume that the triangle is isosceles. In this case, $a = b$ and the triple $(a, h_c, c/2) = (p^\alpha, h_c, 2^{\gamma-1})$ is a Pythagorean triple. Hence,

$$(9.16) \qquad p^{2\alpha} = h_c^2 + 2^{2(\gamma-1)}.$$

Equation (9.16) implies that the triple $(a, h_c, c/2)$ is reduced (i.e., the greatest common divisor of a, h_c, and $c/2$ is 1). From the classical parametrization of all reduced Pythagorean triples, it follows that there exist two coprime integers $m > n$, one of them even and one of them odd, such that

$$(9.17) \qquad p^\alpha = m^2 + n^2, \quad h_c = m^2 - n^2, \quad 2^{\gamma-1} = 2mn.$$

The last equation of (9.17) forces $m = 2^{\gamma-2}$ and $n = 1$. Therefore, the first equation of (9.17) becomes

$$(9.18) \qquad p^\alpha = 2^{2(\gamma-2)} + 1.$$

Notice that equation (9.18) and the fact that p is odd imply $\gamma > 2$. When $\alpha > 1$, equation (9.18) is a particular case of the Catalan equation $x^t = y^s + 1$ with (x, y, t, s) positive integers and $\min(t, s) > 1$ (see Remark 9.3). Lebesgue's result concerning the impossibility of solutions of equation (9.4) shows that α cannot be larger than 1. Hence, equation (9.18) reduces to

$$(9.19) \qquad p = 2^{2(\gamma-2)} + 1.$$

It is well known that the only primes p of the form $2^s + 1$ for some $s \geq 1$ are the Fermat primes, namely, the numbers $p = F_m$ for some $m \geq 0$. Hence, $p = F_m$, F_m is prime, and $\gamma - 2 = 2^{m-1}$. We now get that $(a, b, c) = \left(F_m, F_m, 2^{2^{m-1}+2}\right) = (F_m, F_m, 4(F_{m-1} - 1))$ with F_m prime.

Assume now that the triangle of sides a, b, c is not isosceles. Straightforward computations show that after squaring both sides of formula (9.13) one gets

$$(9.20) \qquad 2a^2c^2 + 2b^2c^2 + 2a^2b^2 - a^4 - b^4 - c^4 = (4S)^2.$$

One may rewrite equation (9.20) as

$$(9.21) \qquad 2c^2(a^2 + b^2) - c^4 - (a^2 - b^2)^2 = (4S)^2.$$

By replacing c with 2^γ in equation (9.21), we obtain

(9.22) $$2^{2\gamma+1}\left(a^2 + b^2\right) - 2^{4\gamma} - \left(a^2 - b^2\right)^2 = (4S)^2.$$

Notice that $a^2 + b^2$ is even. Therefore, we may reduce equation (9.22) modulo $2^{2\gamma+2}$ and get

(9.23) $$-\left(a^2 - b^2\right)^2 \equiv (4S)^2 \pmod{2^{2(\gamma+1)}}.$$

Notice that equation (9.23) forces $2^{\gamma+1} \mid (a^2 - b^2)$. Indeed, assume that this is not the case. It then follows that if $2^\delta \| (a^2 - b^2)$, then $\delta \le \gamma$. Dividing both sides of congruence (9.23) by $2^{2\delta}$, we find that

(9.24) $$-\left(\frac{a^2 - b^2}{2^\delta}\right)^2 \equiv \left(\frac{4S}{2^\delta}\right)^2 \pmod{2^{2(\gamma-\delta+1)}}.$$

However, notice that congruence (9.24) is impossible, because -1 is not a quadratic residue modulo 4. Hence, $2^{\gamma+1}$ divides both $(a^2 - b^2)$ and $4S$. On the one hand, we conclude that $2c = 2^{\gamma+1}$ divides $4S$, and so $c \mid 2S$. On the other hand, we conclude that

$$2^{\gamma+1} \mid (a^2 - b^2) = (a - b)(a + b).$$

Since a and b are both odd, it follows that either $c = 2^\gamma \mid (a - b)$ or $c = 2^\gamma \mid (a + b)$. The case $c \mid (a - b)$ is impossible because of the triangular inequality $0 < |a - b| < c$. Hence, $2^\gamma \mid (a + b)$. Since $\gamma \ge 2$, it follows that one of the numbers a and b is congruent to 1 modulo 4 and the other one is congruent to 3 modulo 4. So, we may assume that $a = p^\alpha \equiv 1 \pmod 4$ and that $b = q^\beta \equiv 3 \pmod 4$. In particular, $q \equiv 3 \pmod 4$ and β is odd. Since formula (9.20) is symmetric in the variables a, b, and c, we may rewrite formula (9.21) as

(9.25) $$2b^2\left(a^2 + c^2\right) - b^4 - \left(a^2 - c^2\right)^2 = (4S)^2.$$

Reducing formula (9.25) modulo b, we get that

(9.26) $$-\left(a^2 - c^2\right)^2 \equiv (4S)^2 \pmod{q^{2\beta}}.$$

One may now use the fact that -1 is not a quadratic residue modulo q (because $q \equiv 3 \pmod 4$) to conclude that equation (9.26) forces $q^\beta \mid (a^2 - c^2)$. In particular, $b = q^\beta \mid 4S$. Hence, $b \mid 2S$. Since $c \mid 2S$, it follows that $bc \mid 2S$. We now get that $S \ge bc/2$. Since

$$S = \frac{bc\sin(A)}{2},$$

it follows that $\sin(A) \ge 1$. This forces $A = \pi/2$, and thus our triangle has a right angle in A. We now get that (a, b, c) is a Pythagorean triple and

(9.27) $$p^{2\alpha} = q^{2\beta} + 2^{2\gamma}.$$

From the standard parametrization of reduced Pythagorean triples it follows that there exist two coprime integers $m > n$, one of them odd and the other one even, such that

(9.28)
$$p^\alpha = m^2 + n^2, \quad q^\beta = m^2 - n^2, \quad 2^\gamma = 2mn.$$

The last formula of (9.28) implies that $m = 2^{\gamma-1}$ and $n = 1$. Now the second formula (9.28) implies that $q^\beta = m^2 - 1 = (m-1)(m+1)$. Since $m-1$ and $m+1$ are coprime odd integers, it follows that $m - 1 = 1$. Therefore, $m = 2$ and $\gamma = 2$. This leads to $q^\beta = 2^2 - 1 = 3$. Thus, $q = 3$ and $\beta = 1$. Finally, the first formula of (9.28) implies that $p^\alpha = 2^2 + 1 = 5$, and so $p = 5$ and $\alpha = 1$. Hence, we obtain $(a, b, c) = (5, 3, 4)$. $\square$

For any positive integer n let $\sigma(n)$ be the sum of its positive divisors. A positive integer n such that $\sigma(n) = 2n$ is called *perfect*. Two distinct positive integers m and n are called *amicable* if $\sigma(m) = \sigma(n) = m + n$.

For instance, 220 and 284 is a pair of amicable numbers, which was already known to the Pythagoreans. Another pair of amicable numbers, namely 17296 and 18416, was discovered by al-Farisi and later independently by Pierre de Fermat.

In the solution to the problem proposed by P. Erdős, that appeared in [Amer. Math. Monthly **62** (1955), 498–499], R. Bojanič proved that

$$\frac{\sigma(F_m)}{F_m} \to 1 \quad \text{and} \quad \frac{\phi(F_m)}{F_m} \to 1 \quad \text{as } m \to \infty,$$

where ϕ is the Euler totient function.

The following theorem appears in [Luca, 2000a].

Theorem 9.8. *A Fermat number is never perfect or part of an amicable pair.*

P r o o f . Assume that

(9.29)
$$\sigma(F_m) = \sigma(x) = F_m + x$$

for some positive integer x. Notice that since F_m is prime for all $m \le 4$, it follows that $\sigma(F_m) - F_m = 1$ for such values of m. In particular, equation (9.29) has no solution for such m's.

Suppose now that $m \ge 5$. Since $\sigma(y) \le y^2$ for all $y \ge 1$, it follows that

$$x^2 \ge \sigma(x) > F_m \ge F_5 > 2^{2^5}.$$

In particular, $x \ge 2^{2^4} + 1 = F_4$.

We further find an upper bound for $\sigma(F_m)/F_m$. We follow the method outlined in [Luca, 2001]. Write

$$F_m = p_1^{\alpha_1} \cdots p_k^{\alpha_k},$$

where $p_1 < p_2 < \cdots < p_k$ are primes. By Lucas's Theorem 6.1, $p_i \equiv 1 \pmod{2^{m+2}}$ for all $i = 1, \ldots, k$. Hence,

$$\log F_m = \sum_{i=1}^{k} \alpha_i \log p_i \ge k \log p_1 \ge k \log(2^{m+2} + 1).$$

Thus,

$$k \log\left(2^{m+2} + 1\right) \leq \log F_m = \log\left(2^{2^m} + 1\right).$$

By Lemma 7.9,

$$\frac{\log(y + 1)}{\log(z + 1)} \leq \frac{\log y}{\log z}, \qquad \text{whenever } y \geq z > 1,$$

and thus

$$k \leq \frac{\log\left(2^{2^m} + 1\right)}{\log\left(2^{m+2} + 1\right)} \leq \frac{\log\left(2^{2^m}\right)}{\log\left(2^{m+2}\right)} = \frac{2^m}{m + 2}.$$

Since

$$\frac{\sigma(y)}{y} \leq \frac{y}{\phi(y)} \qquad \text{for all integers } y \geq 1,$$

where ϕ is the Euler totient function, it follows that

$$(9.30) \qquad 1 + \frac{x}{F_m} = \frac{\sigma(F_m)}{F_m} \leq \frac{F_m}{\phi(F_m)} = \prod_{i=1}^{k}\left(1 + \frac{1}{p_i - 1}\right),$$

or

$$(9.31) \qquad \log\left(1 + \frac{x}{F_m}\right) \leq \sum_{i=1}^{k} \log\left(1 + \frac{1}{p_i - 1}\right) < \sum_{i=1}^{k} \frac{1}{p_i - 1}.$$

Since $p_i \equiv 1 \pmod{2^{m+2}}$, it follows that $p_i \geq 2^{m+2}i + 1$ for all $i = 1, \ldots, k$. Thus,

$$\sum_{i=1}^{k} \frac{1}{p_i - 1} \leq \frac{1}{2^{m+2}} \sum_{i=1}^{k} \frac{1}{i} \leq \frac{1}{2^{m+2}}(1 + \log k) < \frac{1}{2^{m+2}}\left(1 + \log\left(\frac{2^m}{m + 2}\right)\right) < \frac{m \log 2}{2^{m+2}}.$$

Hence,

$$(9.32) \qquad \log\left(1 + \frac{x}{F_m}\right) < \frac{m \log 2}{2^{m+2}}.$$

It now follows that $x < F_m$. Indeed, if $x \geq F_m$, then the above inequality would force

$$\log 2 < \frac{m \log 2}{2^{m+2}},$$

or $2^{m+2} < m$, which is impossible. Therefore, $x < F_m$. Since

$$\log(1 + y) > \frac{y}{2} \qquad \text{for all } y \in (0, 1),$$

it follows that

$$\frac{x}{2F_m} < \frac{m \log 2}{2^{m+2}},$$

or

$$x < \frac{F_m m \log 2}{2^{m+1}}.$$

On the other hand, by [Rosser, Schoenfeld, formulae (2.13), (3.41), and (3.42)], we know that

$$\sigma(y) < \left(1.8 \log \log y + \frac{2.6}{\log \log y}\right) y \qquad \text{for all } y \geq 3,$$

and hence

$$x + F_m = \sigma(x) < \left(1.8 \log \log x + \frac{2.6}{\log \log x}\right) x.$$

Since x is large (at least F_4), it follows that

$$\log \log x > 1.$$

Thus,

$$F_m < 4.4\, x \log \log x < 4.4\, \frac{F_m m \log 2}{2^{m+1}} \log \log \left(\frac{F_m m \log 2}{2^{m+1}}\right)$$

$$< 4.4\, \frac{F_m m \log 2}{2^{m+1}} \log \log F_m,$$

or

$$2^{m+1} < 4.4\, m \log 2 \log \log F_m.$$

However, since $F_m = 2^{2^m} + 1 < 2^{2^{m+1}}$, it follows that

$$2^{m+1} < 4.4\, m \log 2 \log \log 2^{2^{m+1}} = 4.4\, m \log 2((m+1)\log 2 + \log \log 2).$$

But the above inequality is impossible, because the left-hand side of it is larger than the right-hand side of it for all $m \geq 5$. The theorem is therefore proved. $\square$

Remark 9.9. Recall that a *multiply perfect* number is an integer $m > 1$ such that $m \mid \sigma(m)$. From the above arguments it also follows that there are no multiply perfect Fermat numbers. Indeed, from (9.30) and (9.32) in the above proof we know that

$$\log\left(\frac{\sigma(F_m)}{F_m}\right) < \frac{m \log 2}{2^{m+2}} < \log 2.$$

Hence, $\sigma(F_m) < 2F_m$, and therefore F_m cannot be multiply perfect.

10. The Irrationality of the Sum of Some Reciprocals

Satisfaction in science is not associated with
solving a simple problem but comes from understanding
complex situations and enjoying the small steps made
to penetrate the mysteries of nature.

Victor Szebehely

Let $\{n_k\}_{k\geq 1}$ be an increasing sequence of positive integers. In this chapter we investigate some conditions under which the sum of the series

$$(10.1) \qquad \sum_{k\geq 1} \frac{1}{n_k}$$

is an irrational number, and then we apply these results to the case for which the sequence $\{n_k\}_{k\geq 1}$ is the sequence of Fermat numbers.

The first proof of the fact that the sum of the series

$$(10.2) \qquad \sum_{m\geq 0} \frac{1}{F_m}$$

is irrational appears in [Golomb, 1963]. In January 1964, a more general result was given (see [Erdős, Straus]) in terms of the so-called Ahmes series that implies Golomb's result.

For a sequence $\{n_k\}_{k\geq 1}$ of positive integers, let $N_k = \mathrm{lcm}(n_1, n_2, \ldots, n_k)$ be the least common multiple of the first k terms.

Theorem 10.1. *Let $\{n_k\}_{k\geq 1}$ be an increasing sequence of positive integers such that*

(i) $\limsup_{k\to\infty} n_k^2/n_{k+1} \leq 1$,

(ii) $\{N_k/n_{k+1}\}_{k\geq 1}$ *is bounded.*

Then, the sum of the series (10.1) *is rational if and only if $n_{k+1} = n_k^2 - n_k + 1$ for all $k \geq k_0$, in which case we have*

$$(10.3) \qquad \sum_{k\geq 1} \frac{1}{n_k} = \frac{1}{n_1} + \cdots + \frac{1}{n_{k_0-1}} + \frac{1}{n_{k_0} - 1}.$$

P r o o f . We reproduce here the proof from [Erdős, Straus]. Notice that condition (i) ensures that n_k has at least exponential growth. Indeed, first of all, condition (i) shows that $\{n_k\}_{k \geq 1}$ is increasing from some k on. Moreover, for k large enough, we have $n_k^2 / n_{k+1} < 2$. Therefore, $n_{k+1} > n_k^2 / 2$. Choosing k_1 such that $n_k > 4$ for all $k > k_1$, we get that $n_{k+1} > 2n_k$ for all $k > k_1$. Hence, $\{n_k\}_{k \geq 1}$ has at least exponential growth. In particular,

$$(10.4) \qquad \sum_{t \geq k} \frac{1}{n_t} = \mathcal{O}\left(\frac{1}{n_k}\right)$$

holds for all $k \geq 1$.

Assume now that the sum (10.1) is rational and denote it by a/b, where a and b are coprime positive integers. Write $bN_k = c_k n_{k+1} - d_k$, where c_k, d_k are integers with $0 \leq d_k < n_{k+1}$. Then c_k is positive, and condition (ii) ensures that c_k is bounded. Write

$$\sum_{t \geq 1} \frac{1}{n_t} = \sum_{t=1}^{k} \frac{1}{n_t} + \sum_{t \geq k+1} \frac{1}{n_t} = \frac{a}{b}$$

as

$$\frac{M_k}{N_k} + \sum_{t \geq k+1} \frac{1}{n_t} = \frac{a}{b},$$

or

$$(10.5) \qquad bM_k + bN_k\left(\sum_{t \geq k+1} \frac{1}{n_t}\right) = aN_k,$$

for some integer $M_k > 0$. Eliminating the integers from formula (10.5), we get

$$0 \equiv bN_k\left(\frac{1}{n_{k+1}} + \frac{1}{n_{k+2}} + \cdots\right) \pmod{1},$$

or

$$(10.6) \qquad 0 \equiv (c_k n_{k+1} - d_k)\left(\frac{1}{n_{k+1}} + \frac{1}{n_{k+2}} + \cdots\right) \pmod{1},$$

or, by using (10.4) and the fact that c_k is bounded,

$$(10.7) \qquad 0 \equiv -\frac{d_k}{n_{k+1}} + \frac{c_k n_{k+1} - d_k}{n_{k+2}} + \mathcal{O}\left(\frac{n_{k+1}}{n_{k+3}}\right) \pmod{1}.$$

By condition (i) we obtain

$$(10.8) \qquad \frac{n_{k+1}^2}{n_{k+3}} = \frac{n_{k+1}^2}{n_{k+2}} \frac{n_{k+2}^2}{n_{k+3}} \frac{1}{n_{k+2}} = o(1).$$

Using estimate (10.8), formula (10.7) implies

$$(10.9) \qquad d_k \equiv c_k \frac{n_{k+1}^2}{n_{k+2}} - d_k \frac{n_{k+1}}{n_{k+2}} + o(1) \pmod{n_{k+1}}.$$

But certainly,

$$(10.10) \qquad 0 \le c_k \frac{n_{k+1}^2}{n_{k+2}} - \frac{n_{k+1}^2}{n_{k+2}} \le c_k \frac{n_{k+1}^2}{n_{k+2}} - d_k \frac{n_{k+1}}{n_{k+2}} + o(1) \le c_k + o(1).$$

Combining (10.9) with (10.10), we get that

$$(10.11) \qquad\qquad d_k \le c_k \qquad \text{for all sufficiently large } k.$$

Now

$$c_{k+1} n_{k+2} - d_{k+1} = bN_{k+1} \le n_{k+1} bN_k = c_k n_{k+1}^2 - d_k n_{k+1},$$

or

$$(10.12) \qquad\qquad c_{k+1} \le c_k \frac{n_{k+1}^2}{n_{k+2}} + o(1) \le c_k + o(1),$$

so that $c_{k+1} \le c_k$ for all sufficiently large k, which means that $c_k = c = \text{constant}$ for all sufficiently large k. But by condition (i), this is possible only when

$$\lim \frac{n_k^2}{n_{k+1}} = 1.$$

Now (10.9) yields $d_k = c$ for all $k \ge k_2$, and (10.6) becomes

$$\frac{1}{n_{k+1}} = \frac{n_{k+1} - 1}{n_{k+2}} + (n_{k+1} - 1)\left(\frac{1}{n_{k+3}} + \cdots\right), \qquad k \ge k_2,$$

or

$$n_{k+2} = n_{k+1}^2 - n_{k+1} + \frac{n_{k+1}^2}{n_{k+2}} \frac{n_{k+2}^2}{n_{k+3}} + o(1) = n_{k+1}^2 - n_{k+1} + 1 + o(1),$$

so that

$$(10.13) \qquad\qquad n_{k+2} = n_{k+1}^2 - n_{k+1} + 1$$

for all sufficiently large k.

The last statement of Theorem 10.1 holds, since

$$\frac{1}{n_{k_0} - 1} = \frac{1}{n_{k_0}} + \frac{1}{n_{k_0+1}} + \cdots.$$

To see this, note that by induction we have

$$\sum_{k=k_0}^{k_0+\ell} \frac{1}{n_k} = \frac{1}{n_{k_0} - 1} - \frac{1}{n_{k_0} - 1} \frac{1}{n_{k_0}} \frac{1}{n_{k_0+1}} \cdots \frac{1}{n_{k_0+\ell}}$$

and

$$n_{k_0+\ell+1} = (n_{k_0} - 1) n_{k_0} n_{k_0+1} \cdots n_{k_0+\ell} + 1.$$

Theorem 10.1 is therefore established.   □

A series $\sum_{k\geq 1}\frac{1}{n_k}$ such that $n_{k+1} = n_k^2 - n_k + 1$ holds for all sufficiently large k is called an *Ahmes series*. Thus, Theorem 10.1 shows that the sum of a series of the form (10.1) for which the sequence $\{n_k\}_{k\geq 1}$ satisfies (i) and (ii) is rational if and only if the series is Ahmes.

It is easy to see that the irrationality of the sum of the series (10.2) follows immediately from Theorem 10.1. In fact, a more general result can be immediately established.

Corollary 10.2. *Let a and b be positive integers and set $n_k = a^{2^k} + b^{2^k}$ for $k \geq 0$. Then the sum (10.1) is not a rational number.*

P r o o f . Assume first that $a > b$. Since

$$\frac{n_k^2}{n_{k+1}} = \frac{\left(a^{2^k} + b^{2^k}\right)^2}{a^{2^{k+1}} + b^{2^{k+1}}} = \frac{a^{2^{k+1}} + 2a^{2^k}b^{2^k} + b^{2^{k+1}}}{a^{2^{k+1}} + b^{2^{k+1}}} = \frac{1 + 2(b/a)^{2^k} + (b/a)^{2^{k+1}}}{1 + (b/a)^{2^{k+1}}} \to 1,$$

it follows that condition (i) of Theorem 10.1 is fulfilled. Since

$$N_k \mid n_1 n_2 \cdots n_k = (a+b)(a^2+b^2)\cdots\left(a^{2^k}+b^{2^k}\right) = \frac{a^{2^{k+1}} - b^{2^{k+1}}}{a-b},$$

we have

$$\frac{N_k}{n_{k+1}} \leq \frac{1}{a-b}\frac{a^{2^{k+1}} - b^{2^{k+1}}}{a^{2^{k+1}} + b^{2^{k+1}}} < \frac{1}{a-b} \qquad \text{for } k \geq 0.$$

Hence, condition (ii) of Theorem 10.1 is satisfied as well. By Theorem 10.1, it follows now that the sum (10.1) is rational if and only if

(10.14) $$n_{k+1}^2 = n_k^2 - n_k + 1$$

holds for sufficiently large k. Equality (10.14) is equivalent to

$$a^{2^{k+1}} + b^{2^{k+1}} = \left(a^{2^k} + b^{2^k}\right)^2 - \left(a^{2^k} + b^{2^k}\right) + 1 = a^{2^{k+1}} + b^{2^{k+1}} + 2a^{2^k}b^{2^k} - a^{2^k} - b^{2^k} + 1,$$

or

(10.15) $$-(ab)^{2^k} = a^{2^k}b^{2^k} - a^{2^k} - b^{2^k} + 1 = \left(a^{2^k} - 1\right)\left(b^{2^k} - 1\right).$$

Notice now that equality (10.15) is impossible simply because its left-hand side is negative, whereas its right-hand side is nonnegative.

This settles the case $a > b$.

When $a = b$, then $a > 1$ (otherwise, series (10.1) is divergent), and the sum of the series (10.1) is rational if and only if the sum of the series

(10.16) $$\sum_{k\geq 0}\frac{1}{a^{2^k}}$$

is rational. However, if we think of the number given by (10.16) as written in base a, then it has the representation $0.110100010000\ldots 10\ldots$ with 1's only in positions

2^k for $k \geq 0$ and 0's in the other positions. Since this representation has blocks of arbitrary length of consecutive 0's, it follows that this number cannot be rational (otherwise, its base-a representation would be periodic from some point on). □

Corollary 10.2 shows not only that the sum of the series (10.2) is irrational, but even more, that the sum of the reciprocals of all generalized Fermat numbers $F_m(a, b)$ (see Chapter 13) is irrational as well.

In fact, using Theorem 10.1, one can infer the following stronger version of Corollary 10.2 (see [Golomb, 1963]).

Corollary 10.3 (Golomb). *Assume that $a > 1$ is an integer and that $n_k = a^{2^k} + b_k$ for $k \geq 1$, where b_k are integers such that*

$$(10.17) \qquad \sum_{k \geq 1} \frac{|b_k|}{a^{2^k}} < \infty.$$

Then the sum (10.1) is irrational.

P r o o f . Since the series (10.17) is convergent, it follows that its general term b_k/a^{2^k} tends to 0 as $k \to \infty$. Now,
$$(10.18)$$
$$\frac{n_k^2}{n_{k+1}} \leq \frac{\left(a^{2^k} + b_k\right)^2}{a^{2^{k+1}} + b_{k+1}} = \frac{a^{2^{k+1}} + 2a^{2^k} b_k + b_k^2}{a^{2^{k+1}} + b_{k+1}} = \frac{1 + 2\left(b_k/a^{2^k}\right) + \left(b_k/a^{2^k}\right)^2}{1 + \left(b_{k+1}/a^{2^{k+1}}\right)} \to 1$$

so condition (i) of Theorem 10.1 is satisfied. Moreover,

$$(10.19)$$
$$\frac{N_k}{n_{k+1}} \leq \frac{\prod_{t=1}^{k}(a^{2^t} + |b_t|)}{a^{2^{k+1}} - |b_{k+1}|} = \frac{a^2}{1 - |b_{k+1}|/a^{2^{k+1}}} \prod_{t=1}^{k}\left(1 + \frac{|b_t|}{a^{2^t}}\right)$$
$$< \frac{a^2}{1 - |b_{k+1}|/a^{2^{k+1}}} e^{\sum_{t=1}^{k} |b_t|/a^{2^t}}$$

is bounded because series (10.17) is convergent. Therefore, condition (ii) of Theorem 10.1 is satisfied as well.

By Theorem 10.1, it follows that the sum of the series (10.1) is rational if and only if

$$(10.20) \qquad\qquad n_{k+1} = n_k^2 - n_k + 1$$

holds for all sufficiently large k. Equality (10.20) is equivalent to

$$b_{k+1} = 2a^{2^k} b_k + b_k^2 - a^{2^k} - b_k + 1,$$

so that $b_k \neq 0$ for all sufficiently large k implies

$$(10.21) \qquad\qquad |b_{k+1}| > a^{2^k} |b_k|.$$

Since $b_k = 0$ implies $b_{k+1} = -a^{2^k} + 1 \neq 0$, it follows that we may assume that $b_k \neq 0$ for all sufficiently large k. But now (10.21) can be rewritten as

$$\frac{|b_{k+1}|}{a^{2^{k+1}}} > \frac{|b_k|}{a^{2^k}} \qquad \text{for sufficiently large } k,$$

which shows that the general term of the series (10.17) cannot tend to zero. This contradicts the fact that the series (10.17) converges. □

As we stated in previous chapters we do not know whether there are infinitely many Fermat primes. In this section we give two conditions in terms of the rationality of the sum of some reciprocals of the Fermat numbers that are equivalent to the existence of only finitely many Fermat primes.

Theorem 10.4. *Suppose that $0 = m_1 < m_2 < \cdots < m_k < \cdots$ are all the indices m for which F_m is prime (this set of indices might be, of course, finite). Then the sum of the series*

$$(10.22) \qquad \qquad \sum_{k \geq 1} \frac{1}{F_{m_k}} \qquad \qquad .$$

is rational if and only if there are only finitely many Fermat primes.

P r o o f . If there are only finitely many Fermat primes, then the series (10.22) consists of only finitely many terms and is, therefore, rational.

Assume now that there are infinitely many Fermat primes and let $n_k = F_{m_k}$ for all $k \geq 1$. We can use the arguments employed in the proof of Corollary 10.2 to conclude that conditions (i) and (ii) of Theorem 10.1 are satisfied, and therefore series (10.22) cannot be rational unless

$$(10.23) \qquad \qquad n_{k+1} = n_k^2 - n_k + 1$$

holds for all $k \geq k_0$. However,

$$n_k^2 - n_k + 1 = \left(2^{2^{m_k}} + 1\right)^2 - \left(2^{2^{m_k}} + 1\right) + 1 = 2^{2^{m_k+1}} + 2^{2^{m_k}} + 1,$$

which is never equal to $F_{m_{k+1}}$. Hence, equality (10.23) does not hold for any value k. Theorem 10.4 is therefore established. □

Theorem 10.5. *Let $\mathcal{C}$ be the set of all positive integers n for which the regular polygon with n sides can be constructed with ruler and compass. Then, the sum of the series*

$$(10.24) \qquad \qquad \sum_{n \in \mathcal{C}} \frac{1}{n}$$

is rational if and only if there are only finitely many Fermat primes.

P r o o f . Without loss of generality, we may adjoin the numbers 1 and 2 to $\mathcal{C}$. Let $\mathcal{C}_1 \subset \mathcal{C}$ be the subset of all the odd numbers in $\mathcal{C}$. By Gauss's Theorem 4.3, we know that

$$(10.25) \qquad \qquad \mathcal{C} = \left\{2^t n \mid n \in \mathcal{C}_1 \text{ and } t \geq 0\right\},$$

and that

(10.26) $\mathcal{C}_1 = 1 \cup \{n \mid n = F_{m_{i_1}} \cdots F_{m_{i_t}}$ for some $t \geq 1$ and $1 \leq i_1 < \cdots < i_t\}$,

where again $0 = m_1 < \cdots < m_k < \cdots$ are all the indices m for which F_m is prime. Thus, it is easy to see that

(10.27) $$\sum_{n \in \mathcal{C}} \frac{1}{n} = \sum_{t \geq 0} \sum_{n \in \mathcal{C}_1} \frac{1}{2^t n} = 2 \sum_{n \in \mathcal{C}_1} \frac{1}{n} = 2 \prod_{k \geq 1} \left(1 + \frac{1}{F_{m_k}}\right).$$

Hence, it suffices to investigate the irrationality of the product

(10.28) $$\alpha = \prod_{k \geq 1} \left(1 + \frac{1}{F_{m_k}}\right).$$

If there exist only finitely many Fermat primes, then the product (10.28) consists of only finitely many factors, and hence it is rational.

Assume now that there are infinitely many Fermat primes. For $n \geq 1$ we set

$$x_n = \prod_{k=1}^{n} (1 + F_{m_k})$$

and

$$y_n = \prod_{k=1}^{n} F_{m_k}.$$

We first notice that $\alpha < 2$. Indeed,

$$\alpha < \prod_{m \geq 0} \left(1 + \frac{1}{F_m}\right) < \prod_{m \geq 0} \left(1 + \frac{1}{2^{2^m}}\right) = \sum_{j \geq 0} \frac{1}{2^j} = 2.$$

We now find an upper bound for the error of approximating α by x_n/y_n. Notice that

(10.29) $0 < \alpha - \dfrac{x_n}{y_n} = \dfrac{x_n}{y_n} \left(\prod_{k \geq n+1} \left(1 + \dfrac{1}{F_{m_k}}\right) - 1\right) < 2\left(e^{\sum_{k \geq n+1} \frac{1}{F_{m_k}}} - 1\right).$

However,

(10.30) $$\sum_{k \geq n+1} \frac{1}{F_{m_k}} < \sum_{j \geq 2^{m_{n+1}}} \frac{1}{2^j} = \frac{1}{2^{2^{m_{n+1}}-1}} < \frac{3}{F_{m_{n+1}}}.$$

Hence, from formulae (10.29) and (10.30) we get

(10.31) $$0 < \alpha - \frac{x_n}{y_n} < 2\left(e^{\frac{3}{F_{m_{n+1}}}} - 1\right) < \frac{12}{F_{m_{n+1}}}.$$

For the rightmost inequality (10.31) we used the fact that $e^x < 1 + 2x$ for $x \in (0, 1)$. Assume now that α is rational and write it as $\alpha = a/b$, where a and b are coprime positive integers. From inequality (10.31), it follows that

$$(10.32) \qquad 0 < ay_n - bx_n < \frac{12by_n}{F_{m_{n+1}}}.$$

At this point we distinguish two cases:

Case 1. *There exist infinitely many n's with $m_{n+1} > m_n + 1$.*

Suppose that n satisfies the above property. In this case,

$$(10.33) \qquad F_{m_{n+1}} > 2^{2^{m_n+2}} > (2^{2^{m_n+1}} - 1)^2 = \left(\prod_{j=0}^{m_n} F_j \right)^2 \geq y_n^2.$$

With such n's, inequality (10.32) becomes

$$(10.34) \qquad 0 < ay_n - bx_n < \frac{12by_n}{y_n^2} = \frac{12b}{y_n}.$$

If n is large enough, then the right-hand side of inequality (10.34) becomes smaller than 1, which is impossible because $ay_n - bx_n$ is a positive integer. This case is therefore settled.

Case 2. *There exists k_0 such that F_m is prime for all $m > k_0$.*

In this case, there are only finitely many Fermat numbers that are not prime. Since the product

$$(10.35) \qquad \prod_{m \geq 0} \left(1 + \frac{1}{F_m} \right)$$

and the product (10.28) differ only by finitely many rational factors, it suffices to prove that the product given by formula (10.35) is irrational. We keep the previous notation with the convention that $m_k = k$ for all $k \geq 0$. With this notation, inequality (10.32) becomes

$$(10.36) \qquad 0 < ay_m - bx_m < \frac{12by_m}{F_{m+1}}.$$

Since

$$y_m = \prod_{k=0}^{m} F_k = 2^{2^{m+1}} - 1 < F_{m+1},$$

it follows that

$$(10.37) \qquad 0 < ay_m - bx_m < 12b.$$

Let $s = 12b$. By inequality (10.37) and the Dirichlet pigeonhole principle, it follows that there exist infinitely many pairs (m, t) with $m > 0$ and $1 \leq t < s$ such that the equation

$$(10.38) \qquad ay_m - bx_m = ay_{m+t} - bx_{m+t}$$

is satisfied for all such pairs. Equation (10.38) can be rewritten as

$$(10.39) \qquad bx_m \left(\prod_{j=1}^{t} (F_{m+j} + 1) - 1 \right) = a y_m \left(\prod_{j=1}^{t} F_{m+j} - 1 \right).$$

In particular, equality (10.39) implies that

$$(10.40) \qquad F_m + 1 \mid x_m \mid a y_m \left(\prod_{j=1}^{t} F_{m+j} - 1 \right).$$

Since $2^{2^m - 1} + 1 \mid F_m + 1$, it follows that

$$2^{2^m - 1} + 1 \le a \, \gcd\left(2^{2^m - 1} + 1, \ y_m\right) \gcd\left(2^{2^m - 1} + 1, \ \prod_{j=1}^{t} F_{m+j} - 1\right),$$

or

$$(10.41) \quad 2^{2^m - 1} + 1 \le a \prod_{k=0}^{m} \gcd\left(2^{2^m - 1} + 1, \ F_k\right) \gcd\left(2^{2^m - 1} + 1, \ \prod_{j=1}^{t} F_{m+j} - 1\right).$$

By (3.4) and Goldbach's Theorem 4.1, we deduce easily that

$$(10.42) \qquad \gcd\left(2^{2^m - 1} + 1, \ F_k\right) = \begin{cases} 3 & \text{if } k = 0, \\ 1 & \text{if } k > 0. \end{cases}$$

Hence, inequality (10.41) becomes

$$(10.43) \qquad 2^{2^m - 1} + 1 \le 3a \, \gcd\left(2^{2^m - 1} + 1, \ \prod_{j=1}^{t} F_{m+j} - 1\right).$$

Let $D = \gcd\left(2^{2^m - 1} + 1, \ \prod_{j=1}^{t} F_{m+j} - 1\right)$. Since $D \mid 2^{2^m - 1} + 1$, it follows that $2^{2^m} \equiv -2 \pmod{D}$. Hence, $2^{2^{m+j}} \equiv 2^{2^j} \pmod{D}$ for all $j > 0$, and therefore, $F_{m+j} \equiv F_j \pmod{D}$ for all $j > 0$. However, since

$$D \mid \prod_{j=1}^{t} F_{m+j} - 1,$$

it follows that

$$D \mid \prod_{j=1}^{t} F_j - 1.$$

In particular,

$$(10.44) \qquad D \le \prod_{j=1}^{t} F_j - 1 < \prod_{j=0}^{t} F_j = 2^{2^{t+1}} - 1 < 2^{2^s}.$$

Combining inequalities (10.43) and (10.44), we have

$$2^{2^m-1} + 1 < 3a2^{2^s},$$

which is impossible because m can assume infinitely many values. This disposes of Case 2 and concludes the proof of Theorem 10.5.   □

Remark 10.6. In many instances the irrationality of a product of the form (10.28) (or (10.35)) can be inferred from an old criterion from [Brun]. Unfortunately, the convergents x_m/y_m of the products (10.28) or (10.35) do not satisfy the hypothesis from Brun's criterion. However, one can use Brun's criterion to prove the following more general result:

Theorem 10.7. *Let a, b, and q be three integers such that $q \geq 2$, $a > b > 0$, and $(a, b) \neq (2, 1)$, and let $\{m_k\}_{k\geq 1}$ be any increasing sequence of positive integers. Then, the infinite product*

$$\prod_{k\geq 0}\left(1 + \frac{1}{a^{q^k} + b^{q^k}}\right)$$

is irrational.

In what follows we will simply state a few other results concerning sufficient conditions under which the sum of the series (10.1) is irrational. The next theorem is due to [Erdős, Straus].

Theorem 10.8 (Erdős, Straus) *Assume that $\{n_k\}_{k\geq 1}$ is an increasing sequence of positive integers satisfying condition (i) of Theorem 10.1 and the following condition:*

(ii)' $\lim\limits_{k\to\infty} \sup \dfrac{N_k}{n_{k+1}}\left(\dfrac{n_{k+1}^2}{n_{k+2}} - 1\right) \leq 0.$

Then, the sum of the series (10.1) is rational if and only if (10.1) is an Ahmes series.

Notice that conditions (i) and (ii) of Theorem 10.1 imply condition (ii)' of Theorem 10.8, but condition (i) of Theorem 10.1 and condition (ii)' of Theorem 10.8 do not imply condition (ii) of Theorem 10.1.

Theorem 10.9 ([Erdős, 1975]) *Let $n_1 < n_2 < \cdots < n_k < \cdots$ be a sequence of positive integers such that*

$$\lim_{k\to\infty}\sup n_k^{1/2^k} = \infty$$

and

$$n_k > k^{1+\epsilon}$$

for some positive real number ϵ and sufficiently large k. Then, the sum of the series (10.1) is irrational.

The next theorem is due to [Sándor].

Theorem 10.10 (Sándor) *Let $\{n_k\}_{k\geq 1}$ and $\{m_k\}_{k\geq 1}$ be two sequences of positive integers such that*

$$\lim_{k\to\infty}\sup \frac{n_{k+1}}{n_1 n_2 \cdots n_k}\frac{1}{m_{k+1}} = \infty$$

and

$$\liminf_{k \to \infty} \frac{n_{k+1}}{n_k} \frac{m_k}{m_{k+1}} > 1.$$

Then the sum of the series $\sum_{k \geq 1} (m_k / n_k)$ is irrational.

Remark 10.11. Suppose that a and b are positive integers with $a > b$ and assume that $q \geq 3$ is any integer. Theorem 10.9 implies that both the sums

$$\sum_{k \geq 0} \frac{1}{a^{q^k} + b^{q^k}}$$

and

$$\sum_{k \geq 0} \frac{a^{q^{k-1}} - b^{q^{k-1}}}{a^{q^k} - b^{q^k}}$$

are irrational, while Theorem 10.10 implies that

$$\sum_{k \geq 0} \frac{k!}{a^{q^k} + b^{q^k}}$$

is irrational. Moreover, if $s \geq 0$ is an integer, then Theorem 10.10 implies that

$$\sum_{k \geq 0} \frac{k^s}{a^{q^k} + b^{q^k}}$$

is irrational as well.

Remark 10.12. The actual sums of such series and similar series are very hard to compute explicitly in terms of the classical constants. One of the very few examples in which such a sum is explicitly known comes from the Fibonacci sequence (or, slightly more generally, from Lucas sequences of the first kind). Recall that the Fibonacci sequence $\{f_k\}_{k \geq 0}$ is given by $f_0 = 0$, $f_1 = 1$, and $f_{k+1} = f_{k+1} + f_k$ for $k \geq 0$. Then, it has been noted (see [Good] or [Hoggatt, Bicknell]) that

$$(10.45) \qquad \sum_{t \geq 0} \frac{1}{f_{2^t}} = \frac{7 - \sqrt{5}}{2}.$$

In order to prove (10.45), we start with the identity

$$(10.46) \qquad \sum_{t \geq 0} \frac{x^{2^t}}{1 - x^{2^{t+1}}} = \frac{x}{1 - x},$$

which holds for all x with $|x| < 1$. Indeed, to see first why (10.46) holds, notice that every positive integer n can be expressed in a unique way as $n = 2^t(2k + 1)$ where t and k are nonnegative integers. Hence,

$$\frac{x}{1 - x} = \sum_{n \geq 1} x^n = \sum_{\substack{t \geq 0 \\ k \geq 0}} x^{2^t(1+2k)} = \sum_{t \geq 0} x^{2^t} \sum_{k \geq 0} \left(x^{2^{t+1}} \right)^k = \sum_{t \geq 0} \frac{x^{2^t}}{1 - x^{2^{t+1}}},$$

which proves formula (10.46). By replacing x with x^2 in formula (10.46), it follows that

$$(10.47) \qquad \frac{x^2}{1-x^2} = \sum_{t \geq 1} \frac{x^{2^t}}{1 - x^{2^{t+1}}} = \sum_{t \geq 1} \frac{1}{x^{-2^t} - x^{2^t}} \qquad \text{for } 0 < |x| < 1.$$

We now replace x by $\alpha = (\sqrt{5} - 1)/2$ and use the well-known fact that

$$f_{2k} = \frac{\alpha^{-2k} - \alpha^{2k}}{\sqrt{5}} \qquad \text{for } k \geq 0$$

to get that

$$\sum_{t \geq 1} \frac{1}{f_{2^t}} = \sqrt{5} \sum_{t \geq 1} \frac{1}{\alpha^{-2^t} - \alpha^{2^t}} = \sqrt{5} \frac{\alpha^2}{1 - \alpha^2} = \frac{5 - \sqrt{5}}{2},$$

and hence

$$\sum_{t \geq 0} \frac{1}{f_{2^t}} = \frac{1}{f_1} + \sum_{t \geq 1} \frac{1}{f_{2^t}} = 1 + \frac{5 - \sqrt{5}}{2} = \frac{7 - \sqrt{5}}{2},$$

which is exactly formula (10.45).

From formula (10.45), one can see immediately that the sum of the series given in the left-hand side of (10.45) is irrational.

Remark 10.13. In 1980, P. Erdős and R. L. Graham asked about the character of the series (see [Erdős, Graham])

$$(10.48) \qquad \sum_{t \geq 0} \frac{1}{f_{2^t} + 1}.$$

The fact that series (10.48) is irrational is proved in [Badea]. In 1989, André-Jeannin showed that the sum of the series

$$\sum_{n \geq 1} \frac{1}{f_n}$$

of all reciprocals of the members of the Fibonacci sequence is irrational using Padé approximations (see [André-Jeannin]). We mention that earlier, in [Mahler] and [Mignotte], there are two independent proofs of the transcendence (and hence, irrationality) of

$$\sum_{n \geq 1} \frac{1}{n! f_n}.$$

We conclude this chapter by mentioning a few results concerning the character of the sum of the series

$$(10.49) \qquad \sum_{n \geq 1} \frac{1}{q^n + r},$$

where $|q| > 1$ is an integer and r is a nonzero rational number. The first result in this direction was obtained in [Erdős, 1948]. It was shown that the number given by (10.49) is irrational when $q = 2$ and $r = -1$. Erdős's proof first used the fact that

$$(10.50) \qquad \sum_{n \geq 1} \frac{1}{2^n - 1} = \sum_{n \geq 1} \frac{d(n)}{2^n},$$

where $d(n)$ is the divisor function, and then used arithmetic properties of the divisor function to infer the irrationality of the sum of the series occurring in the right-hand side of (10.50). Formula (10.50) follows easily by noticing that if n is a positive integer, then n can be written in exactly $d(n)$ ways as $n = mk$, where m and k are positive integers, and therefore

$$\sum_{k \geq 1} \frac{1}{2^k - 1} = \sum_{k \geq 1} \frac{1}{2^k} \frac{1}{1 - (1/2)^k} = \sum_{k \geq 1} \frac{1}{2^k} \left(\sum_{m \geq 0} \frac{1}{2^{km}} \right) = \sum_{k \geq 1} \sum_{m \geq 1} \frac{1}{2^{km}} = \sum_{n \geq 1} \frac{d(n)}{2^n}.$$

The fact that the sum of the series (10.49) is irrational in general has been finally established in [Borwein]. His proof uses Padé approximations as well.

11. Fermat Primes and a Diophantine Equation

Natural numbers are no doubt from God,
but the devil interspersed them with primes.

Paraphrased quotation of Leopold Kronecker

In this chapter we find all solutions of the Diophantine equation

$$(11.1) \qquad \phi(|x^n - y^n|) = 2^j,$$

where ϕ is the Euler totient function, x and y are integers, and n and j are positive integers such that $n \geq 2$. This problem was first solved in [Luca, 2000c].

We first show that it suffices to find the solutions of the above equation when $n = 4$ and x and y are coprime positive integers. For this last equation we show that aside from a few small solutions, all the others are in one-to-one correspondence with the Fermat primes.

Let $k \geq 3$ be an integer. According to Theorem 4.5, the regular polygon with k sides can be constructed with ruler and compass if and only if $\phi(k)$ is a power of 2. In particular, knowing all solutions of equation (11.1) enables one to find all regular polygons that can be constructed with ruler and compass for which the number of sides is the difference of equal powers of integers. Some equations of a similar flavor as (11.1) were treated in [Luca, 2000b] and in Theorem 8.5. In [Luca, 2000b], all regular polygons were found that can be constructed with ruler and compass whose number of sides is either a Fibonacci or a Lucas number, whereas in Theorem 8.5 we described all regular polygons that can be constructed by ruler and compass and whose number of sides is a binomial coefficient.

Concerning equation (11.1), we first prove the following proposition.

Proposition 11.1. *In order to find all solutions of equation (11.1), it suffices to find only those for which $x > y \geq 1$, $\gcd(x, y) = 1$, and $n = 4$.*

Then we prove the following theorem:

Theorem 11.2. *Assume that (x, y, n, j) is a solution of equation (11.1) satisfying the conditions from Proposition 11.1. Then*

$$(x, y) = \begin{cases} (F_{m-1}, F_{m-1} - 2) & \text{where } m \geq 1 \text{ and } F_m \text{ is a prime number,} \\ (2^{2^m}, 1) & \text{for } m = 0, \ 1, \ 2, \ 3. \end{cases}$$

First we deal with a reduction of the above problem.

In this section we supply a proof of Proposition 11.1.

Let $\mathcal{C} = \{k \geq 1 \mid \phi(k)$ is a power of $2\}$. By the proof of Theorem 4.5, a positive integer k belongs to $\mathcal{C}$ if and only if $k = 2^{\alpha} p_1 \cdots p_t$ for some $\alpha \geq 0$ and $t \geq 0$, where $p_i = 2^{2^{\beta_i}} + 1$ are distinct Fermat primes. In particular, it follows that the elements belonging to the set $\mathcal{C}$ satisfy the following two properties:

(1) *If $a \in \mathcal{C}$ and $b \mid a$, then $b \in \mathcal{C}$.*

(2) *Suppose that $a, b \in \mathcal{C}$. Then, $ab \in \mathcal{C}$ if and only if $\gcd(a,b)$ is a power of 2.*

Let (x, y, n) be such that $|x^n - y^n| \in \mathcal{C}$. We may assume that $x \geq |y| \geq 0$. We first show that it suffices to suppose that $\gcd(x,y) = 1$. Indeed, let $d = \gcd(x,y)$. Write $x = dx_1$ and $y = dy_1$. Then,

$$\left| x^n - y^n \right| = d^n \left| x_1^n - y_1^n \right| \in \mathcal{C}.$$

Since $n \geq 2$, we conclude by (1) and (2) above that $|x_1^n - y_1^n| \in \mathcal{C}$ and that d is a power of 2. Conversely, if (x_1, y_1, n) are such that $|x_1^n - y_1^n| \in \mathcal{C}$ and if d is a power of 2, it follows by (2) that

$$\left| x^n - y^n \right| = d^n \left| x_1^n - y_1^n \right| \in \mathcal{C}$$

as well. Consequently, it suffices to find all solutions of equation (11.1) for which $\gcd(x,y) = 1$.

We assume first that $xy = 0$. It follows that $y = 0$. Since $\gcd(x,y) = 1$ and $x > 0$, we conclude that $x = 1$.

Suppose now that $x = |y|$. Since $\gcd(x,y) = 1$ and ϕ is not defined at 0, it follows that $x = 1$, $y = -1$, and n is odd.

From now on, let $x > |y| > 0$. We first show that we may assume $n > 2$. Indeed, suppose that $n = 2$. Since $n = 2$ is even, we may suppose that $y > 0$. Since

$$(11.2) \qquad\qquad x^2 - y^2 = (x - y)(x + y) \in \mathcal{C},$$

it follows by (1) above that $x - y \in \mathcal{C}$ and $x + y \in \mathcal{C}$. Let $c_1 = x - y$ and $c_2 = x + y$. Since $x > y > 0$, it follows that $c_2 > c_1 > 0$. Moreover, since $\gcd(x,y) = 1$, we conclude that either both c_1 and c_2 are odd and $\gcd(c_1, c_2) = 1$, or both c_1 and c_2 are even, in which case $\gcd(c_1, c_2) = 2$ and one of the numbers c_1 or c_2 is a multiple of 4. Conversely, let $c_2 > c_1$ be any two numbers in $\mathcal{C}$ satisfying one of the above two conditions. Then one can easily see that if we set

$$x = \frac{c_1 + c_2}{2} \quad \text{and} \quad y = \frac{c_1 - c_2}{2},$$

then x and y are positive integers, $x > y$, $\gcd(x,y) = 1$, and $x^2 - y^2 = c_1 c_2 \in \mathcal{C}$. These arguments show that equation (11.1) has an infinity of solutions when $n = 2$ and that all such solutions can be parametrized in terms of two parameters c_1 and c_2 belonging to $\mathcal{C}$ and satisfying certain restrictions.

From now on, let $n > 2$. We first show that n is a power of 2. Assume that this is not the case and let p be an odd prime such that $p \mid n$. Replacing $x^{n/p}$ and $y^{n/p}$ respectively by x and y, we may assume that $|x^p - y^p| \in \mathcal{C}$. From (1) it follows that

$$u_p = \frac{|x^p - y^p|}{|x - y|} \in \mathcal{C}.$$

Since p is odd and $\gcd(x,y) = 1$, it follows that u_p is odd. In particular, u_p is square-free. Let P be a prime dividing u_p. On the one hand, we have $x^p - y^p \equiv 0 \pmod{P}$. On the other hand, since $P \nmid xy$, it follows, by Fermat's Little Theorem 2.9, that $x^{P-1} - y^{P-1} \equiv 1 - 1 \equiv 0 \pmod{P}$. Hence (see also [Carmichael, 1913, p. 38])

$$(11.3) \qquad P \mid \gcd\left(x^p - y^p, x^{P-1} - y^{P-1}\right) = \left|x^{\gcd(p,P-1)} - y^{\gcd(p,P-1)}\right|.$$

Since $P \in \mathcal{C}$, it follows that P is a Fermat prime. Since p is odd, this implies that $\gcd(p, P - 1) = 1$. From formula (11.3), we conclude that $P \mid x - y$. Hence, $x \equiv y \pmod{P}$. It now follows that

$$u_p = \frac{|x^p - y^p|}{|x - y|} \equiv \left|x^{p-1} + x^{p-2}y + \cdots + y^{p-1}\right| \equiv px^{p-1} \pmod{P}.$$

Since $P \mid u_p$, it follows that $p = P$. Since the number u_p is square-free, it follows that $u_p = 1$ or p.

On the other hand, the sequence

$$u_k = \frac{|x^k - y^k|}{|x - y|} \qquad \text{for } k \geq 0$$

is a *Lucas sequence of the first kind*. From [Carmichael, 1913] we know that u_q is divisible by a prime $Q > q$ for any prime $q > 3$. From the above result it follows that $p = 3$ and that $u_3 = 1$ or 3. This leads to the equations

$$x^2 \pm xy + y^2 = 1 \text{ or } 3.$$

The only solution (x,y) of the above equations such that $x > |y| > 0$ is $(2,-1)$ which does not lead to a solution of equation (11.1). Hence, n is a power of 2. Since $n > 2$, it follows that n is a multiple of 4. We may now replace x and y by $x^{n/4}$ and $y^{n/4}$, respectively, and study equation (11.1) only for $n = 4$. Clearly, since $n = 4$ is even, we may assume that $y > 0$. Proposition 11.1 is therefore proved. $\square$

P r o o f   o f   T h e o r e m   1 1 . 2 . Since $x^4 - y^4 = (x-y)(x+y)(x^2+y^2) \in \mathcal{C}$, it follows that $x - y \in \mathcal{C}$, $x + y \in \mathcal{C}$, and $x^2 + y^2 \in \mathcal{C}$. We distinguish two cases:

Case 1. $x \equiv y \equiv 1 \pmod{2}$.

In this case, one of the numbers $x - y$ or $x + y$ is divisible by 4, and the other one is 2 modulo 4. Moreover, since both x and y are odd, it follows that $x^2 + y^2$ is 2 modulo 8. It now follows that there exists $\epsilon \in \{\pm 1\}$ such that

$$x - \epsilon y \equiv 2 \pmod{4},$$
$$x + \epsilon y \equiv 0 \pmod{4}.$$

Write

$$x - \epsilon y = 2 \prod_{i=1}^{I}\left(2^{2^{\alpha_i}} + 1\right),$$

$$(11.4) \qquad x + \epsilon y = 2^s \prod_{j=1}^{J}\left(2^{2^{\beta_j}} + 1\right),$$

$$x^2 + y^2 = 2 \prod_{k=1}^{K}\left(2^{2^{\gamma_k}} + 1\right),$$

where $s \geq 2$, I, J, and K are three nonnegative integers (some of them may be zero), $0 \leq \alpha_1 < \cdots < \alpha_I$, $0 \leq \beta_1 < \cdots < \beta_J$, $0 \leq \gamma_1 < \cdots < \gamma_K$, and $F_\delta = 2^{2^\delta} + 1$ is a Fermat prime whenever $\delta \in \{\alpha_i\}_{i=1}^I \cup \{\beta_j\}_{j=1}^J \cup \{\gamma_k\}_{k=1}^K$.

Notice first that the three sets $\{\alpha_i\}_{i=1}^I$, $\{\beta_j\}_{j=1}^J$, and $\{\gamma_k\}_{k=1}^K$ are pairwise disjoint. Indeed, assume, for example, that $\delta \in \{\alpha_i\}_{i=1}^I \cap \{\beta_j\}_{j=1}^J$. It follows that $2^{2^\delta} + 1 \mid \gcd(x - y, x + y)$, which contradicts the fact that x and y are coprime.

Notice also that $K > 0$ and that $\gamma_1 > 0$. Indeed, if $K = 0$, then $x^2 + y^2 = 2$, which is impossible because $x > y \geq 1$. If $\gamma_1 = 0$, it follows that $3 = 2^{2^0} + 1 \mid x^2 + y^2$, which is impossible because x and y are coprime and -1 is not a quadratic residue modulo 3.

We now use formulae (11.4) and the identity

(11.5) $$2(x^2 + y^2) = (x - y)^2 + (x + y)^2$$

to conclude that

$$4 \prod_{k=1}^K \left(2^{2^{\gamma_k}} + 1\right) = 4 \prod_{i=1}^I \left(2^{2^{\alpha_i}} + 1\right)^2 + 2^{2s} \prod_{j=1}^J \left(2^{2^{\beta_j}} + 1\right)^2,$$

or

(11.6) $$\prod_{k=1}^K \left(2^{2^{\gamma_k}} + 1\right) = \prod_{i=1}^I \left(2^{2^{\alpha_i}} + 1\right)^2 + 2^{2(s-1)} \prod_{j=1}^J \left(2^{2^{\beta_j}} + 1\right)^2.$$

Our main goal is to show that $I = J = 0$.

Suppose that this is not so. In order to achieve a contradiction, we proceed in three steps.

Step 1.I. $0 \in \{\alpha_i\}_{i=1}^I \cup \{\beta_j\}_{j=1}^J$.

Assume that this is not the case. Suppose first that $I > 0$. Hence, $\alpha_1 \neq 0$. Notice first that

(11.7) $$\prod_{i=1}^I \left(2^{2^{\alpha_i}} + 1\right) = \sum_{\mathcal{H} \subseteq \{1,\ldots,I\}} 2^{\sum_{i \in \mathcal{H}} 2^{\alpha_i}},$$

and the sum appearing on the right-hand side of identity (11.7) is precisely the binary expansion of the product appearing on the left-hand side (this is because of the fact that all exponents appearing on the right-hand side of identity (11.7) have distinct binary representations; therefore, they are all distinct). Since $\alpha_1 > 0$, it follows that

(11.8) $$\prod_{i=1}^I \left(2^{2^{\alpha_i}} + 1\right)^2 = 1 + 2^{2^{\alpha_1}+1} + 2^{2^{\alpha_1}+1} + \text{higher powers of 2},$$

where the higher powers of 2 are missing when $I = 1$. From formula (11.6), it follows that

(11.9) $$1 + 2^{2^{\gamma_1}} + \text{higher powers of 2} = \prod_{k=1}^K \left(2^{2^{\gamma_k}} + 1\right).$$

$$= \left(1 + 2^{2^{\alpha_1}+1} + \text{higher powers of } 2\right) + 2^{2(s-1)}(1 + \text{higher powers of } 2).$$

Clearly, the numbers $2^{\alpha_1} + 1$ and $2(s - 1)$ are distinct, because the first is odd and the second is even. On the one hand, from formula (11.9) and the fact that 2^{γ_1} is even we conclude that $2^{\gamma_1} = 2(s - 1)$. On the other hand, since the binary representation of the number given by formula (11.9) has at least three digits of 1, it follows that $K \geq 2$.

If $J = 0$, then formulae (11.6) and (11.9) imply

(11.10)
$$1 + 2^{2^{\gamma_1}} + 2^{2^{\gamma_2}} + \text{higher powers of } 2 = \prod_{k=1}^{K}\left(2^{2^{\gamma_k}} + 1\right)$$

$$= 1 + 2^{2(s-1)} + 2^{2^{\alpha_1}+1} + \text{higher powers of } 2.$$

Formula (11.10) leads to $2^{\gamma_2} = 2^{\alpha_1} + 1$, which is impossible because $\alpha_1 > 0$.

Suppose now that $J > 0$. In this case, $\beta_1 > 0$. Arguments similar to the preceding ones yield that

(11.11)
$$1 + 2^{2^{\gamma_1}} + 2^{2^{\gamma_2}} + \text{higher powers of } 2 = \prod_{k=1}^{K}\left(2^{2^{\gamma_k}} + 1\right)$$

$$= \left(1 + 2^{2^{\alpha_1}+1} + \text{higher powers of } 2\right) + 2^{2(s-1)}\left(1 + 2^{2^{\beta_1}+1} + \text{higher powers of } 2\right).$$

From equation (11.11) and the fact that $2^{\gamma_1} = 2(s - 1)$ it follows that at least one of the following three situations must occur:

(1) $2^{\gamma_2} = 2^{\alpha_1} + 1$. This is impossible because $\alpha_1 > 0$.

(2) $2^{\gamma_2} = 2(s - 1) + 2^{\beta_1} + 1 = 2^{\gamma_1} + 2^{\beta_1} + 1$. This is impossible because both β_1 and γ_1 are positive.

(3) $2^{\alpha_1} + 1 = 2(s - 1) + 2^{\beta_1} + 1$ or $2^{\alpha_1} = 2(s - 1) + 2^{\beta_1} = 2^{\gamma_1} + 2^{\beta_1}$, which is impossible because $\beta_1 \neq \gamma_1$.

This completes the argument in the case $I > 0$.

Assume now that $I = 0$. Hence, $J > 0$ and $\beta_1 > 0$. Arguments similar to the previous ones imply that formula (11.6) reads

(11.12)
$$1 + 2^{2^{\gamma_1}} + \text{higher powers of } 2 = \prod_{k=1}^{K}\left(2^{2^{\gamma_k}} + 1\right)$$

$$= 1 + 2^{2(s-1)}\left(1 + 2^{2^{\beta_1}+1} + \text{higher powers of } 2\right).$$

From equation (11.12) it again follows that $2^{\gamma_1} = 2(s - 1)$ and $K \geq 2$. Formula (11.12) can now be written as

(11.13)
$$1 + 2^{2^{\gamma_1}} + 2^{2^{\gamma_2}} + \text{higher powers of } 2 = \prod_{k=1}^{K}\left(2^{2^{\gamma_k}} + 1\right)$$

$$= 1 + 2^{2(s-1)}\left(1 + 2^{2^{\beta_1}+1} + \text{higher powers of } 2\right).$$

From equation (11.13) it follows that $2^{\gamma_2} = 2(s-1) + 2^{\beta_1} + 1 = 2^{\gamma_1} + 2^{\beta_1} + 1$, which is impossible because both β_1 and γ_1 are positive.

Step 1.I is therefore proved.

Step 1.II. *If $I > 0$, then $\alpha_1 \neq 0$.*

Suppose that this is not the case. Assume that $I > 0$ but $\alpha_1 = 0$. Let $t \geq 1$ be such that $\alpha_i = i - 1$ for $i = 1, \ldots, t$ and either $I = t$ or $\alpha_{t+1} \geq t + 1$. Then by (3.4),

$$(11.14) \quad \prod_{i=1}^{I}\left(2^{2^{\alpha_i}}+1\right) = \prod_{i=1}^{t}\left(2^{2^{i-1}}+1\right) \prod_{i \geq t+1}^{I}\left(2^{2^{\alpha_i}}+1\right) = \left(2^{2^t}-1\right) \prod_{i \geq t+1}^{I}\left(2^{2^{\alpha_i}}+1\right).$$

Hence,

$$(11.15) \quad \prod_{i=1}^{I}\left(2^{2^{\alpha_i}}+1\right)^2 = \left(1 + 2^{2^t+1} + \text{higher powers of } 2\right)\left(1 + \text{higher powers of } 2\right)$$

$$= 1 + 2^{2^t+1} + \text{higher powers of } 2.$$

From formulae (11.6) and (11.15) it follows that

(11.16)
$$1 + 2^{2^{\gamma_1}} + \text{higher powers of } 2 = \prod_{k=1}^{K}\left(2^{2^{\gamma_k}}+1\right)$$

$$= \left(1 + 2^{2^t+1} + \text{higher powers of } 2\right) + 2^{2(s-1)}\left(1 + \text{higher powers of } 2\right).$$

Clearly, $2^t + 1$ and $2(s-1)$ are distinct, because the first number is odd and the other is even. From formula (11.16) it follows that $2^{\gamma_1} = 2(s-1)$ and that $K \geq 2$. Formula (11.6) now becomes

(11.17)
$$1 + 2^{2^{\gamma_1}} + 2^{2^{\gamma_2}} + \text{higher powers of } 2 = \prod_{k=1}^{K}(2^{2^{\gamma_k}}+1)$$

$$= \left(1 + 2^{2^t+1} + \text{higher powers of } 2\right) + 2^{2(s-1)}\left(1 + \text{higher powers of } 2\right).$$

Suppose first that $J = 0$. Then $2^{\gamma_2} = 2^t + 1$, which is false because t is positive.

Suppose now that $J > 0$. Since $\alpha_i = i - 1$ for $i = 1, \ldots, t$, it follows that $\beta_1 \geq t \geq 1$. From the arguments employed in Step 1.I it follows that formula (11.17) can be rewritten as

(11.18)
$$1 + 2^{2^{\gamma_1}} + 2^{2^{\gamma_2}} + \text{higher powers of } 2 = \prod_{k=1}^{K}\left(2^{2^{\gamma_k}}+1\right)$$

$$= \left(1 + 2^{2^t+1} + \text{higher powers of } 2\right) + 2^{2(s-1)}\left(1 + 2^{2^{\beta_1}+1} + \text{higher powers of } 2\right).$$

From equation (11.18) and the fact that $2^{\gamma_1} = 2(s-1)$ it follows that one of the following situations must occur:

(1) $2^{\gamma_2} = 2^t + 1$. This is impossible because $t > 0$.

(2) $2^{\gamma_2} = 2(s-1) + 2^{\beta_1} + 1 = 2^{\gamma_1} + 2^{\beta_1} + 1$. This is impossible because both γ_1 and β_1 are positive.

(3) $2^t + 1 = 2(s-1) + 2^{\beta_1} + 1$ or $2^t = 2^{\gamma_1} + 2^{\beta_1}$, which is impossible because $\gamma_1 \neq \beta_1$.

This completes the proof of Step 1.II.

Step 1.III. *If $J > 0$, then $\beta_1 \neq 0$.*

Notice first that Steps 1.I, 1.II, and 1.III contradict each other.

Assume that the claim made in Step 1.III does not hold. Let $J > 0$ and assume that $\beta_1 = 0$. Let $t \geq 1$ be such that $\beta_j = j - 1$ for $j = 1, \ldots, t$ and either $J = t$ or $J > t$ and $\beta_{t+1} \geq t + 1$. We have

$$(11.19) \quad \prod_{j=1}^{J}\left(2^{2^{\beta_j}}+1\right) = \prod_{j=1}^{t}\left(2^{2^{j-1}}+1\right) \prod_{j \geq t+1}^{J}\left(2^{2^{\beta_j}}+1\right) = \left(2^{2^t}-1\right) \prod_{j \geq t+1}^{J}\left(2^{2^{\beta_j}}+1\right).$$

Hence,
(11.20)
$$\prod_{j=1}^{J}\left(2^{2^{\beta_j}}+1\right)^2 = \left(2^{2^t}-1\right)^2 \prod_{j \geq t+1}^{J}\left(2^{2^{\beta_j}}+1\right)^2 = 1 + 2^{2^t+1} + \text{higher powers of 2}.$$

From formula (11.6) it follows that

$$1 + 2^{2^{\gamma_1}} + \text{higher powers of 2} = \prod_{k=1}^{K}\left(2^{2^{\gamma_k}}+1\right)$$

$$= (1 + \text{higher powers of 2}) + 2^{2(s-1)}\left(1 + 2^{2^t+1} + \text{higher powers of 2}\right).$$

Assume first that $I = 0$. It follows that

$$(11.21) \quad 1 + 2^{2^{\gamma_1}} + \text{higher powers of 2} = \prod_{k=1}^{K}\left(2^{2^{\gamma_k}}+1\right)$$

$$= 1 + 2^{2(s-1)}\left(1 + 2^{2^t+1} + \text{higher powers of 2}\right).$$

From equation (11.21) it follows that $K \geq 2$ and that $2^{\gamma_1} = 2(s-1)$. Formula (11.21) can now be written as

$$(11.22) \quad 1 + 2^{2^{\gamma_1}} + 2^{2^{\gamma_2}} + \text{higher powers of 2} = \prod_{k=1}^{K}\left(2^{2^{\gamma_k}}+1\right)$$

$$= 1 + 2^{2(s-1)}\left(1 + 2^{2^t+1} + \text{higher powers of 2}\right).$$

From equation (11.22) and the fact that $2^{\gamma_1} = 2(s-1)$ it follows that $2^{\gamma_2} = 2(s-1) + 2^t + 1 = 2^{\gamma_1} + 2^t + 1$, which is impossible because both γ_1 and t are positive.

Assume now that $I > 0$. In this case, $\alpha_1 \geq t \geq 1$. From formula (11.6) and the arguments employed at Step 1.I it follows that

(11.23)

$$1 + 2^{2^{\gamma_1}} + \text{higher powers of } 2 = \prod_{k=1}^{K}\left(2^{2^{\gamma_k}} + 1\right)$$

$$= \left(1 + 2^{2^{\alpha_1}+1} + \text{higher powers of } 2\right) + 2^{2(s-1)}\left(1 + 2^{2^t+1} + \text{higher powers of } 2\right).$$

Notice that $2^{\alpha_1} + 1$ and $2(s-1)$ are distinct, because the first number is odd and the other is even. From (11.23) it follows that $2^{\gamma_1} = 2(s-1)$ and that $K \geq 2$. Formula (11.23) can now be rewritten as

(11.24)

$$1 + 2^{2^{\gamma_1}} + 2^{2^{\gamma_2}} + \text{higher powers of } 2 = \prod_{k=1}^{K}\left(2^{2^{\gamma_k}} + 1\right)$$

$$= \left(1 + 2^{2^{\alpha_1}+1} + \text{higher powers of } 2\right) + 2^{2(s-1)}\left(1 + 2^{2^t+1} + \text{higher powers of } 2\right).$$

From equation (11.24) and the fact that $2^{\gamma_1} = 2(s-1)$ it follows that one of the following situations must occur:

(1) $2^{\gamma_2} = 2^{\alpha_1} + 1$. This is impossible because $\alpha_1 > 0$.

(2) $2^{\gamma_2} = 2(s-1) + 2^t + 1 = 2^{\gamma_1} + 2^t + 1$. This is also impossible because both γ_1 and t are positive.

(3) $2^{\alpha_1} + 1 = 2(s-1) + 2^t + 1 = 2^{\gamma_1} + 2^t + 1$. This leads to $\gamma_1 = t$ and $\alpha_1 = t+1$. In this last case, it follows that $\alpha_2 \geq t + 2$ and $\beta_{t+1} \geq t + 2$, whenever they exist. From formulae (11.6) and (11.19) we get

(11.25)

$$1 + 2^{2^t} + 2^{2^{\gamma_2}} + \text{higher powers of } 2 = \prod_{k=1}^{K}\left(2^{2^{\gamma_k}} + 1\right)$$

$$= (2^{2^{t+1}} + 1)^2 \prod_{i \geq 2}^{I}\left(2^{2^{\alpha_i}} + 1\right)^2 + 2^{2^t}\left(2^{2^t} - 1\right)^2 \prod_{j \geq t+1}^{J}\left(2^{2^{\beta_j}} + 1\right)^2$$

$$= \left(2^{2^{t+1}} + 1\right)^2 + 2^{2^t}\left(2^{2^t} - 1\right)^2 + \text{higher powers of } 2$$

$$= 1 + 2^{2^t} + 2^{2^t + 2^{t+1}} + 2^{2^{t+2}} + \text{higher powers of } 2.$$

Equation (11.25) implies $2^{\gamma_2} = 2^t + 2^{t+1}$, which is impossible.

Step 1.III is thus proved.

Steps 1.I, 1.II, and 1.III imply that $I = J = 0$. From formula (11.6), it follows that

(11.26) $$\prod_{k=1}^{K}\left(2^{2^{\gamma_k}} + 1\right) = 1 + 2^{2(s-1)}.$$

From equation (11.26) it follows that $K = 1$ and $2^{\gamma_1} = 2(s - 1)$. Solving the first two equations of system (11.4) for x and y, we get

(11.27) $$x = 2^{2^{\gamma_1 - 1}} + 1 \quad \text{and} \quad y = \epsilon\left(2^{2^{\gamma_1 - 1}} - 1\right),$$

where $2^{2^{\gamma_1}} + 1$ is a Fermat prime and $\epsilon \in \{\pm 1\}$. Since $y > 0$, it follows that $\epsilon = 1$. This belongs to the first family of solutions claimed by Theorem 11.2.

Case 2. $x \not\equiv y \pmod 2$.

In this case all three numbers $x - y$, $x + y$, and $x^2 + y^2$ are odd. Assume that

(11.28)
$$x - y = \prod_{i=1}^{I}\left(2^{2^{\alpha_i}} + 1\right),$$
$$x + y = \prod_{j=1}^{J}\left(2^{2^{\beta_j}} + 1\right),$$
$$x^2 + y^2 = \prod_{k=1}^{K}\left(2^{2^{\gamma_k}} + 1\right),$$

where I, J, and K are three nonnegative integers (some of them may be zero), $0 \le \alpha_1 < \cdots < \alpha_I$, $0 \le \beta_1 < \cdots < \beta_J$, $0 \le \gamma_1 < \cdots < \gamma_K$, and $F_\delta = 2^{2^\delta} + 1$ is a Fermat prime whenever $\delta \in \{\alpha_i\}_{i=1}^{I} \cup \{\beta_j\}_{j=1}^{J} \cup \{\gamma_k\}_{k=1}^{K}$.

Notice again that the three sets $\{\alpha_i\}_{i=1}^{I}$, $\{\beta_j\}_{j=1}^{J}$, and $\{\gamma_k\}_{k=1}^{K}$ are pairwise disjoint, $K > 0$, and $\gamma_1 > 0$. Notice also that $I + J > 0$.

We proceed in four steps.

Step 2.I. $K = J$ and $\gamma_k = \beta_k + 1$ for all $k = 1, \ldots, K$.

From formulae (11.28) and the arguments immediately below formula (11.7) it follows that

(11.29)
$$\lfloor \log_2(x - y) \rfloor = \sum_{i=1}^{I} 2^{\alpha_i},$$
$$\lfloor \log_2(x + y) \rfloor = \sum_{j=1}^{J} 2^{\beta_j},$$
$$\lfloor \log_2(x^2 + y^2) \rfloor = \sum_{k=1}^{K} 2^{\gamma_k}.$$

We now use the following obvious lemma.

Lemma 11.3. (1) If z is a positive number, then

(11.30) $$\lfloor \log_2 z^2 \rfloor \in \{2\lfloor \log_2 z \rfloor,\ 2\lfloor \log_2 z \rfloor + 1\}.$$

(2) If $a > b$ are positive numbers, then

(11.31) $$\lfloor \log_2(a + b) \rfloor \in \{\lfloor \log_2 a \rfloor,\ \lfloor \log_2 a \rfloor + 1\}.$$

From identity (11.5) and Lemma 11.3 it follows that

(11.32)
$$1 + \left\lfloor \log_2(x^2 + y^2) \right\rfloor = \left\lfloor \log_2(2(x^2 + y^2)) \right\rfloor$$
$$= \left\lfloor \log_2((x+y)^2 + (x-y)^2)) \right\rfloor \in \{2\lfloor\log_2(x+y)\rfloor + u \mid u = 0,\ 1,\ 2\}.$$

From formulae (11.29) and (11.32) it follows that

(11.33) $$1 + \sum_{k=1}^{K} 2^{\gamma_k} = u + \sum_{j=1}^{J} 2^{\beta_j + 1} \qquad \text{for some } u \in \{0,\ 1,\ 2\}.$$

Since $\gamma_1 > 0$, it follows that the number appearing on the left-hand side of equation (11.33) is odd. Hence, $u = 1$, $K = J$, and $\gamma_k = \beta_k + 1$ for all $k = 1, \ldots, K$.

Step 2.I is thus proved.

Step 2.II. $0 \in \{\alpha_i\}_{i=1}^{I} \cup \{\beta_j\}_{j=1}^{J}$.

Assume that this is not the case. By Step 2.I, we know that $J > 0$. In particular, $\beta_1 > 0$.

We use formulae (11.28) and identity (11.5) to conclude that

(11.34) $$2 \prod_{k=1}^{K} \left(2^{2^{\gamma_k}} + 1\right) = \prod_{i=1}^{I} \left(2^{2^{\alpha_i}} + 1\right)^2 + \prod_{j=1}^{J} \left(2^{2^{\beta_j}} + 1\right)^2.$$

By the arguments employed in Step 1.I, it follows that

(11.35) $$2 + 2^{2^{\gamma_1}+1} + \text{higher powers of } 2 = 2 \prod_{k=1}^{K} \left(2^{2^{\gamma_k}} + 1\right)$$
$$= \prod_{i=1}^{I} \left(2^{2^{\alpha_i}} + 1\right)^2 + \left(1 + 2^{2^{\beta_1}+1} + \text{higher powers of } 2\right).$$

If $I = 0$, then formula (11.35) becomes

(11.36) $$2 + 2^{2^{\gamma_1}+1} + \text{higher powers of } 2 = 2 \prod_{k=1}^{K} \left(2^{2^{\gamma_k}} + 1\right)$$
$$= 1 + \left(1 + 2^{2^{\beta_1}+1} + \text{higher powers of } 2\right).$$

From formula (11.36) it follows that $2^{\gamma_1} + 1 = 2^{\beta_1} + 1$, or $\gamma_1 = \beta_1$, which is impossible.

Suppose now that $I > 0$. In this case, $\alpha_1 > 0$. By the arguments employed in Step 1.I, it follows that

(11.37) $$2 + 2^{2^{\gamma_1}+1} + \text{higher powers of } 2 = 2 \prod_{k=1}^{K} \left(2^{2^{\gamma_k}} + 1\right)$$

$$= \left(1 + 2^{2^{\alpha_1}+1} + \text{higher powers of 2}\right) + \left(1 + 2^{2^{\beta_1}+1} + \text{higher powers of 2}\right).$$

From equation (11.37), it follows that one of the following situations must occur:

(1) $2^{\gamma_1} + 1 = 2^{\alpha_1} + 1$. This implies $\gamma_1 = \alpha_1$, which is impossible.

(2) $2^{\gamma_1} + 1 = 2^{\beta_1} + 1$. This implies $\gamma_1 = \beta_1$, which is impossible.

(3) $2^{\alpha_1} + 1 = 2^{\beta_1} + 1$. This implies $\alpha_1 = \beta_1$, which is impossible.

Step 2.II is thus proved.

Step 2.III. *If either $I = 0$ or $\alpha_1 \neq 0$, then $x = 2$ and $y = 1$.*

Suppose that either $I = 0$ or $\alpha_1 \neq 0$. By Steps 2.I and 2.II above, it follows that $\beta_1 = 0$ and $\gamma_1 = 1$. We now show that $I = 0$ and $J = 1$. Suppose that this is not the case. Then at least one of the numbers α_1 or β_2 exists. From formula (11.34) and the fact that $\beta_1 = 0$ and $\gamma_1 = 1$, it follows that

$$(11.38) \qquad 2 + 2^3 + \text{higher powers of 2} = 2 \prod_{k=1}^{K} \left(2^{2^{\gamma_k}} + 1\right)$$

$$= \prod_{i=1}^{I} \left(2^{2^{\alpha_i}} + 1\right)^2 + 3^2 \prod_{j \geq 2}^{J} \left(2^{2^{\beta_j}} + 1\right)^2$$

$$= \prod_{i=1}^{I} \left(2^{2^{\alpha_i}} + 1\right)^2 + \left(1 + 2^3\right) \prod_{j \geq 2}^{J} \left(2^{2^{\beta_j}} + 1\right)^2.$$

It follows that $K \geq 2$. Since $J = K$, it follows that $J \geq 2$ as well. Suppose, for example, that $I = 0$. From formula (11.38), it follows that

$$(11.39) \qquad 2 + 2^3 + 2^{2^{\gamma_2}+1} + \text{higher powers of 2} = 2 \prod_{k=1}^{K} \left(2^{2^{\gamma_k}} + 1\right)$$

$$= 1 + \left(1 + 2^3\right)\left(1 + 2^{2^{\beta_2}+1} + \text{higher powers of 2}\right).$$

From equation (11.39) it follows that $2^{\gamma_2} + 1 = 2^{\beta_2} + 1$, which is impossible because $\gamma_2 \neq \beta_2$.

Assume now that $I > 0$. From formula (11.38) it follows that

$$(11.40)$$

$$2 + 2^3 + 2^{2^{\gamma_2}+1} + \text{higher powers of 2} = 2 \prod_{k=1}^{K} \left(2^{2^{\gamma_k}} + 1\right)$$

$$= \left(1 + 2^{2^{\alpha_1}+1} + \text{higher powers of 2}\right) + \left(1 + 2^3\right)\left(1 + 2^{2^{\beta_2}+1} + \text{higher powers of 2}\right).$$

From equation (11.40) it follows that one of the following must occur:

(1) $2^{\gamma_2} + 1 = 2^{\alpha_1} + 1$. This is impossible because $\gamma_2 \neq \alpha_1$.

(2) $2^{\gamma_2} + 1 = 2^{\beta_2} + 1$. This is impossible because $\gamma_2 \neq \beta_2$.

(3) $2^{\alpha_1} + 1 = 2^{\beta_2} + 1$. This is impossible because $\alpha_1 \neq \beta_2$.

Hence, $I = 0$, $J = K = 1$, $\beta_1 = 0$, and $\gamma_1 = 1$. It follows that $x - y = 1$ and $x + y = 3$. Hence, $(x, y) = (2, 1) = \left(2^{2^0}, 1\right)$ which is one of the solutions claimed by Theorem 11.2.

Step 2.III is thus proved.

Assume now that $(x, y) \neq (2, 1)$. By Steps 2.I, 2.II, and 2.III, it follows that $I > 0$ and $\alpha_1 = 0$. The proof of Theorem 11.2 will be completed once we show the following:

Step 2.IV. *If* $\alpha_1 = 0$, *then* $(x, y) = \left(2^{2^m}, 1\right)$ *for some* $m = 1$, 2, 3.

Let $t \geq 1$ be such that $\alpha_i = i - 1$ for $i = 1, \ldots, t$ and either $I = t$ or $I > t$ and $\alpha_{t+1} \geq t + 1$. It now follows that

$$(11.41) \quad \prod_{i=1}^{I}\left(2^{2^{\alpha_i}} + 1\right) = \prod_{i=1}^{t}\left(2^{2^{i-1}} + 1\right) \prod_{i \geq t+1}^{I} \left(2^{2^{\alpha_i}} + 1\right) = \left(2^{2^t} - 1\right) \prod_{i \geq t+1}^{I} \left(2^{2^{\alpha_i}} + 1\right).$$

Hence,

$$(11.42) \quad \prod_{i=1}^{I}\left(2^{2^{\alpha_i}} + 1\right)^2 = 1 + 2^{2^t+1} + \text{higher powers of 2}.$$

From equation (11.34) it follows that

(11.43)

$$2 + 2^{2^{\gamma_1}+1} + \text{higher powers of 2} = 2 \prod_{k=1}^{K}\left(2^{2^{\gamma_k}} + 1\right)$$

$$= \left(1 + 2^{2^t+1} + \text{higher powers of 2}\right) + \left(1 + 2^{2^{\beta_1}+1} + \text{higher powers of 2}\right).$$

From equation (11.43) and the fact that $\gamma_1 = \beta_1 + 1 > \beta_1$ it follows that $2^t + 1 = 2^{\beta_1} + 1$ or $\beta_1 = t$. Hence, $\gamma_1 = t + 1$. Equation (11.34) now becomes

(11.44)

$$2 + 2^{2^{t+1}+1} + \text{higher powers of 2} = 2 \prod_{k=1}^{K}\left(2^{2^{\gamma_k}} + 1\right)$$

$$= \left(2^{2^t} - 1\right)^2 \prod_{i \geq t+1}^{I} \left(2^{2^{\alpha_i}} + 1\right)^2 + \left(2^{2^t} + 1\right)^2 \prod_{j \geq 2}^{J}\left(2^{2^{\beta_j}} + 1\right)^2.$$

We now show that $I = t$ and $J = 1$.

Suppose, for example, that $I > t$ and $J = 1$. Then, from formula (11.44) it follows that $K > 1$, which contradicts the fact that $K = J$.

Suppose now that $I = t$ and $J > 1$. Then $K = J > 1$. Since $\beta_2 \geq t + 2$, it follows by formula (11.44) that

$$(11.45) \quad 2 + 2^{2^{t+1}+1} + 2^{2^{\gamma_2}+1} + \text{higher powers of 2} = 2 \prod_{k=1}^{K}\left(2^{2^{\gamma_k}} + 1\right)$$

$$= \left(2^{2^t} - 1\right)^2 + \left(2^{2^t} + 1\right)^2 + 2^{2^{\beta_2}+1} + \text{higher powers of 2}$$

$$= 2 + 2^{2^{t+1}+1} + 2^{2^{\beta_2}+1} + \text{higher powers of 2}.$$

Equation (11.45) implies that $\gamma_2 = \beta_2$, which is impossible.

Finally, suppose that $I > t$ and $J > 1$. Since $\beta_2 \geq t + 2$ and $\alpha_{t+1} \geq t + 2$, it follows, by formula (11.44), that

(11.46)

$$2 + 2^{2^{t+1}+1} + 2^{2^{\gamma_2}+1} + \text{higher powers of 2} = 2 \prod_{k=1}^{K} (2^{2^{\gamma_k}} + 1)$$

$$= ((2^{2^t} - 1)^2 + 2^{2^{\alpha_{t+1}}+1} + \text{higher powers of 2})$$

$$+ ((2^{2^t} + 1)^2 + 2^{2^{\beta_2}+1} + \text{higher powers of 2})$$

$$= 2 + 2^{2^{t+1}+1} + 2^{2^{\alpha_{t+1}}+1} + 2^{2^{\beta_2}+1} + \text{higher powers of 2}.$$

Equation (11.46) implies that one of the following three situations must occur:

(1) $2^{\gamma_2} + 1 = 2^{\alpha_{t+1}} + 1$. This implies $\gamma_2 = \alpha_{t+1}$, which is impossible.

(2) $2^{\gamma_2} + 1 = 2^{\beta_2} + 1$. This implies $\gamma_2 = \beta_2$, which is impossible.

(3) $2^{\alpha_{t+1}} + 1 = 2^{\beta_2} + 1$. This implies $\alpha_{t+1} = \beta_2$, which is impossible.

The above arguments show that $I = t$, $J = K = 1$, $\alpha_i = i - 1$ for $i = 1, \ldots, t$, $\beta_1 = t$, and $\gamma_1 = t + 1$. It now follows that

(11.47) $$x - y = 2^{2^t} - 1 \quad \text{and} \quad x + y = 2^{2^t} + 1.$$

This implies $x = 2^{2^t}$ and $y = 1$. It remains to show that $t \leq 3$. But this comes from the fact that if $t \geq 4$, then $x^4 - y^4 = 2^{2^{t+2}} - 1$ is divisible by $F_5 = 2^{2^5} + 1$, which is not a prime (in fact, $\phi(2^{2^5} + 1)$ is not a power of 2).

Theorem 11.2 is thus completely proved. $\square$

12. Fermat's Little Theorem, Pseudoprimes, and Superpseudoprimes

Without exception, every prime number measures one
of the powers −1 of any progression whatever, and
the exponent of the said power is a submultiple
of the given prime number −1.

Statement of Fermat's little theorem
by Pierre de Fermat in a letter to
Bernhard Frénicle de Bessy,
October 18, 1640,
[Mahoney, p. 291].

In this chapter we show how to apply Fermat numbers to generate infinitely many pseudoprimes and superpseudoprimes. To define pseudoprimes and superpseudoprimes, we will need to make use of Fermat's little theorem which is a centerpiece of number theory. It gives a fundamental property of primes and is the basis of most tests for primality.

We will make use of Fermat's little theorem in either of its two equivalent forms presented in Chapter 2. The first form states that if p is a prime, then

$$(12.1) \qquad a^p \equiv a \pmod{p}$$

for all integers a. The second form states that if p is a prime and $\gcd(a, p) = 1$, then

$$(12.2) \qquad a^{p-1} \equiv 1 \pmod{p}.$$

However, the converse of Fermat's little theorem is not true. Given any base $a > 1$, there exists a composite integer n coprime to a such that

$$(12.3) \qquad a^n \equiv a \pmod{n}.$$

For example, the composite number $561 = 3 \cdot 11 \cdot 17$ satisfies

$$a^{561} \equiv a \pmod{561}$$

for all integers a (see Example 12.8). Such an integer is called a Carmichael number and will be discussed later.

If $\gcd(a, n) = 1$, then (12.3) holds if and only if

$$(12.4) \qquad\qquad a^{n-1} \equiv 1 \pmod{n}.$$

According to Definition 4.8, a composite integer n is called a pseudoprime to the base a if (12.3) holds. (Note that in [Conway, Guy, Schneeberger, Sloane], the notion of pseudoprime is based on congruence (12.4) in contrast to the congruence given in (12.3), and so produces a smaller class of composite numbers satisfying the converse of Fermat's little theorem.)

If $\gcd(a, n) > 1$, then (12.4) is not satisfied, but (12.3) may be satisfied. For example, the even number 6 is a pseudoprime to the base 4, since

$$4^6 = 4096 \equiv 4 \pmod{6}$$

but

$$4^{6-1} = 1024 \not\equiv 1 \pmod{6}.$$

Pseudoprimes to the particular base 2 have been the earliest and most widely studied. For example, the book [Rotkiewicz, 1972] is primarily concerned with pseudoprimes to the base 2. In keeping with Rotkiewicz, we will refer to pseudoprimes to the base 2 simply as *pseudoprimes*.

Pseudoprimes have also been referred to as *almost primes* (see [Erdős, 1950]) and *Poulet numbers* (see [Duparc]), since they were extensively studied in [Poulet], where all pseudoprimes up to 10^8 are tabulated. See also [Lehmer, 1936].

It has been reported in several sources (see, for example, [Dickson, p. 59] or [Jeans]) that the ancient Chinese as long ago as 500 B.C. believed that

$$(12.5) \qquad\qquad 2^n \equiv 2 \pmod{n}$$

holds if and only if n is a prime number. For a good discussion concerning how this almost certainly false story originated see [Ribenboim, 1996, pp. 103–105]. According to [Mahnke], in September 1680 and also December 1681, Leibniz stated erroneously that congruence (12.5) holds only if n is prime.

The first eight pseudoprimes are 341, 561, 645, 1105, 1387, 1729, 1905, and 2047. The first example of a pseudoprime, namely 341, was found in 1819 by Sarrus (see [Dickson, p. 92]). We will show shortly that there exist infinitely many pseudoprimes. The first even pseudoprime,

$$161\,038 = 2 \cdot 73 \cdot 1103,$$

was found by D. H. Lehmer in 1950 (see [Erdős, 1950]). N. G. W. H. Beeger proved in 1951 that there exist infinitely many even pseudoprimes (see [Beeger]).

There are interesting connections between Fermat numbers and pseudoprimes. It was proved in Theorem 4.13 that all composite Fermat numbers are pseudoprimes. If there are infinitely many composite Fermat numbers, then this would show that there are infinitely many pseudoprimes. Unfortunately, it is not known whether or

not there are infinitely many composite Fermat numbers (cf. Chapter 14). On the
other hand, as we shall see, Fermat numbers have been used by several authors (see
[Cipolla], [Jarden], [Jeans], [Rotkiewicz, 1964a], and [Szymiczek, 1966a]) to generate
infinitely many pseudoprimes. Although pseudoprimes resemble primes by being
infinite in number and by satisfying the assertion of Fermat's little theorem, they
also differ strikingly from primes in various ways. We will utilize Fermat numbers
to illustrate ways in which pseudoprimes are unlike primes and ways in which
they are similar to primes. We will also make use of Fermat numbers to generate
pseudoprimes with additional special properties. We will then explore these themes
further.

The following theorem, proved in [Cipolla] in 1904, shows how Fermat numbers
can be used to generate infinitely many pseudoprimes, including pseudoprimes with
an arbitrary number of prime factors.

Theorem 12.1 (Cipolla). If $a > b > \cdots > s > 1$ and $N = F_a F_b \cdots F_s$, then
N is a pseudoprime if and only if $2^s > a$. In particular, given any positive integer
M, no matter how large, there exist infinitely many pseudoprimes with at least M
distinct prime factors.

P r o o f . Since $\gcd(F_m, F_n) = 1$ if $m \neq n$ by Goldbach's Theorem 4.1, it is easy
to see that

(12.6) $$\operatorname{ord}_N 2 = \operatorname{lcm}(\operatorname{ord}_{F_a} 2, \operatorname{ord}_{F_b} 2, \ldots, \operatorname{ord}_{F_s} 2).$$

By Remark 4.13, $\operatorname{ord}_{F_m} 2 = 2^{m+1}$ for $m \geq 0$. Thus,

(12.7) $$\operatorname{ord}_N 2 = \operatorname{lcm}(2^{a+1}, \ 2^{b+1}, \ldots, \ 2^{s+1}) = 2^{a+1}.$$

Therefore, $2^{N-1} \equiv 1 \pmod{N}$ if and only if $N - 1$ is divisible by 2^{a+1}. However,

(12.8) $$N - 1 = F_a F_b \cdots F_s - 1 = 2^{2^s} K,$$

where K is an odd integer. Hence, N is a pseudoprime if and only if $2^s > a$.

Now let $s > 1$ and let

(12.9) $$N = F_s F_{s+1} F_{s+2} \cdots F_{2^s - 1}.$$

Then N is a pseudoprime by our above arguments. Clearly, given the positive
integer M, we can choose s such that $2^s - s \geq M$. Since $\gcd(F_i, F_j) = 1$ for
$s \leq i < j \leq 2^s - 1$, it follows that N has at least $2^s - s \geq M$ distinct prime factors.
The theorem is thus proved.   □

Remark 12.2. We note that the method given in Theorem 12.1 can be used
to generate pairwise coprime pseudoprimes. Let the pseudoprimes N_1 and N_2 each
be the product of distinct Fermat numbers such that no Fermat number appearing
in N_1 appears in N_2. Since $\gcd(F_m, F_n) = 1$ for $m \neq n$, we see that N_1 and N_2
are coprime. We also note that it was shown in [Erdős, 1949] that for every $M \geq 2$
there exist infinitely many pseudoprimes that are the product of exactly M distinct
primes.

It is of interest that the first proof of the infinitude of pseudoprimes appears in [Jeans] in 1898 and also makes use of Fermat numbers. Specifically, Jeans showed that if $a > b$ and $2^b > a$, then $F_a F_b$ is a pseudoprime. This result is a special case of Theorem 12.1.

The next theorem, which appears in [Sierpiński, 1948] and [Steuerwald], can be used to recursively generate infinitely many pseudoprimes for a given composite Fermat number, say F_5.

Theorem 12.3. *If N is an odd pseudoprime, then $2^N - 1$ is also an odd pseudoprime.*

P r o o f . Since N is composite, there exists an integer c, $1 < c < N$, such that $c \mid N$. Then

$$2^c - 1 \mid 2^N - 1,$$

and $2^N - 1$ is also composite. It now suffices to prove that

$$(12.10) \qquad\qquad 2^N - 1 \mid 2^{2^N - 2} - 1.$$

We prove a stronger result, namely,

$$(12.11) \qquad\qquad 2^N - 1 \mid 2^{(2^N - 2)/2} - 1,$$

which implies (12.10). Noting that N is an odd pseudoprime, we see that

$$\left(2^N - 2\right)/2 = 2^{N-1} - 1 = kN$$

for some integer k. Then

$$2^N - 1 \mid 2^{kN} - 1 = 2^{(2^N - 2)/2} - 1,$$

and (12.11) is satisfied.   □

We note that in 1903, Theorem 12.3 was proved in [Malo] for the case in which N is a prime and $2^N - 1$ is composite.

Remark 12.4. Theorem 12.3 shows in another way, different from Cipolla's Theorem 12.1, that there exist infinitely many pseudoprimes. By Theorem 4.10 we know that if F_m is composite, then F_m is a pseudoprime. In particular, F_5 is composite and thus a pseudoprime. From Theorem 12.3 it follows that for each pseudoprime we can find a greater one. Thus, the set of pseudoprimes is unbounded and hence infinite.

The following theorem, from [Rotkiewicz, 1964a], also uses Fermat numbers to generate infinitely many pseudoprimes and bears a certain resemblance to Theorem 12.1. This theorem generalizes Theorem 12.3 as well in the case for which N is a Fermat number.

Theorem 12.5 (Rotkiewicz). *If $n_1 > n_2 > \cdots > n_s > 1$, $s > 1$, $2^{n_s} > n_1$, and*

$$N = \left(2^{F_{n_1}} - 1\right)\left(2^{F_{n_2}} - 1\right)\cdots\left(2^{F_{n_s}} - 1\right),$$

then N is a pseudoprime. In particular, given any positive integer M, no matter how large, there exist infinitely many pseudoprimes with at least M distinct prime divisors.

Before proving Theorem 12.5, we will need the following lemma (cf. (9.11), (11.3), and Lemma 6.3).

Lemma 12.6. Let a and b be coprime integers with $|a| > |b| \geq 1$ and let m and n be positive integers. Then

$$(12.12) \qquad \gcd\left(a^m - b^m, a^n - b^n\right) = \left|a^{\gcd(m,n)} - b^{\gcd(m,n)}\right|.$$

For a proof of the equality

$$(12.13) \qquad \gcd\left(\frac{a^m - b^m}{a - b}, \frac{a^n - b^n}{a - b}\right) = \left|\frac{a^{\gcd(m,n)} - b^{\gcd(m,n)}}{a - b}\right|,$$

which is equivalent to (12.12), see [Carmichael, 1913, p. 38] or [Williams, 1998, p. 87].

P r o o f   o f   T h e o r e m   1 2 . 5 . Let $d_i = 2^{F_{n_i}} - 1$ for $1 \leq i \leq s$. To prove that N is a pseudoprime, it suffices to show that $\mathrm{ord}_N 2 \mid N - 1$. Since $\gcd(F_{n_i}, F_{n_j}) = 1$ for $1 \leq i < j \leq s$ by Goldbach's Theorem 4.1, it follows by Lemma 12.6 that $\gcd(d_i, d_j) = 1$ if $1 \leq i < j \leq s$. Thus,

$$(12.14) \qquad \mathrm{ord}_N 2 = \mathrm{lcm}(\mathrm{ord}_{d_1} 2, \ \mathrm{ord}_{d_2} 2, \ldots, \ \mathrm{ord}_{d_s} 2).$$

Clearly, the order of 2 modulo $2^{F_{n_i}} - 1$ is equal to F_{n_i} for $1 \leq i \leq s$. Hence, by (12.14),

$$\mathrm{ord}_N 2 = \mathrm{lcm}(F_{n_1}, \ F_{n_2}, \ldots, \ F_{n_s}) = F_{n_1} F_{n_2} \cdots F_{n_s}.$$

We now note that

$$(12.15) \qquad N - 1 = \left(2^{F_{n_1}} - 1\right)\left(2^{F_{n_2}} - 1\right) \cdots \left(2^{F_{n_s}} - 1\right) - 1.$$

Since $\mathrm{ord}_{F_m} 2 = 2^{m+1}$ by Remark 4.13 and since $2^{n_s} > n_1$, it follows that

$$2^{F_{n_i}-1} \equiv 1 \pmod{F_{n_j}}$$

for any i and j such that $1 \leq i, j \leq s$. Hence,

$$(12.16) \qquad 2^{F_{n_i}} - 1 \equiv 2 - 1 \equiv 1 \pmod{F_{n_j}}$$

for $1 \leq i, j \leq s$. By (12.15) and (12.16), we see that

$$N - 1 \equiv (1)(1) \cdots (1) - 1 \equiv 0 \pmod{F_{n_i}}$$

for $1 \leq i \leq s$. Hence,

$$N - 1 \equiv 0 \pmod{F_{n_1} F_{n_2} \cdots F_{n_s}}$$

and $\mathrm{ord}_N 2 \mid N-1$. The rest of the theorem follows by the same arguments as in the proof of Theorem 12.1. $\square$

Even though pseudoprimes are infinite in number, they are much scarcer than primes. For example, it was first proved in [Szymiczek, 1967] that if P_n denotes the nth pseudoprime to the base 2, then

$$\sum_{n=1}^{\infty} \frac{1}{P_n}$$

is convergent, whereas it is well known that if p_n denotes the nth prime, then

$$\sum_{n=1}^{\infty} \frac{1}{p_n}$$

is divergent. However, in [Mąkowski] it was shown that

$$\sum_{n=1}^{\infty} \frac{1}{\log P_n}$$

is divergent.

As usual, let $\pi(x)$ denote the number of primes less than or equal to x, and let $P_a(x)$ denote the number of pseudoprimes to the base a that are less than or equal to x. By the prime number theorem (1.5),

$$(12.17) \qquad\qquad \pi(x) \approx \frac{x}{\log x} \text{ as } x \to \infty.$$

The best known bounds for $P_a(x)$ for x sufficiently large are

$$(12.18) \qquad\qquad e^{(\log x / \log a)^{\alpha}} \le P_a(x) \le \frac{x}{(\ell(x))^{1/2}},$$

where $\ell(x) = e^{\log x \log \log \log x / \log \log x}$ and α may be taken to be $0.68/1.68 > 2/5$ (see [Pomerance, 1981, 1982] and [Ribenboim, 1996, pp. 312–313]). The upper bound in (12.18) is much smaller than $x/\log x$ for large x. We note that by the table in [Pomerance, Selfridge, Wagstaff, p. 1005], $P_2(10^{10}) = 14884$, while by the table in [Ribenboim, 1996, p. 237], $\pi(10^{10}) = 455052511$. Thus if a random number $n \le 10^{10}$ is chosen for which $2^{n-1} \equiv 1 \pmod{n}$, there would be less than a one-in-thirty-thousand chance that the number is a pseudoprime, and not a prime.

The rarity of pseudoprimes to a fixed base a compared to primes provides a rationale for using Fermat's little theorem as the basis for a test for primality. However, as we mentioned earlier, there exist composite integers N, called *Carmichael numbers* or *absolute pseudoprimes,* that are pseudoprimes to every base a, that is,

$$a^N \equiv a \pmod{N}$$

for all integers a. Thus, primality tests using Fermat's little theorem need to be modified. Such modifications will be discussed later.

In 1899 the following necessary and sufficient condition for the composite number N to be an absolute pseudoprime was given (see [Korselt]).

Theorem 12.7 (Korselt) *A composite integer N is an absolute pseudoprime if and only if N is square-free and $p - 1$ divides $N - 1$ for all primes dividing N.*

Example 12.8. For example, it follows immediately from the above assertion that $561 = 3 \cdot 11 \cdot 17$ is an absolute pseudoprime, since $3 - 1 = 2$, $11 - 1 = 10$, and $17 - 1 = 16$ all divide 560.

However, Korselt did not show the actual existence of any absolute pseudoprimes. In 1910, Carmichael independently discovered Korselt's criterion above and reformulated it in the following manner, although he did not use the term "Carmichael number" (see [Carmichael, 1910]):

Theorem 12.9 (Carmichael) *A composite integer N is a Carmichael number if and only if N is square-free and $\lambda(N) \mid N - 1$, where λ is the Carmichael lambda function introduced in Definition 2.20.*

In this paper, Carmichael also showed that a Carmichael number has at least three prime divisors and gave examples of four Carmichael numbers including the smallest such numbers, 561 and 1105. In [Carmichael, 1912] fifteen Carmichael numbers were given.

In [Alford, Granville, Pomerance] it is shown that there exist infinitely many Carmichael numbers. More precisely, it is proven that if $C(x)$ denotes the number of Carmichael numbers less than or equal to x, then for x sufficiently large,

$$C(x) > x^{2/7}.$$

In [Granville] and [Alford, Granville, Pomerance] it is pointed out that the proof of the infinitude of Carmichael numbers depends in part on finding numbers N for which $\lambda(N)$ is especially small compared to N. This is based on the observation that if N is a Carmichael number, then $\lambda(N)$ not only divides $N - 1$, but tends to be small compared to N. For example, let us calculate the value of λ for the Carmichael numbers 1729, 41041, and 825265: $\lambda(1729) = 36$, $\lambda(41041) = 120$, and $\lambda(825265) = 144$.

The following result involving Fermat d-pseudoprimes (see [Somer, 1989], [Carlip, Jacobson, Somer], or [Ribenboim, 1996, p. 117]) provides a rationale for this tendency of $\lambda(N)$ to be small compared to $N - 1$ when N is a Carmichael number.

Let $d \geq 1$. A *Fermat d-pseudoprime* is a composite integer N such that there exists $a \geq 2$, $\gcd(a, N) = 1$, for which $\mathrm{ord}_N a = (N - 1)/d$.

In [Somer, 1989] and [Carlip, Jacobson, Somer] it is proved that for every $d \geq 1$, there exist only finitely many Fermat d-pseudoprimes. A consequence of this result is that

$$(12.19) \qquad \lim_{\substack{N \to \infty \\ N \text{ composite}}} \frac{\gcd(\lambda(N), N - 1)}{N} = 0.$$

Since there are infinitely many Carmichael numbers and $\lambda(N) \mid N - 1$ for the Carmichael number N, (12.19) implies that

$$\lim_{\substack{N \to \infty \\ N \text{ a Carmichael number}}} \frac{\lambda(N)}{N - 1} = 0.$$

Note that (12.19) demonstrates another way in which pseudoprimes differ from primes, since

$$\lim_{\substack{N \to \infty \\ N \text{ prime}}} \frac{\gcd(\lambda(N), N-1)}{N} = \lim_{\substack{N \to \infty \\ N \text{ prime}}} \frac{N-1}{N} = 1.$$

For a generalization of Fermat d-pseudoprimes see [Somer, 1998] and [Carlip, Jacobson, Somer].

We now discuss special types of pseudoprimes to the base a with additional properties such that there exist no absolute pseudoprimes with respect to these pseudoprimes. These pseudoprimes form the basis for probabilistic primality tests.

Definition 12.10. The composite odd integer N is an *Euler pseudoprime to the base a* if $\gcd(a, N) = 1$ and

$$(12.20) \qquad \left(\frac{a}{N}\right) \equiv a^{(N-1)/2} \pmod{N},$$

where $a \geq 1$ and $\left(\frac{a}{N}\right)$ denotes the Jacobi symbol. Euler pseudoprimes to the base 2 will simply be called *Euler pseudoprimes*.

Remark 12.11. It is clear that Euler pseudoprimes to the base a are pseudoprimes to the base a. By Theorem 2.26 (Euler's criterion), if N is an odd prime, then N satisfies (12.20). It is proved in [Solovay, Strassen] that a composite odd integer N can be an Euler pseudoprime for at most $\frac{1}{2}\phi(n)$ bases a, $1 \leq a < n$, $\gcd(a, n) = 1$. Thus, if the base a is chosen randomly, the probability that a composite odd integer N is an Euler pseudoprime to the base a is less than or equal to $\frac{1}{2}$. This forms the basis of the Solovay–Strassen probabilistic primality test:

Let N be a fixed odd integer. Choose a base a such that $\gcd(a, N) = 1$ at random and see whether (12.20) holds. If (12.20) is fulfilled, then repeat the test. The probability that (12.20) will hold for k independent tests if N is composite is less than or equal to 2^{-k}. If N passes k tests, where k is reasonably large, we say that N is a *probable prime*.

Definition 12.12. Let N be a composite odd integer and write $N - 1 = 2^s t$, where t is odd. Let $a \geq 1$ be coprime to N. Then N is a *strong pseudoprime to the base a* if

$$(12.21) \qquad a^t \equiv 1 \pmod{N}$$

or

$$a^{2^r t} \equiv -1 \pmod{N} \quad \text{for some } r, \quad 0 \leq r < s.$$

Strong pseudoprimes to the base 2 will simply be called *strong pseudoprimes*.

Remark 12.13. Strong pseudoprimes to the base a were first defined by Selfridge (see [Williams, 1978, Section 17]). It is evident that strong pseudoprimes to the base a are pseudoprimes to the base a. Note that

$$(12.22) \qquad a^{N-1} - 1 = a^{2^s t} - 1 = \left(a^t - 1\right)\left(a^t + 1\right)\left(a^{2t} + 1\right) \cdots \left(a^{2^{s-1} t} + 1\right).$$

If N is prime, then by Fermat's little theorem,

$$a^{N-1} - 1 \equiv 0 \pmod{N},$$

which implies that one of the factors on the right-hand side of (12.22) is congruent to 0 (mod N); hence, N satisfies (12.21). It is proved in [Rabin] and [Monier] that a composite odd integer $N > a$ can be a strong pseudoprime to the base a for at most $\phi(n)/4$ bases a such that $1 \leq a < n$ and $\gcd(a, n) = 1$. Thus, if the base a is chosen randomly, the probability that N is a strong pseudoprime to the base a is less than or equal to $\frac{1}{4}$. This gives rise to the Miller–Rabin probabilistic primality test, which improves the Solovay–Strassen primality test:

Let N be a fixed odd integer. Choose a base a such that $\gcd(a, N) = 1$ at random and observe whether (12.21) is satisfied. If (12.21) is not satisfied, then N is composite. If (12.21) holds, then repeat the test. The probability that (12.21) will be satisfied for k independent tests if N is composite is less than or equal to 4^{-k}. If N passes k tests, where k is fairly large, we declare N to be a probable prime.

For deterministic primality tests with the best known asymptotic running times, see [Adleman, Pomerance, Rumely] and [Cohen, Lenstra].

The theorem below relates Euler pseudoprimes to the base a and strong pseudoprimes to the base a.

Theorem 12.14. *If N is a strong pseudoprime to the base a, then N is an Euler pseudoprime to the base a. Moreover, if $N \equiv 3 \pmod 4$ and N is an Euler pseudoprime to the base a, then N is a strong psedoprime to the base a.*

The proof of the first assertion is due to Selfridge (see [Williams, 1978, Section 17]). The second assertion was proved in [Malm].

Remark 12.15. In Theorems 4.10 and 5.11 it was proved that both F_m and M_p are pseudoprimes if they are composite. Theorem 12.16 below shows, in addition, that composite Fermat numbers and composite Mersenne numbers are also Euler pseudoprimes and strong pseudoprimes.

Theorem 12.16. *Let F_m be a composite Fermat number and let $M_p = 2^p - 1$ be a composite Mersenne number, where p is a prime. Then F_m and M_p both are Euler pseudoprimes and strong pseudoprimes.*

P r o o f . We note that $m \geq 5$ and $p \geq 11$. We will show that F_m and M_p are both Euler pseudoprimes and strong pseudoprimes without resort to Theorem 12.14. Note that $\left(\frac{2}{F_m}\right) = 1$ and $\left(\frac{2}{M_p}\right) = 1$ by property (vii) of Theorem 2.30.

We first treat the Fermat number F_m. It is clear that

$$(12.23) \qquad\qquad 2^{2^m} \equiv -1 \pmod{F_m},$$

and hence F_m is a strong pseudoprime. Furthermore, for $m \geq 5$,

$$(F_m - 1)/2 = 2^{2^m - 1} > 2^m.$$

Let $k \geq 1$ be such that

$$\frac{(F_m - 1)/2}{2^m} = 2^k.$$

Then by (12.23),

$$2^{(F_m-1)/2} = \left(2^{2^m}\right)^{2^k} \equiv (-1)^{2^k} \equiv 1 \equiv \left(\frac{2}{F_m}\right) \pmod{F_m},$$

and thus F_m is an Euler pseudoprime.

We now consider the Mersenne number M_p. Note that $(M_p - 1)/2 = 2^{p-1} - 1$ is an odd integer. By (5.6) in the proof of Theorem 5.11,

$$2^{(M_p-1)/2} \equiv 1 \equiv \left(\frac{2}{M_p}\right) \pmod{M_p}.$$

Hence, M_p is both a strong pseudoprime and an Euler pseudoprime.   □

The next theorem, which is proved in [Pomerance, Selfridge, Wagstaff, p. 1008], enables us to generate infinitely many strong pseudoprimes and hence by Theorem 12.14 also infinitely many Euler pseudoprimes in a simple manner.

Theorem 12.17. *If N is a pseudoprime, then $2^N - 1$ is a strong pseudoprime.*

P r o o f . Notice that $(2^N - 2)/2 = 2^{N-1} - 1$ is an odd integer. By (12.11) in the proof of Theorem 12.3,

$$2^{(2^N-2)/2} \equiv 1 \pmod{2^N - 1},$$

and hence $2^N - 1$ is a strong pseudoprime.   □

Remark 12.18. Let F_m be a composite Fermat number. By Theorem 12.16, F_m is a strong pseudoprime. Applying Theorem 12.17 recursively, we can generate infinitely many strong pseudoprimes from one strong pseudoprime F_m.

In addition to being infinite in number, another property that pseudoprimes share with primes is that there are infinitely many pseudoprimes in arithmetic progressions. The following result is proved in [Rotkiewicz, 1963, 1967].

Theorem 12.19 (Rotkiewicz). *If $a \geq 1$, $d \geq 1$, and $\gcd(a, d) = 1$, then there exist infinitely many pseudoprimes in the arithmetic progression $\{a + kd \mid k \geq 1\}$.*

We will next discuss superpseudoprimes. To proceed we will first need results on primitive prime divisors.

Let a and b be coprime integers with $|a| > |b| \geq 1$ and let $n \geq 1$ be a positive integer. A prime p is called a *primitive prime divisor* of $a^n - b^n$ if $p \mid a^n - b^n$, but $p \nmid a^m - b^m$ for $1 \leq m < n$. Primitive prime divisors of $a^n + b^n$ are defined similarly. It follows from relation (12.12) that p is a primitive prime divisor of $a^n - b^n$ if and only if $p \nmid a^d - b^d$ for any proper divisor d of n.

Remark 12.20. We make the following observations about primitive prime divisors. If p is a primitive prime divisor of $a^n - 1$ with multiplicity m, then, for $1 \leq i \leq m$,

(12.24) $\mathrm{ord}_{p^i} a = n.$

Furthermore, every prime divisor of the Fermat number $F_m = 2^{2^m} + 1$ is a primitive prime divisor of $2^{2^{m+1}} - 1$.

Later we will need the following two lemmas about primitive prime divisors.

Lemma 12.21. *If the odd prime p is a primitive prime divisor of $a^n - 1$, then*
(i) $p \equiv 1 \pmod{n}$,
(ii) $p \equiv 1 \pmod{2n}$ *if n is odd or if* $\left(\dfrac{a}{p}\right) = 1$.

P r o o f . (i) By Fermat's little theorem and (12.24), $n \mid p - 1$, which implies that

$$p \equiv 1 \pmod{n}.$$

(ii) It follows from (i) that $p \equiv 1 \pmod{2n}$ if n is odd, since p is odd. Now suppose that $\left(\frac{a}{p}\right) = 1$. By Theorem 2.26 (Euler's criterion),

$$a^{(p-1)/2} \equiv 1 \pmod{p},$$

which implies that $n \mid (p - 1)/2$. It then follows that $p \equiv 1 \pmod{2n}$.   □

The following theorem, which was first proved in [Zsigmondy] and independently in [Birkhoff, Vandiver], shows that $a^n - b^n$ always has a primitive divisor if $n \notin \{1, 2, 3, 6\}$.

Theorem 12.22 (Zsigmondy). *Let a and b be coprime integers with $|a| > |b| \geq 1$. Then $a^n - b^n$ has a primitive prime divisor except in the following cases:*

$$n = 1, \quad a - b = 1,$$
$$n = 2, \quad a + b = \pm 2^k \quad (k \geq 1),$$
$$n = 3, \quad a = \pm 2, \ b = \mp 1,$$
$$n = 6, \quad a = \pm 2, \ b = \pm 1.$$

Theorem 12.23 below, from [Schinzel], sharpens the above theorem to give cases in which $a^n - b^n$ has at least two primitive prime divisors. This theorem makes use of the square-free kernel $k(n)$ of n, that is, n divided by its greatest square factor.

Theorem 12.23 (Schinzel). *Let a and b be coprime integers with $|a| > |b| \geq 1$. Let*

$$e = 1 \quad \text{if } k(ab) \equiv 1 \pmod{4},$$
$$e = 2 \quad \text{if } k(ab) \equiv 2 \text{ or } 3 \pmod{4}.$$

If $n/ek(ab)$ is an odd integer, then $a^n - b^n$ has at least two primitive prime divisors, except in the following cases:

$n = 1, \quad a = \pm(2^\alpha + 1)^2, b = \pm(2^\alpha - 1)^2$ or $4a = \pm(p^\alpha + 1)^2, 4b = \pm(p^\alpha - 1)^2$,
     where α is a positive integer and p an odd prime,
$n = 2,$ *same as for $n = 1$ but with $\mp$ for b instead of $\pm$,*
$n = 3, \quad a = \pm 3, b = \mp 1$ or $a = \pm 4, b = \pm 1$ or $a = \pm 4, b = \mp 3$,
$n = 4, \quad |a| = 2, |b| = 1$,
$n = 6, \quad a = \pm 3, b = \pm 1$ or $a = \pm 4, b = \mp 1$ or $a = \pm 4, b = \pm 3$,
$n = 12, \quad |a| = 2, |b| = 1$ or $|a| = 3, |b| = 2$,
$n = 20, \quad |a| = 2, |b| = 1$.

We now define a special type of pseudoprime called a superpseudoprime.

Definition 12.24. A *superpseudoprime N to the base a* is a pseudoprime to the base a all of whose divisors greater than 1 are either primes or pseudoprimes to the base a; that is, if $d \mid N$, then

$$(12.25) \qquad a^{d-1} \equiv 1 \pmod{N}.$$

Superpseudoprimes to the base 2 are simply called *superpseudoprimes*.

Below we give a necessary and sufficient condition for an odd composite integer N to be a superpseudoprime to the base a (see [Somer, 2001]).

Theorem 12.25. *Let p_1, p_2, ..., p_r be distinct odd primes and let $a \geq 2$ be such that $\gcd(a, p_i) = 1$ for every $i = 1, \ldots, r$. Suppose that p_i is a primitive prime divisor of $a^{t_i} - 1$ with multiplicity m_i for $1 \leq i \leq r$. We allow the possibility that $t_i = t_j$ for $1 \leq i < j \leq r$. Let*

$$h = \operatorname{lcm}(t_1, \, t_2, \ldots, \, t_r).$$

Let N be a composite integer such that

$$N = \prod_{i=1}^{r} p_i^{\ell_i},$$

where $1 \leq \ell_i \leq m_i$.

Then N is a superpseudoprime to the base a if and only if for each $i = 1, \ldots, r$ there exists an integer k_i such that

$$(12.26) \qquad p_i = k_i h + 1.$$

P r o o f . Suppose that (12.26) holds for $1 \leq i \leq r$. Let $d = p_1^{g_1} p_2^{g_2} \cdots p_r^{g_r}$ be a composite divisor of N, where $0 \leq g_i \leq \ell_i$ for $i = 1, \ldots, r$. To show that d is a pseudoprime to the base a, it suffices to establish that

$$(12.27) \qquad \operatorname{ord}_d a \mid d - 1.$$

Since $p_i \equiv 1 \pmod{h}$ for $i = 1, \ldots, r$ by (12.26), we have $d \equiv 1 \pmod{h}$, or equivalently,

$$(12.28) \qquad d - 1 \equiv 0 \pmod{h}.$$

Let e_i equal the order of a modulo $p_i^{g_i}$ for $i = 1, \ldots, r$. By Remark 12.20, if $g_i \geq 1$, then $e_i = t_i$; otherwise, $e_i = 1$. Since $\gcd\big(p_i^{g_i}, p_j^{g_j}\big) = 1$ for $1 \leq i < j \leq r$, it follows that

$$(12.29) \qquad \operatorname{ord}_d a = \operatorname{lcm}(e_1, \, e_2, \ldots, \, e_r) \mid \operatorname{lcm}(t_1, \, t_2, \ldots, \, t_r) = h.$$

It now follows from (12.28) and (12.29) that (12.27) holds.

On the other hand, suppose that N is a superpseudoprime to the base a and there exists a prime divisor p_i of N such that $p_i \not\equiv 1 \pmod{h}$ for some i, where $1 \leq i \leq r$. Then there exists a prime q for which $q^m \| h$, but $q^m \nmid p_i - 1$ for some positive integer m. By the definition of h, $q^m \| t_j$ for some j such that $1 \leq j \leq r$. Since p_j is a primitive prime divisor of $a^{t_j} - 1$, we see by Lemma 12.21 that $p_j \equiv 1 \pmod{t_j}$, which implies that

$$p_j \equiv 1 \pmod{q^m}.$$

We claim, however, that $p_i p_j$ is not a pseudoprime to the base a, contradicting the assumption that N is a superpseudoprime to the base a. Since $p_i \not\equiv 1 \pmod{q^m}$ and $p_j \equiv 1 \pmod{q^m}$, we have $p_i p_j \not\equiv 1 \pmod{q^m}$, i.e.,

$$p_i p_j - 1 \not\equiv 0 \pmod{q^m}.$$

Thus,

$$\operatorname{ord}_{p_j} a = t_j \nmid p_i p_j - 1$$

and a fortiori,

$$\operatorname{ord}_{p_i p_j} a \nmid p_i p_j - 1.$$

Hence, $p_i p_j$ is not a pseudoprime to the base a. $\square$

Using Theorem 12.25, we can easily construct superpseudoprimes from tables of primitive prime divisors of $2^n - 1$. For example, using the tables in [Riesel, 1994] or [Brillhart, Lehmer, Selfridge, Tuckerman, Wagstaff], we can generate the superpseudoprime

$$N = 89 \cdot 2113 \cdot 353 \cdot 2931542417.$$

That N is indeed a superpseudoprime follows because each of its four prime factors is congruent to 1 modulo 88, the number 89 is a primitive prime divisor of $2^{11} - 1$, the number 2113 is a primitive prime divisor of $2^{44} - 1$, both 353 and 2931542417 are primitive prime divisors of $2^{88} - 1$, and $\operatorname{lcm}(11, 44, 88) = 88$.

Theorem 12.26 below, which appears in [Somer, 2001], gives a systematic method for producing superpseudoprimes. It is based on Theorem 12.25 and shows how the Fermat numbers can be used to generate infinitely many superpseudoprimes.

Theorem 12.26.

(i) If F_m is composite, then F_m is a superpseudoprime.

(ii) $F_m F_{m+1}$ is a superpseudoprime for all $m \geq 2$. In particular, there exist infinitely many superpseudoprimes.

(iii) Assume that $m \geq 3$ and that each prime divisor of F_m is of the form $k2^{m+3} + 1$, where k is a natural number. Then $F_m F_{m+1} F_{m+2}$ is a superpseudoprime with at least three distinct prime divisors.

(iv) Assume that $p_1, p_2, \ldots, p_s$ are any prime divisors of F_m such that $p_i \equiv 1 \pmod{2^{m+3}}$ for $1 \leq i \leq s$. Then $p_1 p_2 \cdots p_s F_{m+1} F_{m+2}$ is a superpseudoprime with at least three distinct prime divisors.

(v) If N is a pseudoprime with exactly two prime divisors, then N is a superpseudoprime.

(vi) *If the Mersenne number M_p is composite, then M_p is a superpseudoprime.*

(vii) *Let $N = 2^t - 1$ be a composite integer with at least two primitive prime divisors. If p_1, $p_2, \ldots, p_s$ are any primitive prime divisors of N, where $s \geq 2$ and repetitions are allowed up to the multiplicity of a primitive prime divisor, then $p_1 p_2 \cdots p_s$ is a superpseudoprime.*

(viii) *Let $t \geq 5$ be an odd integer and let p_1, $p_2, \ldots, p_s$ be any primitive prime divisors of $2^t - 1$, where $s \geq 1$, and let q_1, $q_2, \ldots, q_j$ be any primitive prime divisors of $2^{2t} - 1$, where $j \geq 1$. Repetitions are permitted for both the p's and q's up to the multiplicity of a primitive prime divisor. Then $p_1 p_2 \cdots p_s q_1 q_2 \cdots q_j$ is a superpseudoprime.*

(ix) *Suppose $8 \mid t$. Let p_1, $p_2, \ldots, p_s$ be any primitive prime divisors of $2^t - 1$, where $s \geq 1$, and let q_1, $q_2, \ldots, q_j$ be any primitive prime divisors of $2^{2t} - 1$, where $j \geq 1$. Then $p_1 p_2 \cdots p_s q_1 q_2 \cdots q_j$ is a superpseudoprime.*

(x) *Any composite divisor d of a superpseudoprime is also a superpseudoprime.*

P r o o f . The assertions (i)–(x) follow from Theorem 12.25, Goldbach's Theorem 4.1, Lucas's Theorem 6.1, Remark 12.20, Lemma 12.21, and Theorem 12.22 upon noting that if $8 \mid t$, then $\left(\frac{2}{p}\right) = 1$ for any primitive prime divisor p of $2^t - 1$.

Remark 12.27. Parts (ii) and (iii) can be found in [Szymiczek, 1966a]. The only known integers m for which $F_m F_{m+1} F_{m+2}$ is a superpseudoprime are $m = 3, 4$, and 8. In order to ascertain whether $M_{2,m}$ is a superpseudoprime we need to examine all the prime factors of F_m for $m \geq 3$ and check whether they are all of the form $k2^{m+3} + 1$. The only Fermat numbers F_m that have been factored completely are those for which $0 \leq m \leq 11$. Of these, we find from Appendix A that the only F_m for which $m \geq 3$ and all of its prime factors are of the proper form are the primes F_3 and F_4 and the composite number F_8.

By using part (iv) rather than part (iii) of Theorem 12.26, we can find many more superpseudoprimes with at least three distinct prime divisors. By tables at the web site [www1], there are over 100 known prime factors of Fermat numbers F_m of the form $k2^{m+3} + 1$ as of the beginning of 2001. By Schinzel's Theorem 12.23, part (vii) holds for any integer t such that $t \geq 28$ and $t \equiv 4 \pmod{8}$.

Part (v) of Theorem 12.26 provides examples of superpseudoprimes N that are products of exactly two primes. These examples are not very interesting, since then the only proper divisors of N greater than 1 are primes. If we knew that there exist infinitely many composite Fermat numbers, then the superpseudoprimes $F_m F_{m+1}$ given in part (ii) of Theorem 12.26 would provide infinitely many superpseudoprimes with at least three prime divisors. The next theorem (see [Somer, 2001]) uses Fermat numbers to generate infinitely many superpseudoprimes that are products of at least three distinct prime divisors.

Theorem 12.28. *Let p be a prime divisor of F_m, where $m \geq 3$. Then $(p-1)/2 > 2^{m+1}$. Let p_1, $p_2, \ldots, p_s$ be distinct primitive prime divisors of $2^{(p-1)/2} - 1$ and let q_1, $q_2, \ldots, q_j$ be distinct primitive prime divisors of $2^{p-1} - 1$, where $j, s \geq 1$. Then $N = p \, p_1 p_2 \cdots p_s q_1 q_2 \cdots q_j$ is a superpseudoprime with at least three distinct prime divisors. In particular, there exist infinitely many superpseudoprimes with at least three distinct primitive prime divisors.*

P r o o f . By Remark 12.20, p is a primitive prime divisor of $2^{2^{m+1}} - 1$. By Zsigmondy's Theorem 12.22, each of the terms $2^{2^{m+1}} - 1$, $2^{(p-1)/2} - 1$, and $2^p - 1$ has primitive prime divisors. By Theorem 6.7, if F_m is composite, then the prime divisor p of F_m is of the form $k2^{m+2} + 1$, where $k \geq 3$. It is now easily seen that if $m \geq 3$, then $(p-1)/2 > 2^{m+1}$ whether F_m is prime or composite. However, we observe that $p \equiv 1 \pmod{16}$, and therefore we obtain $(p-1)/2 \equiv 0 \pmod 8$. According to Lemma 12.21 (i), any primitive prime divisor of $2^{(p-1)/2} - 1$ is congruent to 1 modulo 8. Let p_i, $i \in \{1, \ldots, s\}$, be a primitive prime divisor of $2^{(p-1)/2} - 1$. Then $\left(\frac{2}{p_i}\right) = 1$, and using Lemma 12.21 (ii) again, we see that $p_i \equiv 1 \pmod{2(p-1)/2}$, i.e.,

$$p_i \equiv 1 \pmod{p-1}.$$

Thus, each of the distinct primes p, p_1, $p_2, \ldots, p_s$, q_1, $q_2, \ldots, q_j$ is congruent to 1 $\pmod{p-1}$. Since

$$2^{m+1} \mid p - 1,$$

it follows by Theorem 12.25 that the number N is a superpseudoprime. The theorem is thus proved. □

Remark 12.29. Since any composite divisor of a superpseudoprime is also a superpseudoprime by Theorem 12.26 (x), Theorem 12.28 implies that there exist infinitely many superpseudoprimes with exactly three distinct prime divisors. The papers [Szymiczek, 1965], [Rotkiewicz, 1968], [Fehér, Kiss], and [Somer, 2001] provide other infinite classes of superpseudoprimes to the base a with exactly three distinct prime divisors.

Euler's generalization of Fermat's little theorem given in Theorem 2.17 can be interpreted as saying that for any positive integer n,

$$n \mid a^{\phi(n)} - b^{\phi(n)}$$

for all pairs of integers a, b such that $\gcd(ab, n) = 1$.

Using other number-theoretic functions $f(n)$, we can look for positive integers n having the property

$$n \mid a^{f(n)} - b^{f(n)}$$

for all pairs of integers a, b such that $\gcd(ab, n) = 1$. Then n is called a *pseudoprime for the function $f(n)$*. Rotkiewicz asked which positive integers n are pseudoprimes for the arithmetic function $f(n) = \sigma(n)$, where $\sigma(n)$ is the sum of all positive divisors of n (without multiplicity). We call such positive integers n *Rotkiewicz numbers*. Rotkiewicz also asked whether there are infinitely many such numbers, see [Rotkiewicz, 1972, 1999]. We will see below that Fermat primes can be used to generate Rotkiewicz numbers (see [Luca, Somer]).

Rotkiewicz himself gives all the examples of Rotkiewicz numbers n of the form pq, namely,

(12.30) $2 \cdot 3$, $2 \cdot 7$, $3 \cdot 5$, $5 \cdot 7$, $5 \cdot 13$, $7 \cdot 17$, $13 \cdot 29$.

While we are unable to prove that there exist infinitely many Rotkiewicz numbers, we give examples of families of such numbers.

Lemma 12.30. *A positive integer n is a Rotkiewicz number if and only if $c^{\sigma(n)} \equiv 1 \pmod{n}$ whenever c is an integer such that $\gcd(c, n) = 1$.*

P r o o f . If n is a Rotkiewicz number and c is any integer coprime to n, we can choose $a = c$ and $b = 1$ and obtain that $n \mid c^{\sigma(n)} - 1$.

Conversely, assume that n is an integer satisfying the property from the text of Lemma 12.30 and let a and b be two integers such that ab is coprime to n. Since b is coprime to n, we get that b is invertible modulo n. Let c be an integer such that $c \equiv ab^{-1} \pmod{n}$. Then $c^{\sigma(n)} \equiv 1 \pmod{n}$ implies that $(ab^{-1})^{\sigma(n)} \equiv 1 \pmod{n}$. The last congruence implies that $a^{\sigma(n)} \equiv b^{\sigma(n)} \pmod{n}$, which implies that $n \mid a^{\sigma(n)} - b^{\sigma(n)}$. $\square$

Lemma 12.31. *A positive integer n is a Rotkiewicz number if and only if $\lambda(n) \mid \sigma(n)$, where λ is the Carmichael function from Definition 2.20.*

P r o o f . Let n be a Rotkiewicz number, let p be a prime divisor of n, and let k be such that $p^k \| n$. Assume first that $p > 2$. Let c be a primitive root modulo p^k and let $\gcd(c, n) = 1$. Since $c^{\sigma(n)} \equiv 1 \pmod{n}$, it follows that $c^{\sigma(n)} \equiv 1 \pmod{p^k}$. However, since c is a primitive root modulo p^k, we see that $\lambda(p^k) \mid \sigma(n)$. The case $p = 2$ can be treated similarly (for $k \geq 3$ we can choose $c \equiv 5 \pmod{8}$ and c coprime to all odd prime divisors of n). Since p was an arbitrary prime divisor of n, we have $\lambda(n) \mid \sigma(n)$.

Conversely, suppose that $\lambda(n) \mid \sigma(n)$ and let c be an integer coprime to n. Let p be a prime divisor of n and let k be such that $p^k \| n$. Since c is then also coprime to p, it follows by Carmichael's Theorem 2.22 that $c^{\lambda(p^k)} \equiv 1 \pmod{p^k}$. Since $\lambda(p^k) \mid \lambda(n) \mid \sigma(n)$, we get that $c^{\sigma(n)} \equiv 1 \pmod{p^k}$, and thus $p^k \mid c^{\sigma(n)} - 1$. Since the above divisibility relation holds for all prime factors p of n, we obtain $n \mid c^{\sigma(n)} - 1$. Now Lemma 12.30 implies that n is a Rotkiewicz number. $\square$

Proposition 12.32. *Let $q > 2$ be an integer such that the Mersenne number $M_q = 2^q - 1$ is prime. Then $n = 2^{q-2}M_q$ is a Rotkiewicz number.*

P r o o f . When $q = 3$, we get $n = 2 \cdot 7$, which appears in (12.24). Assume now that $q \geq 5$ and set $n = 2^{q-2}M_q$. Then $\sigma(n) = 2^q(2^{q-1} - 1)$. Moreover, since M_q is prime, it follows that

$$\lambda(n) = \mathrm{lcm}\big(\lambda(2^{q-2}), \phi(M_q)\big) = \mathrm{lcm}\big(2^{q-4}, 2^q - 2\big) = 2^{q-4}\big(2^{q-1} - 1\big).$$

Consequently, we get that $\sigma(n) = 16\,\lambda(n)$, and Lemma 12.31 guarantees that n is a Rotkiewicz number. $\square$

Theorem 12.33. *Let n be a Rotkiewicz number and let $2^\mu \| \sigma(n)$. If F_m is a prime Fermat number not dividing n and*

$$(12.31) \qquad m \leq \frac{\log(\mu + 1)}{\log 2},$$

then nF_m is a Rotkiewicz number as well.

P r o o f . Let the assumptions be satisfied. Then inequality (12.31) is equivalent to $2^{2^m} \mid 2\sigma(n)$. Notice that

$$(12.32) \qquad \lambda(nF_m) = \mathrm{lcm}(\lambda(n), \phi(F_m)) = \mathrm{lcm}\big(\lambda(n), 2^{2^m}\big)$$

and that

$$\sigma(nF_m) = \sigma(n)\sigma(F_m) = 2\sigma(n)\left(2^{2^m-1} + 1\right).$$

In particular, $2\sigma(n) \mid \sigma(nF_m)$. Since n is a Rotkiewicz number, it follows that

(12.33) $\lambda(n) \mid \sigma(n) \mid \sigma(nF_m).$

Moreover,

(12.34) $\phi(F_m) = 2^{2^m} \mid 2\sigma(n) \mid \sigma(nF_m).$

Formulae (12.32)–(12.34) now imply that $\lambda(nF_m) \mid \sigma(nF_m)$, and therefore, by Lemma 12.31, nF_m is a Rotkiewicz number.   □

Theorem 12.33 can be immediately generalized as follows.

Theorem 12.34. *Let n be a Rotkiewicz number and let $2^\mu \| n$. Suppose that $F_{m_1} < F_{m_2} < \cdots < F_{m_j}$ are prime Fermat numbers not dividing n. If*

$$m_j \leq \frac{\log(\mu + j)}{\log 2},$$

then $nF_{m_1}F_{m_2}\cdots F_{m_j}$ is a Rotkiewicz number as well.

In [Luca, Somer] another construction of further Rotkiewicz numbers by means of Fermat primes is described.

13. Generalizations of Fermat Numbers

Man muss jederzeit an Stelle von "Punkten, Geraden, Ebenen"
"Tische, Stühle, Bierseidel" sagen können.

David Hilbert

We will explore generalizations of Fermat numbers that share many of the same properties of the Fermat numbers; these properties were given in earlier chapters. We will also investigate other numbers such as the Cullen numbers, which bear some resemblance to the Fermat numbers.

We define as in [Shanks] the *generalized Fermat numbers*

$$L_{p,m}(a) = \frac{a^{p^{m+1}} - 1}{a^{p^m} - 1},$$

where p is a prime, $a \geq 2$, and $m \geq 0$, which generalize both the Fermat numbers $F_m = L_{2,m}(2) = 2^{2^m} + 1$ and the Mersenne numbers $M_p = L_{p,0}(2) = 2^p - 1$ (see Appendix B). It is proved in [Shanks, 1978, pp. 105–107, Theorem 48] that $p \mid L_{p,m}(a)$ if and only if $p \mid a - 1$. The case $a = 2$ is examined in [Ligh, Jones].

By Goldbach's Theorem 4.1, $\gcd(F_m, F_n) = 1$ for $m \neq n$. It would be desirable to retain this property for another type of *generalized Fermat numbers*,

$$G_{p,m}(a) = \begin{cases} L_{p,m}(a) & \text{if} \quad a \not\equiv 1 \pmod{p}, \\ \frac{1}{p} L_{p,m}(a) & \text{if} \quad a \equiv 1 \pmod{p}. \end{cases}$$

We will demonstrate later that if $0 \leq m < n$, then

$$\gcd(G_{p,m}(a), G_{p,n}(a)) = 1.$$

Recall the definition of the *Möbius function* μ:

$$\mu(n) = \begin{cases} 1 & \text{if } n = 1, \\ (-1)^r & \text{if } n \text{ is the product of } r \text{ distinct primes}, \\ 0 & \text{if } n \text{ is divisible by the square of a prime}. \end{cases}$$

Before proceeding further, we will need the following lemma concerning primitive prime divisors of $a^n - 1$ and strong pseudoprimes to the base a (see Chapter 12 for the definitions of a primitive prime divisor and a strong pseudoprime). If $n \geq 1$ and $\zeta = e^{2\pi i/n}$ is a primitive nth root of unity, then the nth *cyclotomic polynomial* Φ_n is defined by

$$\Phi_n(x) = \prod_{\substack{j=1 \\ \gcd(j,n)=1}}^{n} (x - \zeta^j).$$

Lemma 13.1. *Let $n \geq 1$ and let $\zeta = e^{2\pi i/n}$ be a primitive nth root of unity. Then*

$$(13.1) \qquad \Phi_n(x) = \prod_{d|n}^{n} (x^d - 1)^{\mu(n/d)}.$$

Moreover, if p is the largest prime dividing n, $P_a(n)$ is the product of the primitive prime divisors of $a^n - 1$ counting multiplicity, and it is not the case that $p = n = 2$ and $a \equiv 3 \pmod 4$, then

$$(13.2) \qquad \Phi_n(a) = \lambda P_a(n),$$

where $\lambda = 1$ or $\lambda = p$. Furthermore, $\lambda = p$ if and only if p is a primitive prime divisor of $a^{n/p^k} - 1$, where $\gcd(n/p^k, p) = 1$. If $a \equiv 3 \pmod 4$ and $p = n = 2$, then

$$\Phi_n(a) = a + 1 = \lambda P_a(n),$$

where $\lambda = 2^k$ for some integer k such that $k \geq 2$.

Lemma 13.1 follows from results in [Lidl, Niederreiter, p. 93] and [Birkhoff, Vandiver].

Values of $\Phi_n(x)$, that are primes or probable primes are given in [Morimoto, Kida] for $\phi(n) \leq 100$ and $x \in \{2, 3, \ldots, 1000\}$. Morimoto and Kida further give complete factorizations of the numbers $\Phi_n(x)$ for $\phi(n) \in \{16, 18\}$ and $x \in \{2, 3, \ldots, 1000\}$.

Remark 13.2. By Lemma 13.1, if $m \geq 0$ and if it is not the case that $p = 2$, $m = 0$, and $a \equiv 3 \pmod 4$, then

$$(13.3) \qquad L_{p,m}(a) = \Phi_{p^{m+1}}(a) = \begin{cases} P_a(p^{m+1}) & \text{if } a \not\equiv 1 \pmod p, \\ p P_a(p^{m+1}) & \text{if } a \equiv 1 \pmod p, \end{cases}$$

where the numbers $\Phi_{p^{m+1}}(a)$ and $P_a(p^{m+1})$ are defined as in Lemma 13.1 and $\gcd(p, P_a(p^{m+1})) = 1$. Hence, in this case,

$$(13.4) \qquad G_{p,m}(a) = P_a(p^{m+1}) \quad \text{and} \quad \gcd(p, P_a(p^{m+1})) = 1.$$

If $p = 2$, $m = 0$, and $a \equiv 3 \pmod 4$, then

$$(13.5) \qquad G_{p,m}(a) = \frac{1}{2}\Phi_{p^{m+1}}(a) = \frac{a+1}{2} = 2^c P_a(2),$$

where $c \geq 1$ and $P_a(2)$ is odd.

It now follows from the definition of a primitive divisor that for all primes p and all $a \geq 2$,

(13.6) $$\gcd(G_{p,m}(a), G_{p,n}(a)) = 1$$

if $0 \leq m < n$. Note that

$$F_m = G_{2,m}(2) = L_{2,m}(2)$$

and

$$M_p = G_{p,0}(2) = L_{p,0}(2).$$

We observe that

(13.7) $$G_{p,m}(a) \equiv 1 \pmod{2}$$

if it is not the case that $p = 2$, $m = 0$, and $a \equiv 3 \pmod{4}$.

Lemma 13.3. *Let N be an odd pseudoprime to the base a, where $\gcd(a, N) = 1$. If there exists an integer $k \geq 0$ such that $2^k \| \operatorname{ord}_{p^b} a$ for all prime powers p^b for which $p^b \| N$, then N is a strong pseudoprime to the base a.*

P r o o f . Let $N - 1 = 2^s t$, where t is odd, and let p^b be a prime power such that $b \geq 1$ and $p^b \| N$. Since N is a pseudoprime to the base a and $\gcd(a, N) = 1$, we have

$$\operatorname{ord}_N a \mid N - 1 = 2^s t.$$

Noting that $\operatorname{ord}_{p^b} a \mid \operatorname{ord}_N a$, we see that

$$\operatorname{ord}_{p^b} a = 2^k h,$$

where $0 \leq k \leq s$, h is odd, and $h \mid t$. Therefore,

(13.8) $$a^{2^k t} \equiv 1 \pmod{p^b},$$

but

(13.9) $$a^{2^{k-1} t} \not\equiv 1 \pmod{p^b} \quad \text{if } k \geq 1$$

for all prime powers p^b such that $p^b \| N$. It follows by the proof of the Chinese remainder theorem that

(13.10) $$a^{2^k t} \equiv 1 \pmod{N}.$$

Hence, N is a strong pseudoprime to the base a if $k = 0$.

Now assume that $k \geq 1$. Then by (13.8),

$$p^b \mid \left(a^{2^k t} - 1 \right) = \left(a^{2^{k-1} t} + 1 \right) \left(a^{2^{k-1} t} - 1 \right).$$

If p divides both $a^{2^{k-1}t} + 1$ and $a^{2^{k-1}t} - 1$, then $p \mid 2$, which is impossible, since p is odd. Thus $p \mid a^{2^{k-1}t} + 1$ or $p \mid a^{2^{k-1}t} - 1$, but not both. By (13.9), $p^b \nmid a^{2^{k-1}t} - 1$. Hence,

$$a^{2^{k-1}t} \equiv -1 \pmod{p^b}$$

for all prime powers p^b for which $p^b \| N$. Again using the proof of the Chinese remainder theorem, we obtain

$$a^{2^{k-1}t} \equiv -1 \pmod{N},$$

and N is a strong pseudoprime to the base a.   □

The following theorem gives many properties of the two kinds of generalized Fermat numbers $L_{p,m}(a)$ and $G_{p,m}(a)$ that are possessed by the Fermat numbers.

Theorem 13.4. Let $a \geq 2$ and let $k(a)$ be the square-free kernel of a, that is, a divided by the largest square factor of a. Let p be a prime. Then

(i) $a^{p^{m+1}} - 1 = (a - 1)L_{p,0}(a)L_{p,1}(a) \cdots L_{p,m}(a)$.

(ii) If $0 \leq m < n$, then $\gcd(G_{p,m}(a), G_{p,n}(a)) = 1$.

(iii) If p and q are distinct primes, then $\gcd(G_{p,m}(a), G_{q,n}(a)) = 1$ for all nonnegative integers m and n, not necessarily distinct.

(iv) All positive odd divisors of $G_{p,m}(a)$ are of the form $kp^{m+1} + 1$.

(v) Suppose that p is odd, or $p = 2$ and a is a square, or $p = 2$, $k(a) = 2$, and $m \geq 2$. Then every positive divisor of $G_{p,m}(a)$ is of the form $2kp^{m+1} + 1$.

(vi) If an odd prime number q divides $G_{p,m}(a)$, then

$$q^2 \mid G_{p,m}(a) \quad \Longleftrightarrow \quad a^{q-1} \equiv 1 \pmod{q^2}.$$

(vii) Suppose that it is not the case that $p = 2$, $m = 0$, and $a \equiv 3 \pmod 4$. If $G_{p,m}(a)$ is composite, then $G_{p,m}(a)$ is both a strong pseudoprime to the base a and a superpseudoprime to the base a. In particular, if $G_{p,m}(a)$ is composite, then $G_{p,m}(a)$ is a pseudoprime to the base a and also an Euler pseudoprime to the base a.

(viii) The number $L_{p,m}(2)$ cannot be a perfect power if $p = 2$, $m \geq 0$ (i.e., $L_{2,m}(2) = F_m$), or $m = 0$, $p \geq 2$ (i.e., $L_{p,0}(2) = M_p$), or $m \geq 0$ and $p = 3$. Moreover, $L_{p,m}(a)$ is never a perfect power for any $a \leq 10^4$, any p, and any $m \geq 1$ except for the cases $(a, p, m) = (3, 5, 0)$ and $(a, p, m) = (18, 3, 0)$.

P r o o f . (i) By Remark 13.2, $\Phi_{p^{i+1}}(a) = L_{p,i}(a)$ for i a nonnegative integer. Using the well-known property of the cyclotomic polynomials, namely, that

$$x^n - 1 = \prod_{d \mid n} \Phi_d(x)$$

(the Möbius inverse of (13.1)) for the case $n = p^{m+1}$ and $x = a$, we see that this relation is exactly (i).

(ii)–(iii) These follow from (13.6).

(iv) It suffices to show that $q \equiv 1 \pmod{p^{m+1}}$ for any odd prime divisor q of $G_{p,m}(a)$. Let q be an odd prime divisor of $G_{p,m}(a)$. By (13.4) and (13.5), q is a primitive prime divisor of $a^{p^{m+1}} - 1$. Thus, $\mathrm{ord}_q a = p^{m+1}$, which implies by Fermat's little theorem that $p^{m+1} \mid q - 1$. Hence, $q \equiv 1 \pmod{p^{m+1}}$.

(v) By (13.7), $G_{p,m}(a)$ is odd. Let q be a prime divisor of $G_{p,m}(a)$. It suffices to demonstrate that $q \equiv 1 \pmod{2p^{m+1}}$. First assume that p is odd. By part (iv), $q \equiv 1 \pmod{p^{m+1}}$. Since both q and p are odd, it follows that $q \equiv 1 \pmod{2p^{m+1}}$.

Now suppose that $p = 2$ and a is a square. Then by Theorem 2.26 (Euler's criterion), $a^{(q-1)/2} \equiv 1 \pmod{q}$. By the same argument as in the proof of part (iv), $\mathrm{ord}_q a = p^{m+1}$. Hence, $p^{m+1} \mid (q-1)/2$, which implies that $q \equiv 1 \pmod{2p^{m+1}}$.

Finally, assume that $p = 2$, $k(a) = 2$, and $m \geq 2$. By part (iv), $q \equiv 1 \pmod{2^{m+1}}$, which implies that $q \equiv 1 \pmod 8$. Then by Theorem 2.27 (iii) and (v), $\left(\frac{a}{q}\right) = 1$, where $\left(\frac{a}{q}\right)$ is the Legendre symbol. By using Euler's criterion and the argument given above for the case in which $p = 2$ and a is a square, we see that $q \equiv 1 \pmod{2p^{m+1}}$.

(vi) If $q \mid G_{p,m}(a)$, then by (13.4) and (13.5), q is a primitive prime divisor of $a^{p^{m+1}} - 1$, and hence $\mathrm{ord}_q a = p^{m+1}$. First suppose that $q^2 \mid G_{p,m}(a)$. Then by (13.4) and (13.5), q is a primitive prime divisor of $a^{p^{m+1}} - 1$ with multiplicity at least two. Thus,

$$\mathrm{ord}_{q^2} a = \mathrm{ord}_q a = p^{m+1}.$$

By Fermat's little theorem, $p^{m+1} \mid q - 1$. Therefore,

$$a^{q-1} \equiv 1 \pmod{q^2}.$$

Conversely, assume that $a^{q-1} \equiv 1 \pmod{q^2}$. Thus, $\mathrm{ord}_{q^2} a \mid q - 1$. It is well known and easily proven by use of the binomial theorem (see [LeVeque]) that either $\mathrm{ord}_{q^2} a = \mathrm{ord}_q a$ or $\mathrm{ord}_{q^2} = q\,\mathrm{ord}_q a$. Since $q \nmid \mathrm{ord}_{q^2} a$, we see that

$$\mathrm{ord}_{q^2} a = \mathrm{ord}_q a = p^{m+1},$$

and hence q is a primitive prime divisor of $a^{p^{m+1}} - 1$ with multiplicity at least two. Since $P_a(p^{m+1}) \mid G_{p,m}(a)$ by (13.4) and (13.5), it follows that $q^2 \mid G_{p,m}(a)$.

(vii) Suppose that $G_{p,m}(a)$ is composite. By (13.4), we also have

(13.11) $$G_{p,m}(a) = P_a(p^{m+1}).$$

It now follows from Theorem 12.25 that $G_{p,m}(a)$ is a superpseudoprime to the base a.

We further show that $G_{p,m}(a)$ is a strong pseudoprime to the base a. It will then follow from Remark 12.13 and Theorem 12.14 that $G_{p,m}(a)$ is also both a pseudoprime to the base a and an Euler pseudoprime to the base a. We see by (13.11) that if q is a prime such that $q^b \| G_{p,m}(a)$ for some integer $b \geq 1$, then $\mathrm{ord}_{q^b} a = p^{m+1}$. It now follows from Lemma 13.3 that $G_{p,m}(a)$ is a strong pseudoprime to the base a.

(viii) Let $p = 2$. It was proved in Theorem 9.1 that $L_{2,m}(2) = F_m$ cannot be a perfect power for $m \geq 0$. We now give a simple proof of the known result that $L_{p,0}(2) = M_p$ cannot be a perfect power for any prime p. We prove a stronger result, namely that $2^n - 1$ cannot be a perfect power for any integer $n \geq 2$. Note that $2^n - 1$ cannot be a perfect square, since $2^n - 1 \equiv 3 \pmod 4$. Thus, we can assume that $2^n - 1 = x^r$ for some odd integer r. Then

$$(13.12) \qquad 2^n = x^r + 1 = (x + 1)(x^{r-1} - x^{r-2} + \cdots + x^2 - x + 1),$$

where $x^{r-1} - x^{r-2} + \cdots + x^2 - x + 1 = (x^r + 1)/(x + 1)$ is an odd integer. Since the left-hand side of (13.12) is a power of 2, this forces $(x^r + 1)/(x + 1)$ to be equal to 1. Hence, $r = 1$, and the assertion is proved.

The case for which $p = 3$ and $m \geq 0$ was proved in [Ligh, Jones, p. 15].

Finally, in [Bugeaud, Mignotte] it is shown that the equation

$$\frac{x^n - 1}{x - 1} = y^q$$

has no solutions when $n \geq 3$, $q \geq 2$, and $2 \leq z \leq 10^4$, where $x = z^t$ for some $t \geq 1$, aside from

$$\frac{3^5 - 1}{3 - 1} = 11^2, \quad \frac{7^4 - 1}{7 - 1} = 20^2, \quad \text{and} \quad \frac{18^3 - 1}{18 - 1} = 7^3. \qquad \square$$

Remark 13.5. Part (iv) of Theorem 13.4 was proved in [Shanks, 1978, pp. 105–107]. Part (ii) was proved in [Ligh, Jones] for the case in which $a = 2$. It was further proved in [Ligh, Jones] that $G_{p,m}(2)$ is a pseudoprime to the base 2. In Theorem 1 in [Pomerance, Selfridge, Wagstaff] it was proved that $G_{p,m}(a)$ is a strong pseudoprime to the base a if it is not the case that $p = 2$, $m = 0$, and $a \equiv 3 \pmod 4$. It was proved in [Ligh, Neal] that $2^n - 1$ is never a composite prime power. All parts of Theorem 13.4 were stated in [Keller, 1992, pp. 24–26] for the case in which $a = 2$ except for the result about superpseudoprimes in part (vii).

We now review our earlier results about Fermat numbers, which are special cases of various parts of Theorem 13.4. Part (i) generalizes formula (3.3), parts (ii) and (iii) generalize Goldbach's Theorem 4.1, part (iv) generalizes Euler's Theorem 4.14, part (v) generalizes Lucas's Theorem 6.1, part (vi) generalizes Theorem 6.22, part (vii) generalizes Theorems 4.10, 12.16, and 12.26 (i), and finally, part (viii) generalizes Theorem 9.1.

Of particular interest are the generalized Fermat numbers

$$(13.13) \qquad\qquad F_{a,m} = L_{2,m}(a) = a^{2^m} + 1.$$

We observe that these numbers are the most natural generalizations of the Fermat numbers $F_m = F_{2,m}$. The results we will give regarding $F_{a,m}$ are primarily concerned with the case in which a is even. Note that if a is even, then

$$(13.14) \qquad\qquad F_{a,m} = G_{2,m}(a).$$

The next theorem, which appears in [Szymiczek, 1966b], generalizes Cipolla's Theorem 12.1 and generates infinitely many pseudoprimes to the even base a with each pseudoprime having an arbitrary prescribed number of distinct prime factors.

Theorem 13.6 (Szymiczek). *Suppose that a is a positive even integer such that $2^\alpha \| a$ and $c > d > \cdots > s \geq 0$. Then $N = F_{a,c}F_{a,d} \cdots F_{a,s}$ is a pseudoprime to the base a if and only if $\alpha 2^s > c$.*

P r o o f . Let $a = 2^\alpha M$, where M is an odd integer. Since $\gcd(F_{a,m}, F_{a,n}) = 1$ if $m \neq n$, by (13.14) and Theorem 13.4 (ii), it follows that

$$(13.15) \qquad \mathrm{ord}_N a = \mathrm{lcm}(\mathrm{ord}_{F_{a,c}} a, \mathrm{ord}_{F_{a,d}} a, \ldots, \mathrm{ord}_{F_{a,s}} a).$$

Using completely similar arguments to those given in the proof of Theorem 4.12 and in Remark 4.13, we see that

$$\mathrm{ord}_{F_{a,m}} a = 2^{m+1} \quad \text{for } m \geq 0.$$

Therefore,

$$(13.16) \qquad \mathrm{ord}_N a = \mathrm{lcm}\left(2^{c+1}, 2^{d+1}, \ldots, 2^{s+1}\right) = 2^{c+1}.$$

Thus, $a^{N-1} \equiv 1 \pmod{N}$ if and only if $N - 1$ is divisible by 2^{c+1}. We note that

$$(13.17) \qquad N - 1 = F_c F_d \cdots F_s - 1 = a^{2^s} K = (2^\alpha M)^{2^s} K = 2^{\alpha 2^s} M^{2^s} K,$$

where K is an odd integer, and hence $M^{2^s} K$ is an odd integer. Consequently, N is a pseudoprime to the base a if and only if $\alpha 2^s > c$. $\qquad \square$

An important property that the generalized Fermat numbers $L_{p,m}(a)$ (a even), $G_{p,m}(a)$ ($a \geq 2$ a positive integer), $F_{a,m}$ (a even), share with F_m is that for a fixed a and varying m, some of the values are usually prime, but prime values seem to occur only rarely. We first consider the numbers $L_{p,m}(a)$ and $G_{p,m}(a)$. The numbers $F_{a,m} = L_{2,m}(a)$ will be dealt with separately. Accordingly, we assume that $p \geq 3$ for the numbers $L_{p,m}(a)$ and $G_{p,m}(a)$. The 38 known primes of the form $L_{p,0}(2)$ are the Mersenne primes; they are listed in Appendix B.

In the beginning of the twentieth century, a great project dealing with factorization of the *Cunningham numbers,* i.e., the numbers of the form $a^n \pm 1$ where a is "small" and n "large," was started. The tables for $a \leq 12$ are contained in [Brillhart, Lehmer, Selfridge, Tuckerman, Wagstaff]. The greatest contribution of the Cunningham project is that it stimulated researchers to devise new highly effective methods for primality testing, finding new methods for factorization, etc.

Particularly, in [Brillhart, Lehmer, Selfridge, Tuckerman, Wagstaff], we can find tables of prime factors of $L_{p,m}(a)$ and $G_{p,m}(a)$ for a, p, m such that $2 \leq a \leq 12$, $a \notin \{4, 8, 9\}$, $m \geq 0$, and $p^{m+1} < 1200$ for $a = 2$, $p^{m+1} < 350$ for $a = 3$, $p^{m+1} < 260$ for $a = 5$, $p^{m+1} < 210$ for $a = 6, 7, 10$, and $p^{m+1} < 150$ for $a = 11, 12$. In that range for $m \geq 1$, only the following primes have been found: $L_{3,1}(2) = 73$, $L_{3,2}(2) = 262657$, $L_{7,1}(2) = 4\,432676\,798593$, $L_{3,1}(3) = 757$, $L_{7,1}(5) = 227376\,585863\,531112\,677002\,031251$, $L_{3,1}(11) = 1\,772893$, $L_{3,2}(11) = 5\,559917\,315850\,179173$, $G_{3,1}(10) = 333667$.

In addition, for $p \leq 257$, prime factors of $L_{p,m}(2)$ are listed in Tables 17, 18, and 19 of [Keller, 1992], where $m \leq 300$ for $p = 3$, $m \leq 20$ for $3 < p \leq 43$, and $m \leq 5$ for $43 < p \leq 257$.

The generalized Fermat numbers for which the search for prime factors has been most widely conducted are the numbers $F_{a,m}$ (a even) and $\frac{1}{2}F_{a,m} = G_{2,m}(a)$ (a odd), where $a \geq 3$ and $m \geq 1$. For an interesting result concerning subideals of the Stickelberger ideal (which appears in algebraic number theory) with Mersenne numbers and the generalized Fermat numbers $F_{a,m}$, a even, see [Skula, p. 150]. For additional connections between Fermat numbers and algebraic number theory see [Buchmann, Düllmann], [Watabe], and [Wiedemann]. Further results on generalized Fermat numbers are given in [Gulliver].

In [Morimoto], all primes of the form $F_{a,m}$ are tabulated for a even, $3 \leq a \leq 1000$, and $2 \leq m \leq 8$. In addition, all primes and probable primes of the form $\frac{1}{2}F_{a,m}$ are displayed for a odd and a and m in the same range as above. Dubner lists for each n, $2 \leq n \leq 11$, the smallest integer a such that $F_{a,m}$ is prime (see [Dubner]). Moreover, some very large primes $F_{a,m}$ are found. The two largest are $F_{100964,10}$, with 5125 digits, and $F_{101682,10}$, with 5128 digits. In [Dubner, Keller] prime factors of $F_{a,m}$ are found for $a = 6, 10, 12$ and m as large as 44684. Further, in [Riesel, Björn] and [Björn, Riesel], prime factors of $a^{2^m} + b^{2^m}$ are given for $1 \leq a < b \leq 12$ and $m \leq 1000$. For further results on factors of these generalized Fermat numbers, see [Jiménez Calvo] and [Varshney].

As of the beginning of 2001, well over 165000 primes of the form $F_{a,m}$ are known (see [www16] and [www17]). The largest known primes of the form $F_{a,m}$ as of the beginning of 2001 are $F_{167176,15}$, with 171153 digits, found by Gallot in 2000 and $F_{48594,16}$, with 307140 digits, discovered by Scott and Gallot in 2000. The number $F_{48594,16}$ is the sixth-largest known prime as of the beginning of 2001 and is the largest known prime that is not a Mersenne prime. The integer $F_{167176,15}$ is the ninth-largest prime known as of the beginning of 2001 and is the second largest known prime that is not a Mersenne prime.

It was pointed out in [Riesel, 1969] that a given prime can divide several different generalized Fermat numbers $F_{a,m}$ for different bases a. For example, 641 divides $F_{2,5}$, $F_{10,4}$, $F_{20,5}$, and $F_{40,3}$. In [Dubner, Keller] an explanation is provided for this phenomenon.

Let P be an odd prime of the form $k2^n + 1$, where k is odd. Let a_{tot} denote the total number of bases $a \in \{2, \ldots, P-1\}$ such that $P \mid F_{a,m}$ for some integer m. It is shown in [Dubner, Keller] that

$$(13.18) \qquad \frac{a_{tot}}{P} = \frac{1 - 2^{-n}}{k + 2^{-n}}.$$

The above ratio is very close to $1/k$ for reasonably large n. Based on the equality in (13.18), Dubner and Keller present a heuristic argument that for a given odd prime $P = k2^n + 1$, where k is odd, the probability of P dividing some generalized Fermat number $F_{a,m}$ for a fixed base a and varying m should be $1/k$.

The next theorem appears in [Grytczuk, Luca, Wójtowicz] and generalizes Theorem 7.8.

Theorem 13.7. *Assume that $a \geq 2$ is an even integer and that $m \geq a^{18}$. Then, $P_a(F_{a,m}) > m2^{m-4}$, where $P(F_{a,m})$ denotes the largest prime factor of $F_{a,m}$.*

We now introduce another generalization of the Fermat numbers, the *Ferentinou-Nicolacopoulou numbers* defined by

$$F_{a,a,m} = a^{a^m} + 1,$$

where $a \geq 2$ and $m \geq 1$ (see [Ferentinou-Nicolacopoulou]). Note that $F_m = F_{2,2,m}$. The next theorem, which is proved in [Ribenboim, 1979a], shows that the numbers $F_{a,a,m}$ share many of the same properties as the generalized Fermat numbers $L_{p,m}(a)$ and $G_{p,m}(a)$.

Theorem 13.8. *Let $m \geq 0$ and $a = 2^\ell t$, where $a \geq 2$, $\ell \geq 0$, and t is odd. Let $b_m = a^{a^m}$.*

(i) *Suppose a is even and $k \geq 1$. Then*

$$F_{a,a,m} \mid F_{a,a,m+k} - 2$$

and

$$F_{a,a,m+k} - 2 = F_{a,a,m}\left(b_m^{a^k-1} - b_m^{a^k-2} + \cdots + b_m - 1\right).$$

In particular,

$$\gcd(F_{a,a,m}, F_{a,a,m+k}) = 1.$$

(ii) *Suppose a is odd. Then*

$$F_{a,a,m} \mid F_{a,a,m+1}$$

and

$$F_{a,a,m+1} = F_{a,a,m}\left(b_m^{a-1} - b_m^{a-2} + \cdots + b_m^2 - b_m + 1\right).$$

Moreover, $F_{a,a,m}$ is the product of $m + 1$ mutually coprime factors.

(iii) *Let p be an odd prime dividing $F_{a,a,m}$.*

(1) *If $\left(\frac{a}{p}\right) = 1$, then $p = 1 + k2^{m\ell+2}$ for some $k \geq 1$.*

(2) *If $\left(\frac{a}{p}\right) = -1$, then $p = 1 + k2^{m\ell+1}$ for some $k \geq 1$.*

(3) $a^{t^n(p-1)} \equiv 1 \pmod{F_{a,a,m}}$.

(iv) *If p is an odd prime dividing $F_{a,a,m}$, then*

$$p^2 \mid F_{a,a,m} \quad \Longleftrightarrow \quad a^{p-1} \equiv 1 \pmod{p^2}.$$

(v) *If p is an odd prime and $p^2 \mid F_{a,a,m}$, then $p = k2^r + 1$, where k is odd, $r \geq m\ell + 1$, and*

$$\frac{a^{p-1} - 1}{p} \equiv 1 - r\frac{2^{p-1} - 1}{p} \pmod{p}.$$

(vi) *$F_{a,a,m}$ is not a perfect power.*

Remark 13.9. The last assertion in part (ii) of Theorem 13.8 was stated by J. Ferentinou-Nicolacopoulou. Ribenboim proved part (vi) of Theorem 13.8 by using results about solutions to Catalan's equation (compare with Remark 9.3)

$$x^p - y^q = 1,$$

where x, y, p, and q are positive integers greater than 1 (see [Ribenboim, 1994] for more about Catalan's equation).

We now consider three sequences of numbers related to the Fermat numbers. The first such sequence is defined by

$$G_m = F_m 2^{F_m-1} - 1, \quad m \geq 0.$$

Note that $G_0 = 11$, $G_1 = 79$, and $G_2 = 1114111$ are all prime numbers. It was proved in [Williams, Zarnke] that the 80-digit number

$$G_3 = 2975856693\ 3990262223\ 8577431472\ 3279231829$$
$$0386059069\ 6249581405\ 9909003367\ 4317463551$$

is also a prime. Williams and Zarnke speculated that G_4, which has 19734 digits, and possibly each G_m might also be prime. In [Keller, 1992, pp. 29–35] this conjecture was disproved decisively. It was shown that G_4 is divisible by 16267, G_m is composite for $4 \leq m \leq 9$, and in fact, G_m is composite for infinitely many values of m. As opposed to the Fermat numbers, for which there are no known numbers that are not square-free, Keller showed that infinitely many G_m's are divisible by the square or even a higher power of a prime. For example,

$$23^2 \mid G_m \quad \text{for all } m \equiv 5 \pmod{220},$$
$$47^2 \mid G_m \quad \text{for all } m \equiv 11,\ 2428 \pmod{2530},$$
$$59^2 \mid G_m \quad \text{for all } m \equiv 1135 \pmod{2436},$$
$$149^2 \mid G_m \quad \text{for all } m \equiv 212 \pmod{1332},$$

and

$$23^3 \mid G_m \quad \text{for all } m \equiv 2205 \pmod{5060},$$
$$47^3 \mid G_m \quad \text{for all } m \equiv 42908,\ 116391 \pmod{118910},$$
$$59^3 \mid G_m \quad \text{for all } m \equiv 20623 \pmod{143724},$$
$$149^3 \mid G_m \quad \text{for all } m \equiv 48164 \pmod{198468}.$$

Another sequence of numbers related to the Fermat numbers is the sequence $n^n + 1$, where $n > 1$ is an integer. These numbers have been studied in [Sierpiński, 1958]. It is easy to see that $n^n + 1$ is composite if $n = pK$, where p is an odd prime and $K > 1$, since then $n^K + 1 \mid n^n + 1$. Moreover, if $n = 2^m$, where $m = qt$ for some odd prime q and positive integer t, then

$$2^{t2^{qt}} + 1 \mid 2^{qt2^{qt}} + 1 = n^n + 1,$$

and n is composite again. Hence, $n^n + 1$ can be prime only if

$$n = 2^{2^m}$$

for $m \geq 0$. Thus, $n^n + 1$ can be a prime only if

$$(13.19) \qquad\qquad n^n + 1 = F_{2^m + m}.$$

Since the only Fermat primes known are F_m for $m \in \{0, 1, 2, 3, 4\}$, it follows that the only known primes of the form $n^n + 1$ are $F_1 = 2^2 + 1 = 5$ and $F_3 = 4^4 + 1 = 257$. If there exist infinitely many primes of the form $n^n + 1$, then there are infinitely many Fermat primes. However, there is no convincing evidence for such a conjecture.

The final sequence of numbers related to the Fermat numbers that we will consider are the *Cullen numbers,* defined by

$$C_n = n2^n + 1.$$

In 1905, the Reverend J. Cullen noted that C_n is composite for $n \in \{2, 3, \ldots, 100\}$ with one possible exception, for $n = 53$ (see [Cullen]). A year later, Cunningham stated that 5591 is a prime factor of C_{53} and remarked that C_n is composite for all values of $n \in \{2, 3, \ldots, 200\}$ with a possible exception at $n = 141$ (see [Cunningham]). Slightly over 50 years later, in 1957, Robinson showed that C_{141} is in fact prime (see [Robinson, 1958]). As of the beginning of 2001, only thirteen Cullen primes C_n are known: those with the values

$$n = 1, \ 141, \ 4713, \ 5795, \ 6611, \ 18496, \ 32292, \ 32469,$$
$$59656, \ 90825, \ 262419, \ 361275, \ 4811899$$

(see [www14] and [www15]). The Cullen numbers C_n resemble Fermat numbers in that $C_n - 1$ is divisible by a high power of 2 as n becomes large, and prime Cullen numbers seem to occur very infrequently.

This apparent rarity of Cullen primes is explained in [Keller, 1995]. Keller showed that every prime $p > 3$ leads to four consecutive composite numbers C_n. By Fermat's little theorem, it is easily seen that p divides both C_{p-1} and C_{p-2}. In particular,

$$(p - 1)2^{p-1} + 1 \equiv (-1)1 + 1 \equiv 0 \pmod{p},$$
$$(p - 2)2^{p-2} + 1 \equiv (-2)2^{-1}2^{p-1} + 1 \equiv (-1)1 + 1 \equiv 0 \pmod{p}.$$

Moreover, if p is of the form $6k + 1$, then 3 divides C_p and C_{p+1}, whereas if p is of the form $6k - 1$, then 3 divides C_{p-3} and C_{p-4}. Keller also remarks that if p and $p + 2$ are twin primes, then eight consecutive composite numbers C_n occur.

In [Hooley, p. 119] it was shown that almost all Cullen numbers are composite. Specifically, Hooley proved that

$$\lim_{x \to \infty} \frac{C(x)}{x} = 0,$$

where $C(x)$ denotes the number of positive integers $n \leq x$ for which C_n is prime. Nevertheless, it has been conjectured (see [www14]) that there are infinitely many Cullen primes. It is not yet known whether n and C_n can both be prime (see [www14]). By what was stated earlier, this can occur only if n is a prime of the form $6k - 1$, since if n is a prime of the form $6k + 1$, then $3 \mid C_n$.

14. Open Problems

Several conjectures concerning the Fermat numbers have been resolved. We already know that Fermat's claim that all F_m are primes was disproved by L. Euler.

According to [Dickson, p. 376], an anonymous writer in 1828 conjectured that all numbers of the form

$$2 + 1, \ 2^2 + 1, \ 2^{2^2} + 1, \ 2^{2^{2^2}} + 1, \ldots$$

are primes (cf. also [Sierpiński, 1950, p. 21]). However, in 1953 J. L. Selfridge used a computer to show that $3150 \cdot 2^{18} + 1 \mid F_{16}$; i.e., this old hypothesis was disproved.

In [Riesel, 1994] another open problem was stated: *What is the smallest quadricomposite Fermat number, i.e., a number that has exactly four prime factors?* The answer was shown to be F_{10} by Brent in 1995 [Brent, 1999] (compare with (14.1)).

Let us summarize now several of the most important open problems concerning the Fermat numbers. (Excellent sources of open problems in number theory are [Guy, 1994] and [Shanks].)

(1) *Are there infinitely many prime numbers F_m?*

There is a heuristic probabilistic argument (see [Hardy, Wright, p. 15]) that the number of Fermat primes is finite. According to the prime number theorem (see (1.5)), the number of primes that are less than n is of order $n/\log n$. Thus, the probability that a number n is prime is at most

$$\frac{A}{\log n},$$

where A is a fixed constant, and therefore the total expectation of the number of Fermat primes is at most

$$A \sum_{m=0}^{\infty} \frac{1}{\log\left(2^{2^m} + 1\right)} < A \sum_{m=0}^{\infty} 2^{-m} = 2A.$$

Another heuristic probabilistic argument concerning the expected finitude of Fermat primes, due to H. W. Lenstra, can be found in [Crandall, Pomerance, p. 28].

(2) *Are there infinitely many composite numbers F_m?*

(3) *Are the primes in (1.2) all the Fermat primes?*

(4) *For which regular polygons does there exist a Euclidean construction?*

Tables A.1 and A.3 in Appendix A contain all prime factors of Fermat numbers F_m for $5 \leq m \leq 30$ that were known as of the beginning of 2001. Denoting by Ω_m the number of prime factors of F_m (including multiplicity), we get

(14.1)
$$\Omega_0 = \ldots = \Omega_4 = 1 < \Omega_5 = \ldots = \Omega_8 = 2 < \Omega_9 = 3 < \Omega_{10} = 4 < \Omega_{11} = 5 < 6 < \Omega_{12}.$$

(5) *Is the sequence $\{\Omega_m\}_{m=0}^{\infty}$ monotone?* (See [Křížek].)

(6) *Does there exist a Fermat number with an arbitrary prescribed number of prime factors?* (See [Riesel, 1994].)

(7) *What is the complete factorization of F_{12}, F_{13}, etc., into primes?*

(8) *What is the smallest prime factor of the composite numbers F_{14}, F_{20}, F_{22}, and F_{24}?*

For further open problems we consider factors of the composite Fermat numbers F_m of the form $k2^n + 1$, where k is odd. By Lucas's Theorem 6.1 and by Theorem 6.7, $n \geq m + 2$ and $k \geq 3$.

(9) *Given any odd integer $k \geq 3$ that is not a Sierpiński number, does there exist F_m having a prime factor of the form $k2^n + 1$?*

Recall from Chapter 7 that a Sierpiński number is an odd integer k such that $k2^n + 1$ is not a prime for any value of n. Open problem (9) asks whether for every non-Sierpiński odd integer k there exists a prime divisor of a Fermat number of the form $k2^n + 1$, even though the density of the set of divisors of Fermat numbers is zero in the set of odd integers (see Theorem 7.11). In [Erdős, Odlyzko] it is proved that the density of non-Sierpiński numbers in the set of odd integers is positive. Note that by Theorem 7.12, the set of prime divisors $k2^n + 1$ of Fermat numbers has density zero in the set of all primes, but nothing is known about the density of the k's.

(10) *Does there exist an odd integer $k \geq 3$ such that infinitely many Fermat numbers have a prime factor of the form $k2^n + 1$, where k is fixed and n varies?*

(11) *Given any odd integer $k \geq 3$ that is not a Sierpiński number, do there exist infinitely many Fermat numbers having a prime factor of the form $k2^n + 1$, where n varies?*

(12) *For any integer N, no matter how large, does there exist a Fermat number with at least N prime factors whose orders are all the same?*

By Theorems 6.7 and 6.16, any composite Fermat number has at least two prime factors with the same order n.

(13) *Do there exist infinitely many composite F_m such that all their prime factors have orders $n \geq m + 3$?* (See [Szymiczek, 1966a].)

By Theorem 12.26 (iii), if F_m satisfies the condition given above, then the number $F_m F_{m+1} F_{m+2}$ is a superpseudoprime. We note that by Table A.4 in Appendix A, each of the two prime factors of F_8 has order $n = 11$. Thus, $F_8 F_9 F_{10}$ is a superpseudoprime.

(14) *For any positive integer N, no matter how large, does there exist a composite Fermat number F_m having a prime factor of order n such that $n - m \geq N$?*

For example, F_{635} has a prime factor $4258979 \cdot 2^{645} + 1$, F_{2089} has a prime factor $431 \cdot 2^{2099} + 1$, and F_{18749} has a prime factor $11 \cdot 2^{18759} + 1$, and $n - m = 10$ in these cases (see [www1]). As of the beginning of 2001, the largest known value of $n - m$ is equal to 10.

(15) *Are there infinitely many primes of the form $n^n + 1$?*

If the answer were to be positive, then according to (13.19), there would exist infinitely many Fermat primes.

(16) *Does there exist a Fermat number F_m that is divisible by the square of a prime number?*

There exists a conjecture (see, e.g., [Gostin, McLaughlin]) that all Fermat numbers are square-free. This hypothesis has been verified for almost all known divisors of F_m, and up to now it has not been contradicted (see the end of Chapter 6 for the consequences if Fermat numbers that are not square-free do exist).

(17) *Does there exist for any $m \geq 5$ a positive integer h such that $5h2^{m+2} + 1 \mid F_m$?* (See [Křížek, Chleboun, 1997].)

Let us examine the preceding problem in more detail. We will establish a sufficient condition that determines when a composite Fermat number $F_m = 2^{2^m} + 1$ is divisible by $5h2^{m+2} + 1$.

Remark 14.1. In Theorem 6.7 we proved that any factorization of a composite Fermat number into two nontrivial factors can be expressed in the form

$$(14.2) \qquad F_m = (k2^n + 1)(\ell 2^n + 1)$$

for some odd k and ℓ, $k \geq 3$, $\ell \geq 3$, $n \geq m + 2$, $\max(k, \ell) \geq F_{m-2}$, and

$$(14.3) \qquad \text{either} \quad 3 \mid k \quad \text{or} \quad 3 \mid \ell, \quad \text{but not both.}$$

Lucas's Theorem 6.1, which asserts that every prime divisor of F_m is of the form $k2^{m+2} + 1$, can thus be extended as follows: *For any composite Fermat number there exists at least one h such that $3h2^{m+2} + 1 \mid F_m$.*

Remark 14.2. By Theorem 6.7, k and ℓ from (14.2) are coprime. Hence, if $5 \mid k$, then $5 \nmid \ell$ and vice versa. By (3.5),

$$F_m \equiv 2 \pmod{5}.$$

From this and (14.2) we have

$$k\ell 2^{2n} + (k + \ell)2^n - 1 \equiv 0 \pmod 5.$$

Hence, if $5 \mid k$, then

$$\ell 2^n - 1 \equiv 0 \pmod 5,$$

which implies the conditions

$$\ell = 5r + 1,\ n = 4s \qquad \text{or} \qquad \ell = 5r + 2,\ n = 4s + 3$$
$$\text{or} \qquad \ell = 5r + 3,\ n = 4s + 1 \qquad \text{or} \qquad \ell = 5r + 4,\ n = 4s + 2.$$

Conversely, these conditions evidently imply $5 \mid k$.

Remark 14.3. If $(k2^n + 1)^2 \mid F_m$ (i.e., F_m is not square-free), then the associated cofactor is not divisible by $5h2^n + 1$ for any h (see Theorem 6.16 and use the fact that F_m is not a square, either by Theorem 9.1 or by Remark 6.11).

Theorem 14.4. Let $F_m = \prod_{i=1}^4 f_i$ for some m, where the factors f_i (not necessarily prime) are greater than one. Then there exists a natural number h such that $5h2^{m+2} + 1 \mid F_m$.

P r o o f . By Lucas's Theorem 6.1, any prime factor of the Fermat number F_m is of the form $k2^{m+2} + 1$ for some k. By the principle of induction, we easily find that any factor f_i has also the form $k_i 2^{m+2} + 1$ for $i = 1, 2, 3, 4$.

Suppose that there does not exist any h such that $5h2^{m+2} + 1 \mid F_m$. Let $d(f)$ denote the last digit of a natural number f, i.e., $f \equiv d(f) \pmod{10}$. We see that $d(f_i)$ is always an odd number. Since $d(5h2^{m+2} + 1) = 1$, we find that $d(f_i) \neq 1$ for any $i = 1, 2, 3, 4$. By Goldbach's Theorem 4.1, $d(f_i) \neq 5$, because $F_1 = 5$. The last digit of any f_i thus belongs to the set $\{3, 7, 9\}$.

Further, we may exclude the following possibilities:
(a) $d(f_i) = 3$ and $d(f_j) = 7$ (otherwise, we would have $f_i f_j = 5h2^{m+2} + 1$),
(b) $d(f_i) = d(f_j) = 9$ for $i \neq j$,
(c) $d(f_i) = d(f_j) = 3$ and $d(f_\ell) = 9$ for $i \neq j \neq \ell \neq i$,
(d) $d(f_i) = d(f_j) = 7$ and $d(f_\ell) = 9$ for $i \neq j \neq \ell \neq i$.

Finally, we examine the following two remaining quadruples of $d(f_i)$:

$$(14.4) \qquad\qquad (3, 3, 3, 3), \quad (7, 7, 7, 7).$$

From (14.4) we observe that the last digit of the corresponding product $\prod_{i=1}^4 f_i$ is 1 in both cases. However, this contradicts the fact that $d(F_m) = 7$ for every $m \geq 2$ (see (3.10)). $\square$

Remark 14.5. By the proof of Theorem 14.4, if F_m does not have a factor of the form $5h2^{m+2} + 1$, then F_m has three or fewer prime factors and $d(q) \neq 1$ for each nontrivial factor of F_m. Since $d(F_m) = 7$ for $m \geq 2$ by Remark 3.6, if F_m has prime factors $p_1, p_2, \ldots, p_s$, not necessarily distinct, then

$$\prod_{i=1}^s d(p_i) \equiv 7 \pmod{10}.$$

Since F_0 and F_1 are both primes, it now follows from the excluded cases (a)–(d) in the proof of Theorem 14.4 that F_m does not have a factor of the form $5h2^{m+2} + 1$ only if one of the following possibilities occurs, where p_1, p_2, p_3 are primes such that $d(p_1) \leq d(p_2) \leq d(p_3)$:

(i) F_m is prime,

(ii) $F_m = p_1 p_2$, where $d(p_1) = 3$ and $d(p_2) = 9$,

(iii) $F_m = p_1 p_2 p_3$, where $d(p_1) = d(p_2) = d(p_3) = 3$.

Remark 14.6. Each of the Fermat numbers $F_5, \ldots, F_9$ has fewer than four prime factors. Hence, Theorem 14.4 cannot be used. However, by Appendix A,

$$p_3 \,|F_5 \quad \text{for } p_3 = 641 = 5 \cdot 2^7 + 1,$$

$$p_{14} \,|F_6 \quad \text{for } p_{14} = 67280421310721 = 262814145745 \cdot 2^8 + 1,$$

$$p_{22} \,|F_7 \quad \text{for } p_{22} = 11141971095088142685 \cdot 2^9 + 1,$$

$$p_{62} \,|F_8 \quad \text{for } p_{62} = 456 \ldots 715 \cdot 2^{11} + 1,$$

$$p_7 p_{49} \,|F_9 \quad \text{for } p_7 p_{49} = 2424833 \cdot 364 \ldots 597 = 882 \ldots 485 \cdot 2^{11} + 1,$$

where p_j is a certain prime with j digits. From here, we observe that each of the numbers $F_5, \ldots, F_9$ is divisible by a factor of the form $5h2^{m+2} + 1$.

Let us point out that as late as 1992 the largest known composite Fermat number F_{23471}, discovered by Keller, is divisible by $5 \cdot 2^{23473} + 1$.

Remark 14.7. According to Table A.1 in Appendix A, the Fermat number F_{10} has four prime divisors and F_{11} five prime divisors. By the web site [www1], F_m has at least four factors when $m = 12, 13, 15, 16, 18, 19, 25$ (cf. Tables A.3 and A.4 in Appendix A). Therefore, the assumptions of Theorem 14.4 are satisfied for $F_{10}, F_{11}, F_{12}, F_{13}, F_{15}, F_{16}, F_{18}, F_{19}$, and F_{25}.

Theorem 14.4 and Remarks 14.5, 14.6, and 14.7 lead us to open problem (17).

Remark 14.8. The following question appears in [Jones, Pearce, p. 96]: *Does there exist, for each positive integer n, an odd natural number k such that $k2^n + 1$ is prime?* The answer is yes. The point is that p is a prime of the form $k2^n + 1$ for some odd k and fixed n if and only if p is a prime of the form

$$t2^{n+1} + (2^n + 1).$$

However, the arithmetic progression with constant difference 2^{n+1} and initial term $2^n + 1$ contains infinitely many primes by Dirichlet's Theorem 2.23 on primes in arithmetic progressions.

Remark 14.9. The sum of the reciprocals of the smallest prime divisors of Fermat numbers converges, since by Lucas's Theorem 6.1,

$$\sum_{m=2}^{\infty} \frac{1}{p(F_m)} \leq \sum_{m=2}^{\infty} \frac{1}{2^{m+2} + 1} < \infty,$$

where $p(F_m)$ denotes the smallest prime divisor of F_m. The following question appears in [Golomb, 1955, p. 273]: *Does the sum of the reciprocals of all prime divisors of all Fermat numbers converge?* The answer is yes. Here is a more general result, which is given in [Křížek, Luca, Somer].

Theorem 14.10. *Let $\mathbb{P}$ and $\mathbb{D}$ be the sets appearing in Theorems 7.12 and 7.11, respectively, and let $\lambda > \frac{1}{2}$. Then, both the series*

$$\sum_{p \in \mathbb{P}} \frac{1}{p^\lambda}$$

and

$$\sum_{d \in \mathbb{D}} \frac{1}{d^\lambda}$$

are convergent. Moreover, for any prime number $p \in \mathbb{P}$, let $\alpha(p)$ be the power of p appearing in the prime factorization of the (unique) Fermat number F_m that is a multiple of p. Then, the series

$$\sum_{p \in \mathbb{P}} \frac{\alpha(p)}{p}$$

is convergent.

Remark 14.11. According to Remark 5.10, elite primes are connected with Fermat numbers. It is unknown whether there exist infinitely many elite primes. However, it has been shown in [Křížek, Luca, Somer] that the relative density of the set $\mathbb{E}$ of elite primes in the set of all primes is equal to zero and

$$\sum_{p \in \mathbb{E}} \frac{1}{p}$$

is convergent.

Remark 14.12. There is an interesting relation (see [Křížek, Křížek]) between Fermat primes and stereometry, in particular the Platonic polyhedra. It is well known that there are just five convex regular polyhedra: the tetrahedron, the cube, the octahedron, the dodecahedron, and the icosahedron (see Figure 14.1).

Theorem 14.13. *If $n + 4$ is the number of faces of a regular polyhedron, then $2^n + 1$ is a prime number.*

P r o o f . The number of faces of a regular polyhedron belongs to the set $\{4, 6, 8, 12, 20\}$. Hence, $n \in \{0, 2, 4, 8, 16\}$, and we immediately see that $2^0 + 1$, $2^2 + 1$, $2^4 + 1$, $2^8 + 1$ and $2^{16} + 1$ are primes. $\square$

There is a conjecture based on computer factorizations (see (1.4) and Appendix A) that

$$(14.5) \qquad\qquad F_m \text{ is prime} \implies m \leq 4.$$

Under this hypothesis we can prove a converse theorem.

Theorem 14.14. *If (14.5) holds and $2^n + 1$ is a prime number, then for $n \neq 1$ there exists a regular polyhedron whose number of faces is $n + 4$.*

P r o o f . For $n = 0$ the theorem obviously holds. Assume that $n > 1$ is not a power of 2. Then by (1.3) the number $2^n + 1$ is composite. Now the rest of the proof follows from (14.5) for $n = 2^m$ and $m = 1, 2, 3, 4$.   □

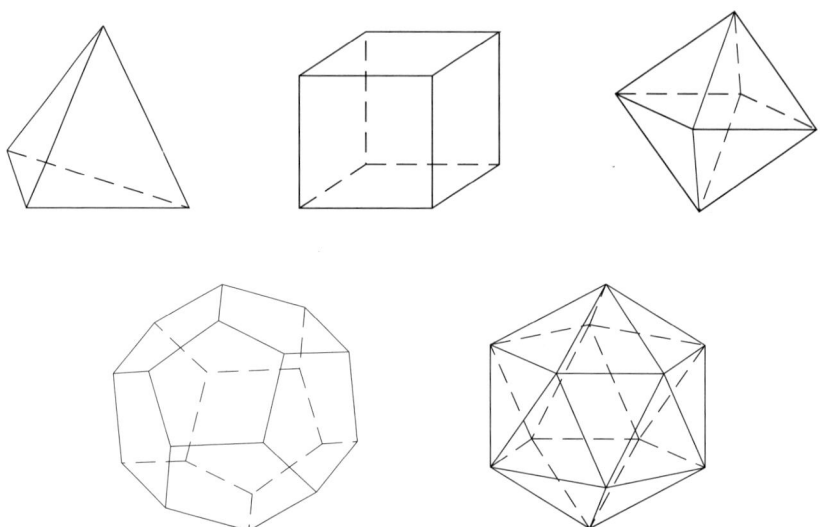

Figure 14.1. Regular polyhedra: tetrahedron, cube, octahedron, dodecahedron, and icosahedron.

Remark 14.15. If we replace in the above Theorems 14.13 and 14.14 the term "faces" by "vertices," the theorems are again true.

Gauss's Theorem 4.3 establishes a remarkable connection between Fermat primes and regular polygons. Theorems 14.13 and 14.14 express another relation of Fermat primes with the Platonic polyhedra. It may happen that Theorems 14.13 and 14.14 hold just due to the strong law of small numbers [Guy, 1988, 1990]. But it is an open problem whether this is indeed the case, or whether there is some deeper principle explaining this property.

15. Fermat Number Transform and Other Applications

The advantage of mathematics as a science is that you can check whether you are right or wrong.

Norman Macrae: *John von Neumann,* 1999, p. 139.

Up to now we have presented several useful applications of the Fermat numbers in number theory, e.g., in proving that there exist infinitely many primes and pseudoprimes, and in establishing the existence of Sierpiński numbers (see Remark 4.2, Theorems 12.1 and 7.4). However, there are more practical applications of F_m, as we shall see in this chapter. In particular, we introduce the use of Fermat numbers in number-theoretic transforms; in binary arithmetic modulo F_m, which leads to fast multiplication of large numbers; in pseudorandom number generators; in hashing schemes; in the chiral Potts model; and in an analysis of the logistic equation by means of divisors of Fermat numbers.

Fermat Number Transform

Transforms having the cyclic convolution property modulo a Fermat number were first defined in [Schönhage, Strassen] and applied to fast multiplication of very large numbers. Later they were used in digital signal processing.

To introduce the main idea of the Fermat number transform, suppose for simplicity that F_m is prime. Let $\alpha \in \{2, 3, \ldots, F_m - 1\}$ be given and let N be chosen such that $N \mid \mathrm{ord}_{F_m} \alpha$. The number N is called a *transformation length,* and $\mathrm{ord}_{F_m} \alpha$ the *maximum transformation length.* For instance, if $m \geq 1$ and $\alpha = 3$, then the maximum transformation length is 2^{2^m} due to Pepin's test (5.1). If $\alpha = 2$, then clearly $\mathrm{ord}_{F_m} \alpha = 2^{m+1}$ for $m = 0, 1, \ldots$; see Remark 4.13 and Figure 15.1.

Given the vector $x = (x(0), x(1), \ldots, x(N-1))^{\top}$ of integers such that $x(k) \in \{0, 1, \ldots, F_m - 1\}$ for $k = 0, 1, \ldots, N - 1$, the one-dimensional *Fermat number transform* (FNT) and its inverse are defined by the formulae

(15.1)
$$X(j) \equiv \sum_{k=0}^{N-1} x(k)\alpha^{jk} \pmod{F_m}, \qquad j = 0, 1, \ldots, N - 1,$$

$$x(k) \equiv N^{-1} \sum_{j=0}^{N-1} X(j)\alpha^{-jk} \pmod{F_m}, \quad k = 0, 1, \ldots, N - 1,$$

respectively, where $X(j) \in \{0, 1, \ldots, F_m - 1\}$ for all $j \in \{0, 1, \ldots, N - 1\}$ and N^{-1} denotes the integer such that $NN^{-1} \equiv 1 \pmod{F_m}$. In Proposition 15.5 below, we prove that the second formula of (15.1) really represents the inverse.

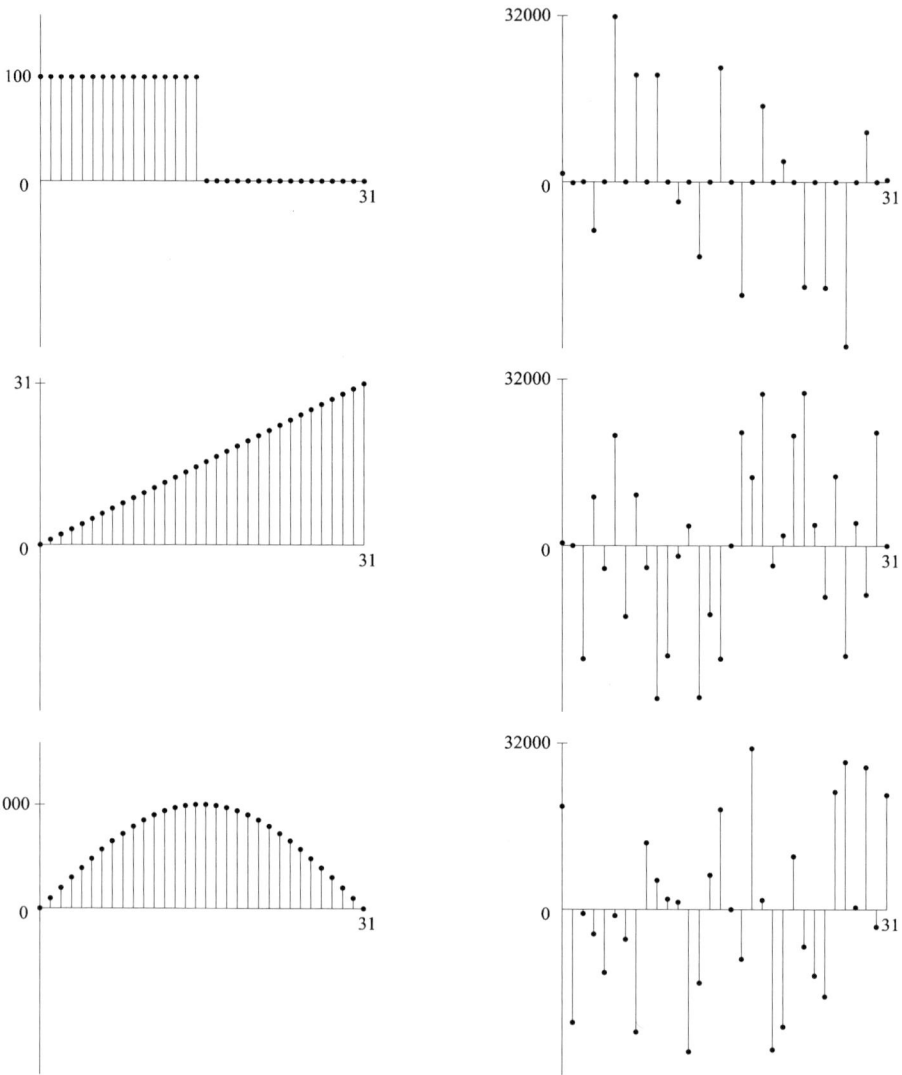

Figure 15.1. Signals (on the left) and their images (on the right) by the FNT for $m = 4$, $\alpha = 2$, and $N = 32$.

Remark 15.1. The case for which F_m is composite is treated, e.g., in [Creutzburg, Grundmann, 1985], [Creutzburg, Tasche], and [Nussbaumer, 1981].

Remark 15.2. Fourier and Laplace transforms are associated with continuous-time signals. For computer implementation they have to be discretized. Let us recall that the *discrete Fourier transform* (DFT) and its inverse are defined by the formulae (compare with (15.1))

$$Y(j) = \sum_{k=0}^{N-1} y(k)\, e^{-2\pi ijk/N}, \qquad j = 0, 1, \ldots, N-1,$$

$$y(k) = N^{-1} \sum_{j=0}^{N-1} Y(j)\, e^{2\pi ijk/N}, \quad k = 0, 1, \ldots, N-1,$$

respectively, i.e., DFT operates on signals in the complex number field. Contrary to (15.1), here $N^{-1} = 1/N$ is a real number. The DFT requires N^2 multiplications of complex numbers. However, if N is a power of 2, then we get the so-called *fast Fourier transform*, where the number of multiplications can be substantially reduced to $(N/2)\log_2 N$ (see [Cooley, Tukey]). It has a number of applications in such diverse disciplines as spectral analysis, image processing, medicine, acoustics, data filtering, convolution and correlation studies, structural vibrations, and pattern recognition.

Remark 15.3. The Fermat number transform for fast digital convolution was first developed in [Agarwal, Burrus, 1973]. From (15.1) we see that the FNT is truly a digital transform, taking into account the quantization in amplitude and the finite precision of digital signals. It is ideally suited for digital computation, since it requires only $\mathcal{O}(N \log N)$ additions, subtractions, and bit shifts, but no multiplications. Note that multiplications in all stages of the FNT are replaced by shifts of a special form, as we shall see later (cf. also [Creutzburg], [Creutzburg, Grundmann, 1985], [Gorshkov, Kravchenko], [Schönhage, Strassen]). Moreover, the FNT avoids complex arithmetic and evaluates convolutions exactly, i.e., there are no round-off errors, since all calculations are performed in the ring of integers modulo F_m. On the other hand, the fast Fourier transform requires $\mathcal{O}(N \log N)$ multiplications and cannot avoid round-off errors in complex arithmetic. In contrast to Fourier transform solution methods, calculation of the inverse transformation matrix for the FNT is immediate (see Example 15.4 below). Thus the Fermat number transform is considerably simpler than the fast Fourier transform. By the use of appropriate bit coding, the algorithm of the FNT is substantially faster than that of the fast Fourier transform. For instance, by [Creutzburg, Grundmann, 1983b], a reduction of 90% is realized for $N = 32$. A detailed comparison of the complexity of arithmetic operations for the FNT and the fast Fourier transform is given in [Agarwal, Burrus].

Define the finite discrete convolution

$$z = x' * x$$

of two vectors $x' = (x'(0), x'(1), \ldots, x'(N-1))^\top$ and $x = (x(0), x(1), \ldots, x(N-1))^\top$ by the relation

(15.2) $$z(k) = \sum_{j=0}^{N-1} x'(k-j)x(j), \quad k = 0, 1, \ldots, N-1,$$

where $x'(k - j) = x'(k - j + N)$ if $k < j$ (i.e., the argument is evaluated modulo N). We could also assume that the sequences are periodically extended with the period N, and then we speak about *cyclic convolutions* (see, e.g., [Kučera]).

The FNT shares many properties with the DFT, in particular, the following convolution identity (see, e.g., [Agarwal, Burrus, 1974] for the proof):

$$z = x' * x = \mathrm{FNT}^{-1}(\mathrm{FNT}(x') \circ \mathrm{FNT}(x)),$$

where $\circ$ denotes the pointwise product

(15.3) $X' \circ X = (X'(0)X(0), X'(1)X(1), \ldots, X'(N - 1)X(N - 1))^{\mathsf{T}}$

and where capital letters denote the transformed sequences. Therefore, the convolution "$*$" of two periodic signals of length N can be performed by taking the inverse transform of the pointwise product "$\circ$" of the transformed signals.

In many practical applications it is advantageous to adopt the following balanced representation of the signal x,

(15.4) $x(k) \in \left\{ -\dfrac{F_m - 1}{2}, \ldots, -2, -1, 0, 1, 2, \ldots, \dfrac{F_m - 1}{2} \right\}$

instead of $x(k) \in \{0, 1, \ldots, F_m - 1\}$. We shall also assume (15.4) in the following example, presented in [Agarwal, Burrus, 1974] (cf. also [Creutzburg, Grundmann, 1983b], [Elliott, Rao]).

Example 15.4. Consider two sequences $x = (2, -2, 1, 0)^{\mathsf{T}}$ and $x' = (1, 2, 0, 0)^{\mathsf{T}}$, whose convolution is desired. Let $m = 2$, $\alpha = 4$, and let $N = \mathrm{ord}_{F_m} \alpha$, i.e., $N = 4$. Since only linear invertible transforms are considered, they can be represented by an $N \times N$ nonsingular matrix T. In our case, the transformation matrix $T \equiv (\alpha^{jk})_{j,k=0}^{N-1}$ (mod 17) is given by

$$T \equiv \begin{pmatrix} 1 & 1 & 1 & 1 \\ 1 & 4 & 4^2 & 4^3 \\ 1 & 4^2 & 4^4 & 4^6 \\ 1 & 4^3 & 4^6 & 4^9 \end{pmatrix} \equiv \begin{pmatrix} 1 & 1 & 1 & 1 \\ 1 & 4 & -1 & -4 \\ 1 & -1 & 1 & -1 \\ 1 & -4 & -1 & 4 \end{pmatrix} \equiv \begin{pmatrix} 1 & 1 & 1 & 1 \\ 1 & 4 & 16 & 13 \\ 1 & 16 & 1 & 16 \\ 1 & 13 & 16 & 4 \end{pmatrix} (\mathrm{mod}\ 17).$$

Since $4^{-1} \equiv 13$ (mod 17), we easily find that the inverse transformation matrix $T^{-1} \equiv N^{-1}(\alpha^{-jk})_{j,k=0}^{N-1}$ (mod 17) is defined by

$$T^{-1} \equiv 4^{-1} \begin{pmatrix} 1 & 1 & 1 & 1 \\ 1 & 4^{-1} & 4^{-2} & 4^{-3} \\ 1 & 4^{-2} & 4^{-4} & 4^{-6} \\ 1 & 4^{-3} & 4^{-6} & 4^{-9} \end{pmatrix} \equiv 13 \begin{pmatrix} 1 & 1 & 1 & 1 \\ 1 & -4 & -1 & 4 \\ 1 & -1 & 1 & -1 \\ 1 & 4 & -1 & -4 \end{pmatrix}$$

$$\equiv 13 \begin{pmatrix} 1 & 1 & 1 & 1 \\ 1 & 13 & 16 & 4 \\ 1 & 16 & 1 & 16 \\ 1 & 4 & 16 & 13 \end{pmatrix} (\mathrm{mod}\ 17).$$

Hence, the transformation of x is given by

$$X \equiv Tx \equiv \begin{pmatrix} 1 & 1 & 1 & 1 \\ 1 & 4 & 16 & 13 \\ 1 & 16 & 1 & 16 \\ 1 & 13 & 16 & 4 \end{pmatrix} \begin{pmatrix} 2 \\ 15 \\ 1 \\ 0 \end{pmatrix} \equiv \begin{pmatrix} 18 \\ 78 \\ 243 \\ 213 \end{pmatrix} \equiv \begin{pmatrix} 1 \\ 10 \\ 5 \\ 9 \end{pmatrix} \pmod{17}.$$

Similarly, $X' \equiv Tx' \equiv (3, 9, 16, 10)^\top \pmod{17}$ and thus by (15.3) we get

$$Z \equiv X' \circ X \equiv (3, 90, 80, 90)^\top \equiv (3, 5, 12, 5)^\top \pmod{17},$$

since the transform of the cyclic convolution of two sequences is equal to the pointwise product of their transforms. Taking now the inverse of Z, we get

$$z \equiv T^{-1}Z \equiv 13 \begin{pmatrix} 1 & 1 & 1 & 1 \\ 1 & 13 & 16 & 4 \\ 1 & 16 & 1 & 16 \\ 1 & 4 & 16 & 13 \end{pmatrix} \begin{pmatrix} 3 \\ 5 \\ 12 \\ 5 \end{pmatrix}$$

$$\equiv 13 (25, 280, 175, 280)^\top \equiv (2, 2, 14, 2)^\top \equiv (2, 2, -3, 2)^\top \pmod{17}.$$

The vector $(2, 2, -3, 2)^\top$ thus satisfies assumption (15.4).

The following proposition can be found in [Agarwal, Burrus]. However, we give another proof.

Proposition 15.5. *Let F_m be a prime, $\alpha \in \{2, 3, \ldots, F_{m-1}\}$, and let $N \mid$ $\mathrm{ord}_{F_m}\alpha$. If $T \equiv (\alpha^{jk})_{j,k=0}^{N-1} \pmod{F_m}$ and $T^{-1} \equiv N^{-1}(\alpha^{-jk})_{j,k=0}^{N-1} \pmod{F_m}$ are the matrices corresponding to transformations (15.1), then*

$$TT^{-1} \equiv T^{-1}T \equiv I \pmod{F_m},$$

where I is the identity matrix.

P r o o f . We show that

$$N^{-1} \sum_{k=0}^{N-1} \alpha^{ik}\alpha^{-jk} \equiv \delta_{ij} \pmod{F_m},$$

where $0 \leq i, j < N$ and δ_{ij} is the Kronecker delta.

First assume that $i = j$. Then immediately

$$N^{-1} \sum_{k=0}^{N-1} \alpha^0 = N^{-1}N \equiv 1 \pmod{F_m}.$$

Further, let $p = i - j > 0$ (the case $i < j$ can be treated similarly) and let q be a positive integer such that $qN = \mathrm{ord}_{F_m}\alpha$. Then $\alpha^{pq} \not\equiv 1 \pmod{F_m}$, since $p < N$. Hence, using the formula for the sum of a geometric progression, we obtain

$$(\alpha^{pq} - 1) \sum_{k=0}^{N-1} \alpha^{pqk} = \alpha^{pqN} - 1 \equiv 0 \pmod{F_m}.$$

Finally, multiplying this congruence by $N^{-1}(\alpha^{pq} - 1)^{-1}\alpha^{-q}$, we get

$$N^{-1} \sum_{k=0}^{N-1} \alpha^{(i-j)k} \equiv 0 \pmod{F_m}. \qquad \square$$

The following proposition can be found, e.g., in [Creutzburg, Grundmann, 1985].

Proposition 15.6. *Let* $\| \cdot \|_p$ *be the standard* ℓ_p*-norm, i.e.,*

$$\|x\|_p = \begin{cases} \left(\frac{1}{N} \sum_{j=0}^{N-1} |x(j)|^p \right)^{1/p} & \text{for} \quad p \in [1, \infty), \\ \max_j |x(j)| & \text{for} \quad p = \infty. \end{cases}$$

Then for $z(k)$ *given by* (15.2) *we have*

$$|z(k)| \leq N \|x'\|_p \|x\|_q$$

for $p, q \geq 1$ *satisfying* $1/p + 1/q = 1$.

The convolution (15.2) is used for filtering signals x by the *filter* x', see [Creutzburg, Grundmann, 1985]. In particular, by Proposition 15.6,

$$|z(k)| \leq \max_j |x'(j)| \sum_{j=1}^{N-1} |x(j)| \leq \frac{F_m - 1}{2}.$$

Remark 15.7. In [Nussbaumer, 1977, 1982] pseudo-Fermat number transforms are discussed, which provide much more flexibility in selecting word lengths than the conventional FNT described above. For another generalization of the FNT, see [Dimitrov, Cooklev, Donevsky].

Remark 15.8. Reed–Solomon error-correcting codes are used to protect standard compact disks (CD's) against the effects of minor damage (e.g., scratching). They enable missing information to be reconstructed. Error-correcting codes are also employed in many other fields, for instance, to correct errors in signals (data) coming from interplanetary vehicles. The fast decoding of Reed–Solomon codes utilizing the Fermat number transform is introduced in [Reed, Scholtz, Truong, Welch] and also in [Reed, Truong, Welch].

The Fermat number transform and its modified versions are applied to digital filtering, e.g., in [Agarwal, Burrus, 1974], [Creutzburg, Grundmann, 1983a], [Lee, Min, Suk], [Li, Peterson].

The paper [Leibowitz] presents a simplified binary arithmetic for the FNT. In [Agarwal, Burrus] an implementation of the FNT is presented, and in [McClellan] hardware for the FNT is described. A parallel implementation of the FNT is developed in [Lucká, Vajteršic, Creutzburg, Grundmann]. It is used for fast coding and decoding. A parallel algorithm for computing digital convolutions using the FNT is given in [Lucká, Creutzburg, Grundmann, Vajteršic].

In [Jiang, Yu] the FNT of hyperlarge-scale two-dimensional cyclic convolutions is introduced. Two-dimensional convolutions are also mentioned in [Agarwal, Burrus]. A k-dimensional error-free deconvolution using the Fermat number transform is suggested in [Morháč, 1989] and [Yang]. In [Gorshkov, Kravchenko, Rvachev, Rvachev] the Chinese remainder theorem is employed to reduce larger power-of-two lengths to lower dimensions. For the use of the Chinese remainder theorem in connection with number-theoretic transforms see also [Creutzburg, Tasche] and [Truong, Chang, Hsu, Pei, Reed].

The paper [Crandall, Fagin] extends the idea of using Fourier transforms for long-integer arithmetic to a new concept, the discrete weighted transform based on arithmetic modulo a Fermat or Mersenne number. In [Morikawa, Hamada, Yamane] a fast Fourier transform algorithm using the Fermat number transform is introduced.

Remark 15.9. According to formula (6.3), we can define an analogue of the square root of 2 in a ring of integers modulo the Fermat prime F_m for $m > 1$. Indeed, denoting by $\sqrt{2}$ an element b for which $b^2 \equiv 2 \pmod{F_m}$ and $0 < b < F_m/2$, we have

$$\sqrt{2} \text{ "=" } b = 2^{2^{m-2}}\left(2^{2^{m-1}} - 1\right).$$

For instance, $6 \equiv \sqrt{2} \pmod{17}$, since $6^2 \equiv 2 \pmod{17}$. Let us note that there are at least two square roots of 2 $\pmod{F_m}$ (more than two if F_m is composite.)

Remark 15.10. Complex convolutions via Fermat number transforms are examined in [Elliott, Rao], [Labunets], [Nussbaumer, 1976, 1982], and [Truong, Chang, Hsu, Pei, Reed]. All operations are again performed modulo a Fermat number. For instance, the imaginary unit $i = \sqrt{-1}$ is represented by the element $a = 2^{2^{m-1}}$, since $a^2 = 2^{2^m} \equiv -1 \pmod{F_m}$. Note that the filtering of complex signals can be applied in a number of cases concerning, e.g., radar, sonar, and modems.

In [Morháč, 1986] a new algorithm to solve convolution systems of linear equations using modular arithmetic and the FNT is described. This algorithm allows elimination of round-off errors in the solution of the deconvolution problem due to integer arithmetic.

Binary Arithmetic Modulo Fermat Numbers

Formulae (15.1) can be efficiently evaluated using a relatively simple binary arithmetic modulo a Fermat number $F_m = 2^n + 1$, where $n = 2^m$. Any integer x defined modulo F_m can be realized as an $(n + 1)$-bit word (see, e.g., [Creutzburg, Grundmann, 1985])

$$(15.5) \qquad x = \sum_{i=0}^{n-1} x_i 2^i + x_n 2^n + 1,$$

where

$$x_0, x_1, \ldots, x_n \in \{0, 1\} \quad \text{and} \quad x_i x_n = 0 \text{ for } i < n.$$

For instance, if $n = 4$, then

$$1 \simeq 0\ 0000,\ 2 \simeq 0\ 0001,\ 3 \simeq 0\ 0010, \ldots,\ 16 \simeq 0\ 1111,\ 0 = 17 \simeq 1\ 0000,$$

where the leading binary digit x_n is followed by a space and the symbol $\simeq$ expresses the relationship between the integer x and its binary coefficients x_i.

For a given x we define $\overline{x}$ by

$$x + \overline{x} \equiv 0 \pmod{F_m}.$$

We see that $\overline{x} = 0$ for $x = 0$. Negation for $x \neq 0$ can be simply realized by complementing all bits x_i $(i < n)$ except for x_n, i.e.,

$$\overline{x} = \sum_{i=0}^{n-1} \overline{x}_i 2^i + \overline{x}_n 2^n + 1,$$

where

$$\overline{x}_i = 1 - x_i \text{ for } i < n \quad \text{and} \quad \overline{x}_n = x_n.$$

Indeed, from here, (15.5), and the fact that $x_n = 0$ for $x \neq 0$, we obtain

$$x + \overline{x} = \sum_{i=0}^{n-1} 2^i + 2 = 2^n - 1 + 2 = 2^n + 1 \equiv 0 \pmod{F_m}.$$

Example 15.11. Let $n = 4$. Since $3 \simeq 0\ 0010$, we obtain

$$\overline{3} \simeq \overline{0\ 0010} = 0\ 1101 \simeq 14 \equiv -3 \pmod{17}.$$

The addition of two integers modulo F_m is described, e.g., in [Creutzburg, Grundmann, 1983b, 1985], [Nussbaumer, 1982]. Subtraction is done with help of negation and addition. Here we will deal only with multiplication by powers of 2 to show the main advantage of this special binary arithmetic.

Multiplication of an integer $x \neq 0$ by 2 yields

$$2x = 2\left(\sum_{i=0}^{n-1} x_i 2^i + 1\right) = x_{n-1} 2^n + \sum_{i=0}^{n-2} x_i 2^{i+1} + 1 + 1.$$

If $x_{n-1} = 1$, then

$$2x = (2^n + 1) + \sum_{i=0}^{n-2} x_i 2^{i+1} + 1 \equiv \sum_{j=1}^{n-1} x_{j-1} 2^j + 0 \cdot 2^0 + 1 \pmod{F_m},$$

and for $x_{n-1} = 0$ we find that

$$2x = \sum_{i=0}^{n-2} x_i 2^{i+1} + 1 + 1 \equiv \sum_{j=1}^{n-1} x_{j-1} 2^j + 1 \cdot 2^0 + 1 \pmod{F_m}.$$

In both cases, we see that multiplication of $x \neq 0$ by 2 reduces to one bit shift to the left such that the first leading binary digit remains zero and the new value of the last binary digit is the complementary value of x_{n-1}.

Multiplication of an integer x by 2^d (cf. (15.1)) requires one to perform the foregoing procedure d times.

Example 15.12. Let $n = 4$, $x = 10$, and $d = 3$. Since

$$2^3 \cdot 10 \equiv 12 \pmod{17},$$

the multiplication proceeds as follows:

$$10 \simeq 0\ 1001,$$
$$2 \cdot 10 \simeq 0\ 0010,$$
$$2^2 \cdot 10 \simeq 0\ 0101,$$
$$2^3 \cdot 10 \simeq 0\ 1011 \simeq 12.$$

Multiplication by negative powers 2^{-d} (cf. (15.1)) is based on the relation

$$2^d\left(-2^{n-d}\right) = -2^n \equiv 1 \pmod{F_m},$$

which yields

$$2^{-d} \equiv -2^{n-d} \pmod{F_m}.$$

For multiplication of x by a general integer see [Creutzburg, Grundmann, 1983b, 1985].

Recall that binary arithmetic modulo a Fermat number is used in [Schönhage, Strassen] for fast multiplication of large ñumbers. Their fast algorithm requires only $\mathcal{O}(N \log N \log \log N)$ bit operations for two N-digit binary numbers, whereas the number of bit operations for standard straightforward multiplication is proportional to N^2. This algorithm was used in [Young] for fast squaring in order to find prime factors of Fermat numbers. A method for a fast multiplication modulo Fermat primes in presented also in [Gorshkov, 1994a]. For a computer implementation of fast mutliplication modulo $2^n + 1$ see, e.g., [Crandall, Pomerance, p. 457].

Generators of Pseudorandom Numbers

A sequence of *pseudorandom* numbers $\{X_i\}$ can be characterized as a deterministic sequence of numbers in $[0, 1]$ having the same relevant statistical properties as a sequence of random numbers. Recall that pseudorandom numbers are used in primality testing and in factoring large numbers, in simulation of physical processes, in solving partial differential equations by the Monte Carlo method (especially in higher dimensions), in cryptography to produce pseudorandom sequences of bits, in various statistical tests, in most computer games, etc. They are usually generated with the help of congruences. For example, in the popular home microcomputer ZX Spectrum, which was developed in the 1980s, pseudorandom numbers X_i are defined by means of remainders obtained when the sequence of powers 75, 75^2, $75^3, \ldots$ is divided by the Fermat prime $F_4 = 2^{16} + 1$. The remainders are then divided by $F_4 - 1$. Specifically,

$$(15.6) \qquad X_i = \frac{r_i - 1}{F_4 - 1} \quad \text{for } i = 1, 2, \ldots,$$

where $r_i \in \{1, \ldots, F_4 - 1\}$ is the remainder such that

(15.7) $r_i \equiv 75r_{i-1} \pmod{F_4}$ with $r_0 = 1$.

This yields a periodic sequence $\{X_i\}$ of pseudorandom numbers in the range from 0 to 1 (excluding 1) with the maximum possible period 65 536, since 75 is a primitive root modulo F_4. To see this, we first apply Theorem 2.27 and (5.2) to get

$$\left(\frac{75}{F_4}\right) = \left(\frac{3}{F_4}\right)\left(\frac{5^2}{F_4}\right) = \left(\frac{3}{F_4}\right) = -1,$$

and by Definition 2.25, we find that 75 is a quadratic nonresidue modulo F_4. Now by Theorem 5.28 we obtain that 75 is a primitive root modulo F_4, that is, $\mathrm{ord}_{F_4} 75 = F_4 - 1$.

The sequence (15.6) looks fairly random and the values X_i are uniformly distributed in the interval $[0,1)$. For some weaknesses of this pseudorandom number generator, see [Ripley, p. 40].

Minimal Perfect Hashing Scheme

The problem of how to store data so as to facilitate efficient searching is a basic problem of computer science. Hashing refers to an effective means of searching for information. We will briefly describe a hashing scheme introduced in [Chang], which is based on Euler's theorem and Fermat numbers.

A hashing function can be considered to be a mapping of a given set of records into a set of addresses in computer memory. Let each record R be uniquely identified by its key k. Besides k, the record contains some additional information INFO(k) as depicted in Figure 15.2. For instance, k can be the license plate number of an automobile and INFO(k) can contain other data about this car such as its color, type, size, engine, owner. We will order all records and define a hashing function in such a way that for a given key k, we can directly find the associated record without going through all the keys.

Specifically, a *hashing function* is a function h defined on a key set

$$K = \{k_1, k_2, \ldots, k_m\}$$

into the set $\{1, 2, \ldots, n\}$ of locations (addresses). Assume for simplicity that K is a subset of natural numbers and $m \leq n$. For instance, we can set $h(k) = k$ (mod n). However, this simple hashing function may lead to duplications, i.e., two keys could correspond to the same address. Thus, the definition of h has to be modified somehow. To avoid duplications, suppose that h is an injective mapping. Such a transformation h is called a *perfect hashing function*. Moreover, if the cardinality of the set of records is equal to the cardinality of addresses (i.e., $n = m$), a perfect hashing function is called a *minimal perfect hashing function*. In this case, h is a one-to-one mapping and requirements on computer memory are minimal (see Figure 15.2).

The set of keys with additional information			addresses
k_1,	$\mathrm{INFO}(k_1)$	$\mapsto$	1
k_2,	$\mathrm{INFO}(k_2)$	$\mapsto$	2
k_3,	$\mathrm{INFO}(k_3)$	$\mapsto$	3
$\vdots$			$\vdots$
k_n,	$\mathrm{INFO}(k_n)$	$\mapsto$	n

Figure 15.2. The hashing scheme for $h : k_j \in K \mapsto j \in \{1, \ldots, n\}$.

A transformation function $T : K \rightarrow \{1, 2, \ldots, n\}$ is called *pairwise coprime* if $\gcd(T(k), T(\ell)) = 1$ for $k \neq \ell$ and $k, \ell \in K$.

Example 15.13. Let $K = \{3, 4, 6, 8\}$ and let $T(k) = k + 1$. Then $T(3) = 4$, $T(4) = 5$, $T(6) = 7$, and $T(8) = 9$, which means that T is a pairwise coprime transformation function on K.

The following assertion appears in [Chang].

Lemma 15.14. Let $K = \{k_1, k_2, \ldots, k_n\}$ be a finite set of positive integers. The hashing function $h(k) =$ the remainder of $C/T(k)$ is a minimal perfect hashing function if

$$C = \sum_{j=1}^{n} j \left(\prod_{i \neq j}^{n} T(k_i) \right)^{\phi(T(k_j))},$$

where T is a pairwise coprime transformation function on K.

P r o o f . For $j = 1, \ldots, n$ put

$$M_j = \left(\prod_{i \neq j}^{n} T(k_i) \right)^{\phi(T(k_j))}.$$

Since T is a pairwise coprime transformation function on K, we get

$$\gcd\left(\prod_{i \neq j}^{n} T(k_i), T(k_j) \right) = 1,$$

and thus, by Euler's Theorem 2.17,

(15.8) $M_j \equiv 1 \pmod{T(k_j)}$ for $j \in \{1, \ldots, n\}$.

Moreover, it is clear that

(15.9) $M_j \equiv 0 \pmod{T(k_i)}$ for $i \neq j$.

Let

$$C = \sum_{j=1}^{n} j M_j$$

and let k_i be given. Then, by (15.8) and (15.9),

$$(15.10) \qquad\qquad C \equiv \sum_{j=1}^{n} jM_j \equiv i \pmod{T(k_i)}$$

and $i \leq T(k_i)$. This implies that $h(k)$ = the remainder of $C/T(k)$ is a minimal perfect hashing function.   □

Remark 15.15. If

$$C^* \equiv \sum_{j=1}^{n} jM_j \left(\bmod \ \prod_{i=1}^{n} T(k_i) \right)$$

and $0 \leq C^* < \prod_{i=1}^{n} T(k_i)$, then obviously $h(k)$ = the remainder of $C^*/T(k)$ is also a minimal perfect hashing function.

An *ordered minimal perfect hashing function* is a minimal perfect hashing function that preserves the numerical order of the elements of the key set. From (15.10) we observe that Lemma 15.14 enables us to store keys in K in ascending order.

Example 15.16. Let $K = \{3, 4, 6\}$ and let $T(k) = k + 1$ again. Then, by Lemma 15.14,

$$C = (T(4)T(6))^{\phi(T(3))} + 2(T(3)T(6))^{\phi(T(4))} + 3(T(3)T(4))^{\phi(T(6))}.$$

From this and (2.14) we can find that $C = 193\,230\,537$. Therefore, by Remark 15.15, $C^* = 17$, since $T(3)T(4)T(6) = 140$ and $C^* \equiv C \pmod{140}$. Consequently,

$$h(k) = \text{ the remainder of } \frac{17}{k+1}$$

is an ordered minimal perfect hashing function. Indeed, the remainder of $17/4$ is 1, the remainder of $17/5$ is 2, and the remainder of $17/7$ is 3.

The following theorem can also be found in [Chang].

Theorem 15.17 (Chang). *Let K be a finite set of positive integers $m_1 < m_2 < \cdots < m_n$. The hashing function $h(m)$ = the remainder of C upon division by F_m for $m \in K$ is an ordered minimal perfect hashing function if*

$$C = \sum_{j=1}^{n} j \Big(\prod_{i \neq j}^{n} F_{m_i} \Big)^{\phi(F_{m_j})}.$$

P r o o f . According to Goldbach's Theorem 4.1, Fermat numbers are pairwise coprime. Setting $T(m) = F_m$ in Lemma 15.14, we obtain the assertion of Theorem 15.17.   □

Remark 15.18. For large n the applicability of Chang's theorem could be rather difficult. A theoretical foundation of finding a simpler minimal perfect hashing function is described in [Jaeschke, 1981].

Chiral Potts Model

In [Bellon, Maillard, Rollet, Viallet] Fermat numbers are used to examine deformations of dynamics associated with the chiral (asymmetric) Potts model. One of the deformation parameters can be seen to be the number n of states of this model. The problem in question requires one to find such n for which the values of $\frac{1}{2}(-1 \pm \sqrt{n})$ can be realized as a sum of nth roots of unity. This can be realized only by the Fermat primes. In particular, for $n = 17$ we have (compare with Remark 17.4)

$$z^{3^0} + z^{3^2} + z^{3^4} + z^{3^6} + \cdots + z^{3^{14}} = \frac{1}{2}(-1 + \sqrt{17}),$$

$$z^{3^1} + z^{3^3} + z^{3^5} + z^{3^7} + \cdots + z^{3^{15}} = \frac{1}{2}(-1 - \sqrt{17}),$$

where $z = \cos\frac{2\pi}{17} + i\sin\frac{2\pi}{17}$.

Logistic Equation and Chaos

Now we introduce an interesting connection between the logistic equation and the Fermat numbers (and also the Mersenne numbers). The theory of chaos was significantly advanced by observations made by the young physicist M. J. Feigenbaum when he began studying with a pocket calculator the bifurcations of the following nonlinear difference equation:

$$(15.11) \qquad x_{n+1} = \lambda x_n(1 - x_n), \quad \lambda, x_n \in \mathbb{R}, \quad n \in \mathbb{N}.$$

For a fixed control parameter λ and a given initial value $x_1 \in [0, 1]$, the sequence $\{x_n\}$ may have a different number of accumulation points. By (15.11), we see that the sequence $\{x_n\}_{n=1}^{\infty}$ is bounded for every $\lambda \in [0, 4]$.

Equation (15.11) is called the *logistic equation*. It describes, for instance, the evolution of a biological dynamical system. The linear term λx_n characterizes an increase of some population, whereas the quadratic term $(-\lambda x_n^2)$ represents a decrease of this population due to some process.

If $\lambda \in [0, 1]$, then the sequence $\{x_n\}$ has zero as a limit for $n \to \infty$, i.e., zero is an attractor. (This represents the extinction of the population.) We see that $x_n = 0$ is the solution of (15.11) for any λ. However, for a given fixed $\lambda > 3$ and $x_1 \in (0, 1]$ the sequence $\{x_n\}$ may have more accumulation points.

The set of all accumulation points of x_n defined by (15.11) for various values of the control parameter $\lambda \in [0, 4]$ is illustrated in Figure 15.3. We observe that solutions of a simple nonlinear mathematical model may have a very complicated behavior. The "solution" branches have bifurcation points. The first bifurcation point is at $\lambda_1 = 1$. The corresponding solution branch is given by the function $f(\lambda) = 1 - 1/\lambda$ for $\lambda \in [1, 4]$, which expresses a stable equilibrium state until the second bifurcation point, which is at $\lambda_2 = 3$. Further bifurcation points are at $\lambda_3 \approx 3.449490$, $\lambda_4 \approx 3.544090$, $\lambda_5 \approx 3.564407$, etc. They correspond to the so-called *doubling of the period* (see Figure 15.3).

For $\lambda > \lambda_2$ there exist constant (but unstable) solutions whose initial values $x_1 = x_1(\lambda)$ lie on the graph of the function f, e.g.,

$$(15.12) \qquad x_n = 0.7 \text{ for all } n \in \mathbb{N} \text{ with } \lambda = \frac{10}{3}.$$

Similarly, for $\lambda > \lambda_3$ there are other unstable solutions, e.g., $x_n = \frac{6}{7}$ for n odd, $x_n = \frac{3}{7}$ for n even, and $\lambda = \frac{7}{2}$. The unstable solutions are indicated by broken lines in Figures 15.3 and 15.4.

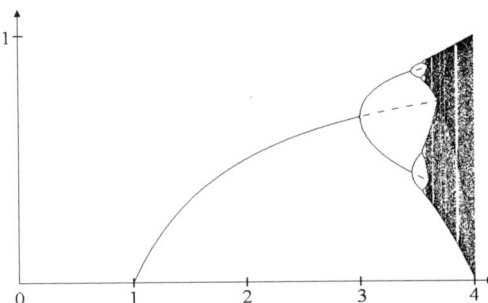

Figure 15.3. Feigenbaum's transition to chaos of dynamical system (15.11). For the increasing control parameter $\lambda \in [0, 4]$, the figure illustrates the set of associated accumulation points of all sequences $\{x_n\}$ for all initial values $x_1 \in [0, 1]$.

If the initial value $x_1 \in (0, 1]$ does not lie on the broken lines of Figure 15.4, we get two accumulation points for $\lambda \in (\lambda_2, \lambda_3]$, four accumulation points for $\lambda \in (\lambda_3, \lambda_4]$, etc. As we shall see later, these accumulation points will correspond to nonconstant periodic solutions. For instance, if $\lambda \in (\lambda_2, \lambda_3]$ and if we choose the initial value x_1 on the upper or lower branch, then the associated sequence $\{x_n\}$ oscillates between these two values (compare with Figures 15.4 and 15.7).

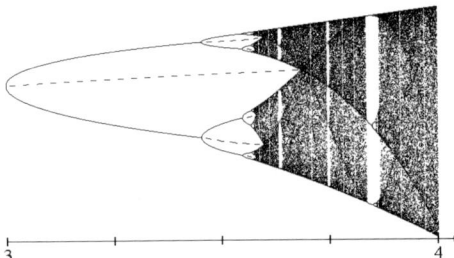

Figure 15.4. A more detailed view of Feigenbaum's transition to chaos of dynamical system (15.11) for $\lambda \in [3, 4]$ illustrates the existence of the so-called windows.

Feigenbaum (see [Feigenbaum]) has found an interesting property, namely, that

$$\frac{\lambda_n - \lambda_{n-1}}{\lambda_{n+1} - \lambda_n} \to 4.66920160910299067\ldots \quad \text{as } n \to \infty,$$

i.e., the distances $\lambda_{n+1} - \lambda_n$ form almost a geometric progression. The limit is called the *Feigenbaum number*, and the repeated doublings of the periods at λ_1, $\lambda_2, \lambda_3, \ldots$, are called *Feigenbaum cascades*.

For λ exceeding the value $\lambda_\infty = \lim_{n\to\infty} \lambda_n \approx 3.5699456$, we can obtain chaos, but also finitely many accumulation points (see Figure 15.3 and 15.4). This means that for specific ranges of values $\lambda \in [\lambda_\infty, 4]$ the so-called windows appear, where the chaotic behavior of the solution branches disappears and is replaced by regular behavior (see Figure 15.4 for a detailed diagram).

Further, we briefly describe an algorithm from [Antonyuk, Stanyukovich, 1990a] for determination of periodic solutions of (15.11). Using the transformation

$$x_n = \frac{1}{2} - \frac{y_n}{\lambda}, \quad y_n \in \mathbb{R},$$

logistic equation (15.11) can be rewritten in the canonical form

$$(15.13) \qquad\qquad y_{n+1} = y_n^2 + b,$$

where $b = \lambda/2 - \lambda^2/4$. In Figure 15.5 we again see the associated bifurcation points.

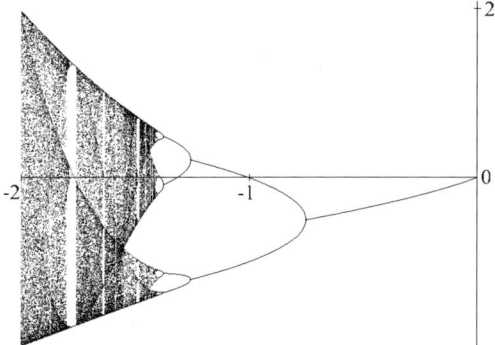

Figure 15.5. For an increasing value of the control parameter $b \in [-2, 0]$, the figure illustrates the set of associated accumulation points of all sequences $\{y_n\}$ for all initial values $y_1 \in [-2, 2]$. Unstable solutions are not indicated.

In our further analysis we will examine equation (15.13) in the complex plane $\mathbb{C}$, i.e., $y_n \in \mathbb{C}$ and $b \in \mathbb{C}$. This approach has several indisputable advantages as we shall see later. Now we show that some fractal objects in the complex plane are described by means of the same equation as that for chaos in the real variable, namely the equation (15.13).

Remark 15.19. Let us indicate the dependence of sequence (15.13) on b in the usual way: $y_n = y_n(b)$. Then for the initial conditions $y_1(b) = 0$ for all $b \in \mathbb{C}$, the set M of all bounded sequences $\{y_n(b)\}_{n=1}^\infty$ is nothing else but the well-known *Mandelbrot set* presented in [Mandelbrot], i.e.,

$$M = \{b \in \mathbb{C} \mid \exists c \in \mathbb{R} \; \forall n \in \mathbb{N} : |y_n(b)| \leq c\}$$

(see Figure 15.6). Various color pictures of a neighborhood of this fractal set are obtained in the following way: To each point $b \in \mathbb{C} \setminus M$ we assign a certain color shade according to the rate of divergence of the sequence $\{y_n(b)\}_{n=1}^\infty$.

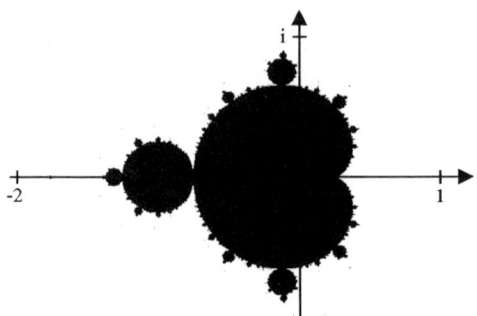

Figure 15.6. The Mandelbrot set of all complex numbers b for which the magnitude of terms in the sequence $y_1(b) = 0$, $y_2(b)$, $y_3(b)$, ..., given by (15.13) remains bounded.

Definition 15.20. A solution $\{y_n\}_{n=1}^{\infty}$ of (15.13) is said to be *periodic* if there exists $p \in \mathbb{N}$ such that

$$y_{n+p} = y_n \text{ for all } n \in \mathbb{N} \quad \text{and} \quad y_{n+r} \neq y_n \text{ for all } r \in \{1, \ldots, p-1\}.$$

The number p is then called the *minimal period*, and any positive integer multiple of p is called a *period*. All other solutions are called *nonperiodic*.

Similarly to Definition 15.20, periodic and nonperiodic solutions of the logistic equation (15.11) are defined. For instance, the solution in (15.12) is periodic with $p = 1$, since $x_{n+1} = x_n$ (see the intersection of the dashed line with the parabola $f(x) = \frac{10}{3}x(1-x)$ in Figure 15.7).

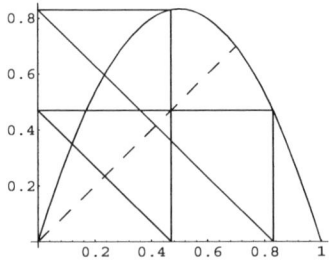

Figure 15.7. Geometric interpretation of the periodic solution of equation (15.11) with the minimal period $p = 2$ for $\lambda = \frac{10}{3}$ and $x_1 = (13 - \sqrt{13})/20 \approx 0.4697224$, $x_2 = (13 + \sqrt{13})/20 \approx 0.8302776$, $x_3 = x_1$, $x_4 = x_2$, etc. Starting at x_1, the associated value is $x_2 = f(x_1)$. Then $x_3 = f(x_2) = x_1$.

For $\lambda \geq \lambda_{\infty}$ we observe various small periods that are not powers of 2 in regions of the graph called windows, e.g., $p = 3$ for $\lambda = 3.83$, $p = 5$ for $\lambda = 3.74$, or $p = 7$ for $\lambda = 3.702$ (see Figure 15.4).

First consider (15.13) in the simplest case, when $b = 0$, i.e.,

(15.14) $$y_{n+1} = y_n^2.$$

Then, by (15.14), all solutions (periodic or nonperiodic) have the form

$$y_n = a^{2^n}, \quad a \in \mathbb{C}.$$

All periodic solutions with period p (and also with period q such that $q \mid p$) are defined by

$$y^{2^p} - y = 0.$$

This equation has a trivial solution $y_n = 0$, and the other periodic solutions lie on the unit circle and satisfy

$$y^{2^p - 1} - 1 = \prod_{d \mid (2^p - 1)} \Phi_d(y) = 0,$$

where Φ_d is the cyclotomic polynomial of degree $\phi(d)$. Recall that

$$\Phi_1(y) = y - 1, \qquad\qquad \Phi_2(y) = y + 1,$$
$$\Phi_3(y) = y^2 + y + 1, \qquad\qquad \Phi_4(y) = y^2 + 1,$$
$$\Phi_5(y) = y^4 + y^3 + y^2 + y + 1, \quad \Phi_6(y) = y^2 - y + 1,$$

etc., (compare with (13.1)). Hence, all periodic solutions of (15.14) with period $p = 1$ are given by the equations $\Phi_0(y) \equiv y = 0$ and $\Phi_1(y) = 0$, and all periodic solutions with the minimal period $p > 1$ are given by the equation $\Phi_d(y) = 0$, where $d \mid (2^p - 1)$ and $d \nmid (2^q - 1)$ for $q \mid p$ such that $q < p$.

Consider the divisors of $2^p - 1$. If p is composite, then so is $2^p - 1$ (cf. (B.1)). If $q \mid p$, then also $(2^q - 1) \mid (2^p - 1)$. On the other hand, if p is prime, then $M_p = 2^p - 1$ is a Mersenne number (cf. Appendix B). The periodic solution of (15.14) with the minimal period p is then given by the equation

$$\Phi_{M_p}(y) \equiv y^{M_p - 1} + y^{M_p - 2} + \cdots + y + 1 = 0$$

if and only if M_p is prime.

Example 15.21. If $p = 3$, then $M_3 = 7$ and $\Phi_7(y) = y^6 + y^5 + \cdots + y + 1 = 0$. For

$$y_k = \cos\frac{2^k\pi}{7} + i\sin\frac{2^k\pi}{7}, \quad k = 1, 2, \ldots,$$

we get by (15.14) that $y_4 = y_1$, $y_5 = y_2$, etc., which confirms that the period is 3.

Using the prime factorizations of $2^p - 1$ for $p \leq 9$, the associated periodic solutions can be expressed by the equations

$$
\begin{aligned}
p &= 1, \quad \Phi_0(y) = 0, \quad \Phi_1(y) = 0, \\
p &= 2, \quad \Phi_3(y) = 0, \\
p &= 3, \quad \Phi_7(y) = 0, \\
p &= 4, \quad \Phi_5(y) = 0, \quad \Phi_{15}(y) = 0, \\
p &= 5, \quad \Phi_{31}(y) = 0, \\
p &= 6, \quad \Phi_9(y) = 0, \quad \Phi_{21}(y) = 0, \quad \Phi_{63}(y) = 0, \\
p &= 7, \quad \Phi_{127}(y) = 0, \\
p &= 8, \quad \Phi_{17}(y) = 0, \quad \Phi_{51}(y) = 0, \quad \Phi_{85}(y) = 0, \quad \Phi_{255}(y) = 0, \\
p &= 9, \quad \Phi_{73}(y) = 0, \quad \Phi_{511}(y) = 0.
\end{aligned}
$$

Remark 15.22. Solutions of (15.13) with a small $b \neq 0$ can be computed by perturbation theory. All periodic solutions with period p (and also with period q such that $q \mid p$) are defined by the equation

$$y^{2^p} - y = bP(y, b),$$

where P is a polynomial in two variables with integer coefficients that can be obtained from the definition of periodicity. Searching for a periodic solution of (15.13) leads to factoring of the last equation into a set of algebraic equations of the form

$$\Phi(y) = bQ(y, b),$$

where Φ is the product of finitely many cyclotomic polynomials arising in solving the equation (15.14), and Q is a polynomial in two variables with rational coefficients that can be computed by perturbation theory. Let

$$Q(y, b) = Q_k(y)b^k + \cdots + Q_1(y)b + Q_0(y),$$

where the polynomials $Q_0, Q_1, \ldots, Q_k$ are calculated successively for $|b| \ll 1$. However, according to [Antonyuk, Stanyukovich, 1990a], this solution will be exact even for large $|b|$. Every periodic solution of (15.13) is, of course, bounded (cf. Remark 15.19).

Remark 15.23. Equation (15.11) with $\lambda = 4$ is equivalent to (15.13) with $b = -2$. Let us point out that the general form of solutions of (15.13) for $b = -2$ has the form

$$(15.15) \qquad\qquad y_n = a^{2^n} + a^{-2^n}, \quad a \in \mathbb{C}.$$

This enables us to express explicitly a subsequence of the Lucas sequence ($L_k = a^k + a^{-k}$) defined recurrently by

$$(15.16) \qquad\qquad S_{n+1} = S_n^2 - 2, \quad S_1 = 4$$

(compare with the Lucas–Lehmer test in Theorem B.2). From (15.15) we get

$$S_n = \left(2 + \sqrt{3}\right)^{2^{n-1}} + \left(2 - \sqrt{3}\right)^{2^{n-1}} = \left\lfloor \left(2 + \sqrt{3}\right)^{2^{n-1}} \right\rfloor + 1,$$

where the last equality holds only for $n \geq 2$.

Further, we show how the sequence of period-doubling bifurcations of equation (15.11) is related to the Fermat numbers. This result is based on the paper [Antonyuk, Stanyukovich, 1990b]. Before we start to examine a connection between the Fermat numbers and the logistic equation (15.11), we will again deal with the simpler auxiliary equation (15.14) and look for its periodic solutions.

Let p be a period of a periodic solution of equation (15.14). Every periodic solution is determined by an appropriate initial condition $y_1 \in \mathbb{C}$. All periodic solutions with period $p = 2^m$, $m = 0, 1, 2, \ldots$, or periods q such that $q \mid p$ and $q < p$ are defined by the equation

$$y^{2^{2^m}} - y = 0.$$

Excluding the trivial solution $y_n = 0$, we get the equation

$$y^{2^{2^m}-1} - 1 = 0.$$

Again using the cyclotomic polynomials Φ_d of degree $\phi(d)$, we get

$$y^{2^{2^m}-1} - 1 = \prod_{d|(2^{2^m}-1)} \Phi_d(y) = 0.$$

The polynomial on the left-hand side is thus factored into irreducible lower-order polynomials.

Finding the divisors of $2^{2^m} - 1$ is connected with factoring the Fermat numbers F_m into prime factors, since by formula (3.4) we have

$$F_m - 2 = 2^{2^m} - 1 = \prod_{k=0}^{m-1}\left(2^{2^k} + 1\right) = \prod_{k=0}^{m-1} F_k.$$

In this way periodic solutions of the minimal period $p = 2^m$ are defined by the equation

$$\left(\prod_{d|\prod_{k=0}^{m-1} F_k} \Phi_d(y)\right) \bigg/ \left(\prod_{d|\prod_{k=0}^{m-2} F_k} \Phi_d(y)\right) = 0,$$

which for $p = 2^m$ gives the set of solutions

(15.17)
$$\Phi_0(y) \equiv y = 0, \quad \Phi_1(y) = 0 \text{ for } m = 0,$$
$$\Phi_3(y) = 0 \text{ for } m = 1,$$
$$\Phi_d(y) = 0 \text{ for } d \mid \prod_{k=0}^{m-1} F_k, \quad d \nmid \prod_{k=0}^{m-2} F_k, \text{ and } m \geq 2.$$

Example 15.24. Let $m = 2$. Then, by (15.17), $d \mid F_0 F_1$ and $d \nmid F_0$. Hence, we have to consider the cyclotomic polynomials $\Phi_5(y) = y^4 + y^3 + y^2 + y + 1$ and $\Phi_{15}(y) = y^8 - y^7 + y^5 - y^4 + y^3 - y + 1$. We observe that the associated two sequences

$$y_k = \cos\frac{2^k\pi}{5} + i\sin\frac{2^k\pi}{5}, \quad y_k' = \cos\frac{2^k\pi}{15} + i\sin\frac{2^k\pi}{15}, \quad k = 1, 2, \ldots,$$

yield periodic solutions with the minimal period $p = 4$.

Let $w_{m,m+1}$ denote the multiple by which the number of polynomials Φ_d defining solutions of (15.17) increases when the period $p = 2^m$ is replaced by $p = 2^{m+1}$. Clearly, $w_{0,1} = \frac{1}{2}$ for $m = 0$. If $m \geq 1$, then

(15.18)
$$w_{m,m+1} = \frac{\tau(F_{m-1})(\tau(F_m) - 1)}{\tau(F_{m-1}) - 1},$$

where $\tau(\cdot)$ denotes the number of positive divisors. The relation (15.18) follows from the fact that $\tau(\cdot)$ is multiplicative and the fact that the Fermat numbers are pairwise coprime (see Goldbach's Theorem 4.1). In Theorem 15.25 below, we show that a doubling of the number of polynomials Φ_d arises if and only if F_{m-1} and F_m are prime numbers for $m \geq 1$ ($w_{m,m+1} = 2 \iff \tau(F_{m-1}) = \tau(F_m) = 2$). Starting from $p = 2$, the number of polynomials increases when the period is doubled ($w_{m,m+1} > 1$ for all $m \in \mathbb{N}$).

Further, we describe the process of doubling the period generated by (15.11) for $\lambda < \lambda_\infty$. Using solutions (15.17), we may find by perturbation theory all periodic solutions with period $p = 2^m$ of equation (15.11). The algorithm for finding such a solution is given in [Antonyuk, Stanyukovich, 1990a]. Let $x_1 = x \in \mathbb{R}$. All periodic solutions with the period $p = 2^m$ are defined by a finite number of irreducible (over the field $\mathbb{Q}$ of rational numbers) algebraic equations $\{X(x, \lambda) = 0\}$. If one polynomial Φ_d is employed for finding a solution of the equation $X(x, \lambda) = 0$, then we indicate this by the index d such that $X_d(x, \lambda) = 0$. For instance, periodic solutions for the first two periods are given by the equations (compare with Figure 15.3)

$$X_0(x, \lambda) \equiv x = 0, \quad X_1(x, \lambda) \equiv x\lambda - \lambda + 1 = 0 \text{ for } p = 1,$$
$$X_3(x, \lambda) \equiv x = x^2\lambda^2 - x\lambda^2 - x\lambda + \lambda + 1 = 0 \text{ for } p = 2.$$

Every equation $X(x, \lambda) = 0$ yields a planar algebraic curve. The plane (λ, x) with these curves (for all periods) is called a *bifurcation diagram,* and the associated set of curves is called a *bifurcation tree.* According to [Feigenbaum], for any period $p = 2^m$, $m \neq 0$, there exist 2^{m-1} curves. In doubling the period, i.e., when the period $p = 2^m$ is changed to $p = 2^{m+1}$, the number of curves is also doubled: This leads to a bifurcation when the period doubles.

Now we briefly introduce the tree of cyclotomic polynomials. At present we know five Fermat primes: $F_0 = 3, \ldots, F_4 = 65537$ ($\tau(F_0) = \tau(F_1) = \tau(F_2) = \tau(F_3) = \tau(F_4) = 2$), which yield four doublings of the number of polynomials Φ_d,

$$w_{1,2} = w_{2,3} = w_{3,4} = w_{4,5} = 2,$$

corresponding to four bifurcations for the values $p = 4, 8, 16, 32$. The numbers F_5, F_6, F_7, F_8 are known to have just two prime factors, i.e., $\tau(F_5) = \tau(F_6) = \tau(F_7) = \tau(F_8) = 4$. Therefore, by (15.18),

$$w_{5,6} = 6 \quad \text{and} \quad w_{6,7} = w_{7,8} = w_{8,9} = 4.$$

For $m = 9, 10$, and 11, we have the following values of $w_{m,m+1}$ based on the complete factorizations of F_m given in Appendix A:

$$w_{9,10} = \frac{28}{3}, \quad w_{10,11} = \frac{120}{7}, \quad \text{and} \quad w_{11,12} = \frac{496}{15}.$$

For $m \geq 12$ we cannot at present compute $w_{m,m+1}$, since the complete factorizations of F_m are not known. The following result appears in [Antonyuk, Stanyukovich, 1990b].

Theorem 15.25 (Antonyuk, Stanyukovich). *Let $m \geq 1$. Then*

(a) $w_{m,m+1} > 2 \iff F_m$ *is composite,*

(b) $w_{m,m+1} = 2 \iff F_{m-1}$ *and* F_m *are prime,*

(c) $w_{m,m+1} < 2 \iff F_{m-1}$ *is composite and* F_m *is prime.*

P r o o f . (a) If $w_{m,m+1} > 2$, then $\tau(F_{m-1})\tau(F_m) - \tau(F_{m-1}) > 2\tau(F_{m-1}) - 2$ due to (15.18). Therefore,

$$\tau(F_m) > 3 - \frac{2}{\tau(F_{m-1})} \geq 2,$$

since always $\tau(F_{m-1}) \geq 2$. Consequently, F_m is composite.

Conversely, if F_m is composite, then by (15.18) we have

$$w_{m,m+1} > \tau(F_m) - 1 \geq 2.$$

(b) If $w_{m,m+1} = 2$, then by (15.18),

$$\tau(F_m) = 3 - \frac{2}{\tau(F_{m-1})}.$$

This may happen only if $\tau(F_{m-1}) = \tau(F_m) = 2$, i.e., F_{m-1} and F_m are prime.

The converse implication is evident.

(c) This case is complementary to cases (a) and (b). The proposition is therefore proved. $\square$

Remark 15.26. Pairs of Fermat numbers satisfying (c) are not known.

Remark 15.27. From Theorem 15.25 we see that if $w_{m,m+1} \neq 2$, then the number of curves will increase by a factor different from two.

If $w_{m,m+1} > 2$, then there occurs an additional (in comparison with the case $w_{m,m+1} = 2$) splitting of the curves into individual branches, which can be characterized once the complete factorization of F_m into primes is known. Then the additional splitting is called the *fine structure of the curve*. The fine structure may appear starting from the period $p = 2^6 = 64$ if F_m is composite for $m \geq 5$. It is possible that the fine structure is characterized by the appearance not only of attractor-curves, but also repeller-curves, and this is what makes it very difficult to find such cases by computer.

If $5 \leq m \leq 30$, then F_m is composite and thus $w_{m,m+1} > 2$ (see Appendix A). For the composite Fermat number F_{382447}, we have $w_{382447,382448} > 2$ and $w_{382448,382449} \neq 2$.

The order of splitting of the cyclotomic polynomials Φ_d, in the case when the period is doubled, is defined by the following rule (its formulation is given only for the case $w_{m,m+1} = 2$): In changing from $p = 2^m$ to $p = 2^{m+1}$, the index d of the cyclotomic polynomial Φ_d is replaced by two indices d_1 and d_2 of two new cyclotomic polynomials Φ_{d_1} and Φ_{d_2}, and we have as in [Antonyuk, Stanyukovich, 1990b]

$$d = F_{k_1} \cdots F_{k_s} F_{m-1} \rightarrow \begin{cases} d_1 = F_{k_1} \cdots F_{k_s} F_m, \\ d_2 = F_{k_1} \cdots F_{k_s} F_{m-1} F_m, \end{cases}$$

where $F_{k_1} < \cdots < F_{k_s} < F_{m-1} < F_m$.

We assign to the bifurcation tree of curves a tree of cyclotomic polynomials, which are constructed with respect to the above splitting. This correspondence is defined via the algorithm for obtaining all periodic solutions of equation (15.11). For periods $p = 2^m$ such that $1 \leq m \leq 4$, there is a one-to-one correspondence between curves and cyclotomic polynomials.

For periods $p = 2^m$ such that $m \geq 5$, this one-to-one correspondence can be violated (a single curve can correspond to several cyclotomic polynomials), but in all cases, for a given period, the number of curves can never be greater than the number of cyclotomic polynomials. It is also shown in [Antonyuk, Stanyukovich, 1990b] that the range in which d varies (for a fixed m) is given by the inequalities

$$F_{m-1} \leq d \leq F_m - 2 = (F_{m-1} - 2)F_{m-1}, \quad m \neq 0.$$

A Negative Property of Fermat Primes

As opposed to our previous applications of Fermat numbers based on their special properties, we show below that Fermat primes are the only odd primes that cannot be used in a particular application because of those same properties. In [Schroeder, pp. 164–167] there is a proposal to improve concert hall acoustics by a method involving primitive roots modulo an odd prime p. If $p - 1$ has two coprime factors $r > 1$ and $s > 1$, then by using the Chinese remainder theorem, one can form a two-dimensional $r \times s$ rectangular array that can be employed in the enhancement of concert hall acoustics. The Fermat primes are the only odd primes p for which $p - 1$ has only one prime divisor, namely 2, thus making the construction of such an $r \times s$ rectangular array impossible.

16. The Proof of Gauss's Theorem

Recall Gauss's Theorem 4.3: *There exists a Euclidean construction (i.e., by ruler and compass) of the regular n-gon if and only if*

$$n = 2^i p_1 p_2 \cdots p_j,$$

where $n \geq 3$, $i \geq 0$, $j \geq 0$, *and* $p_1, p_2, \ldots, p_j$ *are distinct Fermat primes.*

The original proof due to Gauss covers more than 50 pages [Gauss, Sect. VII], even though the necessity was not completely shown (the proof was given by [Wantzel] and later also by [Pierpont]). In this section we give a shorter proof of Gauss's theorem using the modern machinery of field extensions and Galois theory. Throughout this section all numbers are complex. We recall the following definition:

Definition 16.1. *Let* $K \not\subseteq \{0\}$ *be a subset of the complex numbers* $\mathbb{C}$. *Then,* K *is called a field if whenever* $x, y \in K$ *with* $y \neq 0$, *both* $x - y$ *and* x/y *belong to* K.

Let us notice that if K is a field, then both 0 and 1 are elements of K. Indeed, it suffices to choose $x = y \neq 0$ in the above definition. Moreover, if we choose $x = 0$ or $x = 1$, we get that whenever y is a nonzero element of K, both $-y$ and $1/y$ belong to K. Finally, assume that x and y are elements of K with $y \neq 0$. Since both $-y$ and $1/y$ are nonzero elements of K, it follows, by the above definition, that both $x + y = x - (-y)$ and $x \cdot y = x/(1/y)$ are in K.

We also notice that the rational numbers $\mathbb{Q}$ form a field, the real numbers $\mathbb{R}$ form another field, and that the complex numbers $\mathbb{C}$ form yet another field. Moreover, every field K contains $\mathbb{Q}$. Indeed, since $1 \in K$, it follows by the above remarks that $2 = 1 + 1$, $3 = 2 + 1$, and so on, are also elements of K. By induction, it follows that any positive integer m belongs to K. By Definition 16.1, we now get that every positive rational number m/n belongs to K. Finally, since K contains 0 and since once $x \in K$ then $-x \in K$ as well, it follows that K contains $\mathbb{Q}$.

We just showed that $\mathbb{Q} \subset K \subset \mathbb{C}$ for any field K. Whenever we have two fields K and L with $K \subset L$ we say that K is a *subfield* of L or that L is a *field extension* of K. So, with our terminology, $\mathbb{Q}$ is a subfield of every field K and every $K \subseteq \mathbb{C}$ is a subfield of $\mathbb{C}$.

Remark 16.2. In general, the notion of a field is defined in a much more general context. For a thorough account of the theory of fields the reader may refer to any book in abstract algebra such as [Hungerford]. In this chapter, we intend to supply just enough background so that we can prove Gauss's theorem.

It also follows easily from the above remarks that if K is a field, $x \in K$, and q is a rational number, then $qx \in K$. Indeed, this follows simply because K contains $\mathbb{Q}$. At this step we recall the definition of a linear space over a field.

Definition 16.3. *Let $K \subset \mathbb{C}$ be a field. A nonempty subset V of the complex numbers is a linear space over K if whenever $v, w \in V$ and $x \in K$ both xv and $v - w$ are in V.*

Let us notice that $0 \in V$ (pick $v = w$ in Definition 16.3). Moreover, if we let $v = 0$, we get that $-w \in V$ whenever $w \in V$. Finally, since $-w \in V$ whenever $w \in V$, it follows that $v + w = v - (-w) \in V$ whenever v and w are in V.

Notice also that if $K \subset L$ is a field extension and V is a linear space over L, then V is a linear space over K as well.

Let V be a linear space over a field K. A subset $I \subset V$ is called a *linearly independent set* if the relation

$$\lambda_1 v_1 + \lambda_2 v_2 + \cdots + \lambda_n v_n = 0$$

with $v_1, v_2, \ldots, v_n$ distinct elements of I and $\lambda_1, \lambda_2, \ldots, \lambda_n$ elements of K holds only for $\lambda_1 = \cdots = \lambda_n = 0$. A subset $B \subset V$ is called a *basis* of V if B is a linearly independent set and every element $v \in V$ can be written in the form

$$v = \lambda_1 v_1 + \lambda_2 v_2 + \cdots + \lambda_n v_n$$

for some distinct elements $v_1, v_2, \ldots, v_n \in B$ and some $\lambda_1, \lambda_2, \ldots, \lambda_n \in K$.

It is well known (see [Hungerford]), although not entirely trivial to prove, that any linear space V over a field K has a basis B and that any two bases B_1 and B_2 have the same cardinality. In particular, if V has a basis B with n elements, then all other bases of V have also exactly n elements. In this case, the number n is called the *dimension* of V over K, and we also say that V is *finite-dimensional*. Of course, when B happens to have infinite cardinality we say that V is *infinite-dimensional*.

Assume now that $K \subset L$ is a field extension. In this case, L is a linear space over K. We denote the dimension of L over K by $[L : K]$. We recall the following transitivity property of field extensions:

Proposition 16.4. *Assume that $K \subset L$ and $L \subset M$ are field extensions with both $[L : K]$ and $[M : L]$ finite. Then $[M : K] = [M : L] \cdot [L : K]$.*

P r o o f . Let $n = [L : K]$, $m = [M : L]$, and let $\{e_1, e_2, \ldots, e_n\}$ be a basis of L over K and $\{f_1, f_2, \ldots, f_m\}$ be a basis of M over L. It is now easy to show that $\{e_i f_j\}_{\substack{1 \leq i \leq n \\ 1 \leq j \leq m}}$ is a basis of M over K. $\square$

Suppose that K is a field and that $\zeta \in \mathbb{C}$ is a complex number. We use the notation $K(\zeta)$ to denote the smallest field containing both K and ζ (compare with Figures 16.1 and 16.2).

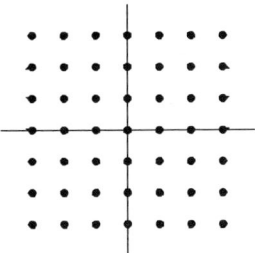

Figure 16.1. The Gaussian integers $a + ib$ of $K\left(\sqrt{-1}\right)$ for $a, b \in \mathbb{Z}$.

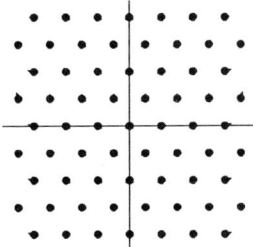

Figure 16.2. The integers $a + \frac{1}{2}(1 + i\sqrt{3})b$ of $K(i\sqrt{3})$ for $a, b \in \mathbb{Z}$.

One of the most natural ways of constructing finite-dimensional fields over $\mathbb{Q}$ is by using *algebraic numbers*. We recall that an algebraic number α is a complex number that is a root of some polynomial of the form

$$F(X) = X^d + a_1 X^{d-1} + \cdots + a_d,$$

with rational coefficients $a_1, a_2, \ldots, a_d$. Recall that a polynomial like the one above (that is, with the leading coefficient 1) is called *monic*. Notice that every rational number q is the root of the polynomial $X - q$; hence, it is algebraic. For an algebraic number α let f be a monic polynomial of minimal degree d having α as a root. By use of the division algorithm for polynomials over a field, it is easy to show that f is unique with the above property. Then f is called the *minimal polynomial of α*, and the degree d of f is called the *degree of α*. We recall the following properties of algebraic numbers.

Proposition 16.5. *Assume that α is an algebraic number of degree d. Then $[\mathbb{Q}(\alpha) : \mathbb{Q}] = d$ and $\{1, \alpha, \ldots, \alpha^{d-1}\}$ is a basis of $\mathbb{Q}(\alpha)$ over $\mathbb{Q}$.*

P r o o f . Notice first of all that $B = \{1, \alpha, \ldots, \alpha^{d-1}\}$ is a linearly independent set over $\mathbb{Q}$. Indeed, if not, then there exists a relation of the form

$$\lambda_1 \alpha^{d-1} + \lambda_2 \alpha^{d-2} + \cdots + \lambda_{d-1}\alpha + \lambda_d = 0,$$

with all λ_i rational but not all of them zero, which implies that α is a root of the nonzero polynomial with rational coefficients

$$\lambda_1 X^{d-1} + \cdots + \lambda_{d-1}X + \lambda_d,$$

which has degree less than d. This contradicts the definition of d.

Denote by V the linear space over $\mathbb{Q}$ having the basis $B = \{1, \alpha, \ldots, \alpha^{d-1}\}$. It suffices to show that V is a field. Let

$$f(X) = X^d + a_1 X^{d-1} + \cdots + a_d$$

be the minimal polynomial with rational coefficients having α as a root. Since

$$\alpha^d = -a_1 \alpha^{d-1} - \cdots - a_d,$$

it follows that $\alpha^d \in V$. One may further use induction to show that $\alpha^n \in V$ for all integers $n \geq 0$. Now let $\gamma \in V$ and let $\beta \in V$ be any nonzero element. Notice that since $\alpha^n \in V$ for all $n \geq 0$, it follows that $\gamma\beta \in V$. Since $\gamma - \beta \in V$ (because V is a linear space), it follows that it suffices to show that $1/\beta \in V$. Write

$$\beta = b_1 \alpha^{d-1} + \cdots + b_{d-1}\alpha + b_d,$$

where b_i are rational numbers. Since $\alpha^n \in V$ for all $n \geq 0$, it follows that $\beta^n \in V$ for all $n \geq 0$ as well. Since V has dimension d, it follows that the set $\{1, \beta, \ldots, \beta^d\}$ cannot be a linearly independent set. It follows that there exist $\lambda_1, \ldots, \lambda_d$ rational numbers not all zero such that

$$\lambda_1 \beta^d + \cdots + \lambda_{d-1}\beta + \lambda_d = 0.$$

We may assume that $\lambda_d \neq 0$ because otherwise we may simplify the above relation by dividing each term by β and get a similar relation with d replaced by $d - 1$. Since $\lambda_d \neq 0$, it follows that

$$\frac{1}{\beta} = -\frac{\lambda_1}{\lambda_d}\beta^{d-1} - \cdots - \frac{\lambda_{d-1}}{\lambda_d}.$$

Since $\beta^n \in V$ for all $n \geq 0$, it follows that $1/\beta \in V$ as well, which finishes the proof of Proposition 16.5. $\qquad\square$

We are now well equipped to deal with the proof of Gauss's theorem. We start with the following definition.

Definition 16.6. *A complex number z is called constructible if the point P representing the complex number z in the complex plane $\mathbb{C}$ can be constructed with ruler and compass starting from the points of integer coordinates.*

In other words, we say that a complex number $z = x + iy$, where both x and y are real, is constructible if and only if both x and y are constructible. It is easily seen that if both

$$z_1 = x_1 + iy_1 \qquad \text{and} \qquad z_2 = x_2 + iy_2$$

are constructible with $z_2 \neq 0$, then both

$$z_1 - z_2 = (x_1 - x_2) + i(y_1 - y_2)$$

and

$$\frac{z_1}{z_2} = \frac{(x_1 x_2 + y_1 y_2) + i(x_2 y_1 - x_1 y_2)}{x_2^2 + y_2^2}$$

are constructible. Indeed, this last assertion follows from the fact that if a and b are constructible lengths of segments, then $a + b$, $|a - b|$, ab, and a/b if $b \neq 0$ are also lengths of segments that can be constructed via the use of similar triangles.

In order to prove Gauss's theorem, we need to understand better the properties of constructible numbers.

Propositon 16.7. *Assume that α is a constructible number. Then α is algebraic and its degree is a power of 2.*

P r o o f . Assume that we can construct α by a construction involving at most r intermediary points obtained as intersections line–circle or circle–circle. We then show that α is algebraic and that the dimension of the smallest field containing both α and all the intermediary points is a divisor of 2^{r+1}. In particular, the degree of α is a divisor of 2^{r+1}. We do this by induction on r.

We begin with a useful observation. Suppose that K is a field and that l_1 and l_2 are two lines each one passing through a point P_1 and P_2, respectively, such that both P_1 and P_2 have both coordinates in K. Assume also that the slopes of both l_1 and l_2 are in K as well. Finally, let P be the point of intersection of l_1 and l_2. Then, both coordinates of P are in K as well.

Assume now that $r = 0$. In this case, α can be constructed using only intersections of straight lines with the first few lines passing through two points having integer values for both their coordinates. By the preceding observation, it follows that both coordinates of α are rational. Hence, $\alpha = p + iq$ with p and q rational. Since α is a root of $(X - p)^2 + q^2 = 0$, it follows that α is algebraic of degree at most 2.

Assume now that one can construct α by a construction involving exactly r intermediary points obtained as intersection line–circle or circle–circle. Let $P_1, \ldots, P_r$ be these intermediate points. Let α_i be the complex number representing P_i. By the induction hypothesis, the smallest field containing $\alpha_1, \alpha_2, \ldots, \alpha_{r-1}$ has degree a divisor of 2^r. We denote this field by K_1. From the above observation, it follows that $\alpha_r = x_r + iy_r$ satisfies a system of equations

$$(x - u_1)^2 + (y - v_1)^2 = r_1,$$
$$u_2 x + v_2 y = r_2,$$

or

$$(x - u_1)^2 + (y - v_1)^2 = r_1,$$
$$(x - u_2)^2 + (y - v_2)^2 = r_2,$$

where $u_i, v_i, r_i \in K_1$. It now follows easily that $[K_1(\alpha_r) : K_1] = 1$ or 2. By the preceding arguments, it also follows that $\alpha \in K_1(\alpha_r)$. Now Proposition 16.4 completes the induction step. $\square$

From Proposition 16.7 we conclude that if α is constructible, then it is algebraic and its degree is a power of 2.

Finally, let us return to the constructibility with ruler and compass of the regular polygon with n sides. With our formalism, this is equivalent to the constructibility of the nth root of unity, namely, the complex number

$$\zeta_n = \cos \frac{2\pi}{n} + i \sin \frac{2\pi}{n}.$$

This complex number is a root of the polynomial $X^n - 1$. Its minimal polynomial is the nth cyclotomic polynomial of degree $\phi(n)$, where ϕ is the Euler totient function. The nth cyclotomic polynomial is denoted by Φ_n (see Lemma 13.1, where Φ_n is introduced) and can be defined recursively using the formulae $\Phi_1 = X - 1$ and

$$X^n - 1 = \prod_{d \mid n} \Phi_d.$$

For a proof that Φ_n is irreducible the reader is referred to [Hungerford].

From Proposition 16.7 and the preceding discussion, it follows that if the regular polygon is constructible with ruler and compass, then $\phi(n)$ is a power of 2. But this is equivalent, by the proof of Theorem 4.5, to the fact that $n = 2^i p_1 p_2 \cdots p_j$, where $p_1, p_2, \ldots, p_j$ are distinct Fermat primes.

For the converse statement we need some Galois theory.

Definition 16.8. *Let L be a field extension. An automorphism of L is a nonzero map $\gamma : L \to L$ such that $\gamma(x + y) = \gamma(x) + \gamma(y)$ and $\gamma(xy) = \gamma(x)\gamma(y)$ for all $x, y \in L$. If $K \subset L$ is a field extension, then an automorphism γ of L is called a Galois automorphism of L over K if $\gamma(x) = x$ for all $x \in K$. The set of all Galois automorphisms of L over K is denoted by $\mathrm{Gal}(L/K)$.*

Suppose that $K \subset L$ is a finite extension, i.e. $[L : K]$ is finite. It is then easily seen that every element of $\mathrm{Gal}(L/K)$ is both one-to-one and onto as a function from L to L. Moreover, $\mathrm{Gal}(L/K)$ forms a group together with the usual composition law. The inverse of the element $\gamma \in \mathrm{Gal}(L/K)$ is exactly the inverse map of γ as a map from L to L. Moreover, since $\mathrm{Gal}(L/K)$ is a subgroup of the group of permutations of the set with $[L : K]$ elements, it follows that $\mathrm{Gal}(L/K)$ is a finite group.

Proposition 16.9. *Suppose that ζ_n is a primitive root of unity of order n. Let $L = \mathbb{Q}(\zeta_n)$. Then, $\mathrm{Gal}(L/\mathbb{Q})$ is cyclic with $\phi(n)$ elements.*

For the proof of Proposition 16.9 the reader may consult [Hungerford].

Finally, we need another result about field extensions with abelian Galois groups.

Proposition 16.10. *Suppose that L is a field such that $\mathrm{Gal}(L/\mathbb{Q})$ has the same cardinality as $[L : \mathbb{Q}]$. Moreover, assume that $G = \mathrm{Gal}(L/\mathbb{Q})$ is commutative. Then, there exists a one-to-one inclusion-reversing correspondence between the intermediary subfields $K \subset L$ and the subgroups $H \subset G$.*

The proof of Proposition 16.10 can be found in [Hungerford], too.

We can now deal with the converse of Gauss's theorem. Assume that $\phi(n) = 2^r$ for some integer $r \geq 1$ and let $L = \mathbb{Q}(\zeta_n)$. Since $|G| = |\mathrm{Gal}(L/\mathbb{Q})| = [L : \mathbb{Q}] = 2^r$, it follows that we may apply Proposition 16.10 to get a one-to-one inclusion-reversing correspondence between the subgroups of the cyclic group with 2^r elements and the intermediate subfields of L. Since G has a chain of subgroups $1 < H_1 < \cdots < H_r = G$ with $[H_{i+1} : H_i] = 2$ for all i, it follows that there exists a chain of subfields $\mathbb{Q} \subset K_1 \subset K_2 \subset \cdots \subset K_r = L$. Moreover, by Proposition 16.4 and the fact that $[L : \mathbb{Q}] = 2^r$, it follows that $[K_{i+1} : K_i] = 2$ for all i. It now follows easily, by induction, that every real positive element from K_{i+1} can be constructed with ruler and compass starting from elements of K_i (simply because this construction reduces to solving an equation of the second degree with coefficients in K_i). In particular, ζ_n is constructible, which ends the proof of Gauss's theorem.

17. Euclidean Construction of the Regular Heptadecagon

*On that same occasion I have lifted the mystery which
had rested over Gauss's theory of the division of the circle;
I now see as clear as daylight how he has been led to it.*

Niels Henrik Abel

Figure 17.1. Young Carl Friedrich Gauss.

Figure 17.2. Gauss's home in Gotmarstrasse 11, Göttingen.

Gauss's theorem from the previous chapter establishes a remarkable necessary and sufficient condition for a Euclidean construction of regular polygons. Euclid ($\approx 330 - \approx 260$ B.C.) already knew how to construct regular polygons with n sides by ruler and compass for

$$n = 2^i 3^j 5^k,$$

where $n \geq 3$ and $i \geq 0$ are integers and $j, k \in \{0, 1\}$. However, he did not know whether it is possible to construct the regular heptagon or nonagon. C. F. Gauss (see Figure 17.1) gave the answer to this question 2000 years later. When he was only nineteen (cf. Figure 17.2), he wrote a short treatise about the division of a circle into 17 equal parts by geometric means (see Figure 17.3 and (17.3)). Here he essentially used the fact that 17 is a Fermat prime. This fundamental discovery is presented on the base of his statue in Brunswick (in German *Braunschweig*), where he was born (see Figure 17.4).

Several authors incorrectly place the 17-gon on Gauss's tombstone in Göttingen (see Figure 17.5), where he requested it to be put. However, in [Reid] the circumstances concerning the representation of Gauss's 17-gon are correctly explained. The statue of Gauss in Brunswick is shown in Figure 17.6 and details of its base in Figure 17.7. Since the regular 17-gon would look like a circle (compare with Figure 4.1 or 17.8), the 17-pointed star is illustrated there.

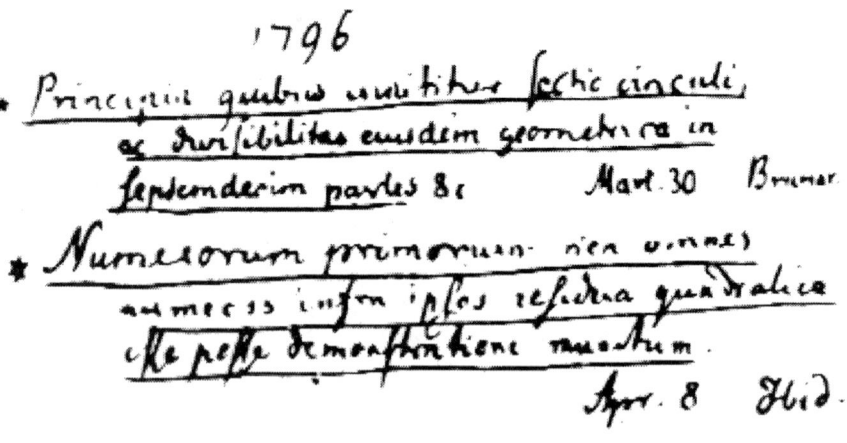

Figure 17.3. The upper part of a page of Gauss's notebook, where the first entry concerns the epochal Euclidean construction giving the division of a circle into 17 equal parts.

Figure 17.4. Memorial plaque of Gauss's birthplace in Brunswick.

Figure 17.5. Gauss's tombstone in Göttingen.

Figure 17.6. Statue of Carl Friedrich Gauss in his native Brunswick. On the left-hand side of its base the 17-pointed golden star shines in honor of Gauss's fundamental discovery.

Figure 17.7. Details of the base. Sides of the 17-pointed star are parts of the longest diagonals of the regular 17-gon.

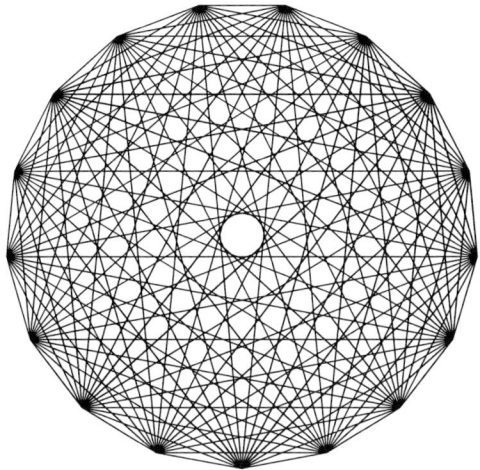

Figure 17.8. All diagonals of the regular heptadecagon.

Remark 17.1. Gauss's construction is based on the fact that if a, b, c are lengths of three straight line segments, then it is possible to construct by ruler and compass straight line segments of the lengths $a \pm b$, ab/c, and $\sqrt{ab}$, and thus also $\sqrt{a}$ for $b = 1$. To this end we can apply, e.g., similarities of triangles (see Figure 17.9) and Euclid's theorem for a right triangle (see Figure 17.10).

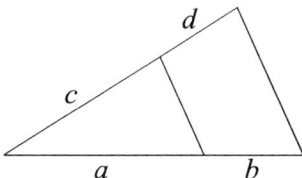

Figure 17.9. Illustration of the construction of the length $d = ab/c$ using similar triangles.

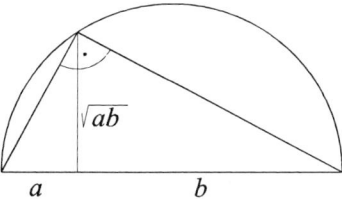

Figure 17.10. Euclid's theorem.

In this way we are able to construct a straight line segment whose length is expressed by a finite number of square roots. In particular, in the case of the regular triangle, pentagon, and heptadecagon (whose numbers of sides are F_0, F_1, and F_2, respectively) we have the following theorem:

Theorem 17.2. *The following formulae hold true:*

(17.1)
$$\cos\frac{2\pi}{3} = -\frac{1}{2},$$

(17.2)
$$\cos\frac{2\pi}{5} = \frac{\sqrt{5}-1}{4},$$

(17.3)
$$\cos\frac{2\pi}{17} = \frac{1}{16}\left(-1+\sqrt{17}+\sqrt{2(17-\sqrt{17})}\right.$$
$$\left. + 2\sqrt{17+3\sqrt{17}-\sqrt{2(17-\sqrt{17})}-2\sqrt{2(17+\sqrt{17})}}\right).$$

P r o o f . By the Fundamental theorem of algebra the binomial equation $z^n - 1 = 0$ has just n roots in the complex plane. They are uniformly distributed on the unit circle with center at the origin and have the form (see Figure 17.11 for $n = 17$)

(17.4) $z_k = e^{ik\alpha} = \cos k\alpha + i\sin k\alpha$ for $k = 0, \ldots, n-1,$

where

$$\alpha = \frac{2\pi}{n}.$$

We have $z_0 = 1$, and the sum of all the roots is zero:

(17.5) $z_{n-1} + z_{n-2} + \cdots + z_1 + 1 = 0.$

Moreover, by (17.4)

(17.6) $z_k + z_{n-k} = 2\cos k\alpha,$ $k = 1, \ldots, n-1.$

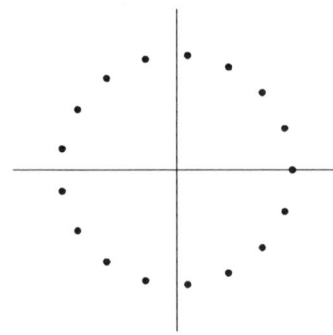

Figure 17.11. Roots of the equation $z^n - 1 = 0$.

Choosing first $n = 3$, we see from (17.5) and (17.6) that $-1 = z_1 + z_2 = 2 \cos \alpha$, which implies (17.1).

Further, let $n = 5$. Setting

(17.7) $$x_1 = z_1 + z_4,$$
(17.8) $$x_2 = z_2 + z_3,$$

we easily find that $x_1 > 0 > x_2$. From this and (17.5) we have that

(17.9) $$x_1 + x_2 = -1.$$

Using (17.7), (17.8), (17.6), the relation

(17.10) $$2 \cos k\alpha \cos \ell\alpha = \cos(k + \ell)\alpha + \cos(k - \ell)\alpha$$

for any integers k, ℓ, (17.6) again, and (17.5), we obtain

$$x_1 x_2 = 4 \cos \alpha \cos 2\alpha = 2(\cos 3\alpha + \cos \alpha) = z_3 + z_2 + z_1 + z_4 = -1.$$

From this and (17.9) we find that x_1 and x_2 are roots of the quadratic equation with integer coefficients
$$x^2 + x - 1 = 0,$$

i.e.,
$$x_1 = \frac{-1 + \sqrt{5}}{2}.$$

The last equation, (17.7), and (17.6) yield (17.2).

Finally, let $n = 17$ and let

(17.11) $$x_1 = z_1 + z_2 + z_4 + z_8 + z_9 + z_{13} + z_{15} + z_{16},$$
(17.12) $$x_2 = z_3 + z_5 + z_6 + z_7 + z_{10} + z_{11} + z_{12} + z_{14}.$$

Due to (17.5), we have

(17.13) $$x_1 + x_2 = -1.$$

Using (17.11), (17.12), (17.6), (17.10), (17.6) again, and (17.5), we obtain

$$
\begin{aligned}
x_1 x_2 = {}& 4(\cos \alpha + \cos 2\alpha + \cos 4\alpha + \cos 8\alpha)(\cos 3\alpha + \cos 5\alpha + \cos 6\alpha + \cos 7\alpha) \\
= {}& 2(\cos 4\alpha + \cos 2\alpha + \cos 6\alpha + \cos 4\alpha + \cos 7\alpha + \cos 5\alpha + \cos 8\alpha + \cos 6\alpha \\
& + \cos 5\alpha + \cos \alpha + \cos 7\alpha + \cos 3\alpha + \cos 8\alpha + \cos 4\alpha + \cos 9\alpha + \cos 5\alpha \\
& + \cos 7\alpha + \cos \alpha + \cos 9\alpha + \cos \alpha + \cos 10\alpha + \cos 2\alpha + \cos 11\alpha + \cos 3\alpha \\
& + \cos 11\alpha + \cos 5\alpha + \cos 13\alpha + \cos 3\alpha + \cos 14\alpha + \cos 2\alpha + \cos 15\alpha + \cos \alpha) \\
= {}& 4(z_1 + z_2 + \cdots + z_{16}) = -4.
\end{aligned}
$$

From this and (17.13) we find that x_1 and x_2 are roots of the quadratic equation with integer coefficients

$$x^2 + x - 4 = 0.$$

Since $x_2 < z_3 + z_7 + z_{10} + z_{14} < 0$, we have

$$x_1 = \frac{-1 + \sqrt{17}}{2}, \quad x_2 = \frac{-1 - \sqrt{17}}{2}.$$

Setting

$$y_1 = 2(\cos \alpha + \cos 4\alpha),$$
$$y_2 = 2(\cos 2\alpha + \cos 8\alpha),$$
$$y_3 = 2(\cos 3\alpha + \cos 5\alpha),$$
$$y_4 = 2(\cos 6\alpha + \cos 7\alpha),$$

we see that $y_1 > y_2$ and $y_3 > y_4$. By (17.11) and (17.6),

$$y_1 + y_2 = x_1,$$

and using (17.6) again and (17.5), we get

$$\begin{aligned}
y_1 y_2 &= 4(\cos \alpha + \cos 4\alpha)(\cos 2\alpha + \cos 8\alpha) \\
&= 2(\cos 3\alpha + \cos \alpha + \cos 9\alpha + \cos 7\alpha + \cos 6\alpha + \cos 2\alpha + \cos 12\alpha + \cos 4\alpha) \\
&= z_3 + z_{14} + z_1 + z_{16} + z_9 + z_8 + z_7 + z_{10} + z_6 + z_{11} + z_2 + z_{15} \\
&\quad + z_{12} + z_5 + z_4 + z_{13} = -1.
\end{aligned}$$

From this we come to another quadratic equation,

$$y^2 - x_1 y - 1 = 0,$$

i.e.,

(17.14)
$$y_1 = \frac{x_1 + \sqrt{x_1^2 + 4}}{2} = \frac{-1 + \sqrt{17} + \sqrt{34 - 2\sqrt{17}}}{4},$$
$$y_2 = \frac{x_1 - \sqrt{x_1^2 + 4}}{2}.$$

Similarly, we get

$$y_3 + y_4 = x_2,$$
$$y_3 y_4 = 4(\cos 3\alpha + \cos 5\alpha)(\cos 7\alpha + \cos 6\alpha) = -1.$$

This again yields the quadratic equation for y_3 and y_4:

$$y^2 - x_2 y - 1 = 0,$$

i.e.,

$$(17.15) \qquad y_3 = \frac{x_2 + \sqrt{x_2^2 + 4}}{2} = \frac{-1 - \sqrt{17} + \sqrt{34 + 2\sqrt{17}}}{4},$$

$$y_4 = \frac{x_2 - \sqrt{x_2^2 + 4}}{2}.$$

We know that

$$y_1 = 2\cos\alpha + 2\cos 4\alpha,$$
$$y_3 = 2(\cos 5\alpha + \cos 3\alpha) = 4\cos\alpha\cos 4\alpha,$$

where the last equality follows from (17.10). Therefore,

$$w_1 = 2\cos\alpha \quad \text{and} \quad w_2 = 2\cos 4\alpha$$

$(w_1 > w_2)$ are the zeros of the quadratic equation

$$w^2 - y_1 w + y_3 = 0.$$

Therefore,

$$w_1 = \frac{y_1 + \sqrt{y_1^2 - 4y_3}}{2} = 2\cos\frac{2\pi}{17}.$$

Substituting from (17.14) and (17.15), we see that

$$2\cos\frac{2\pi}{17} = -\frac{1}{8} + \frac{1}{8}\sqrt{17} + \frac{1}{8}\sqrt{34 - 2\sqrt{17}}$$
$$+ \frac{1}{4}\sqrt{17 + 3\sqrt{17} - \sqrt{34 - 2\sqrt{17}} - 2\sqrt{34 + 2\sqrt{17}}},$$

hence (17.3) holds.   □

Remark 17.3. The previous proof is based on ideas from [Klein], [Rademacher], and [Stewart, I., 1989].

Remark 17.4. Decompositions (17.11) and (17.12) are due to Gauss. He ordered the roots z_k in a cycle according to powers of 3 modulo 17 as follows:

$$(17.16) \qquad z_3, z_9, z_{10}, z_{13}, z_5, z_{15}, z_{11}, z_{16}, z_{14}, z_8, z_7, z_4, z_{12}, z_2, z_6, z_1,$$

since

$$3^1 \equiv 3 \ (\text{mod } 17), \quad 3^2 \equiv 9 \ (\text{mod } 17), \quad 3^3 \equiv 10 \ (\text{mod } 17), \dots$$
$$3^{14} \equiv 2 \ (\text{mod } 17), \quad 3^{15} \equiv 6 \ (\text{mod } 17), \quad 3^{16} \equiv 1 \ (\text{mod } 17).$$

Recall that 3 is a primitive root modulo 17; i.e., the minimum solution of the congruence

$$3^j \equiv 1 \ (\text{mod } 17)$$

for $j > 0$ is $17 - 1 = 16$. Namely, for $j = 1, \ldots, 16$ we obtain 16 different remainders $r_j \in \{1, \ldots, 16\}$ such that

$$3^j = 17q_j + r_j$$

for appropriate integers q_j. Notice that

$$z_{r_j} = z_1^{r_j} = z_1^{3^j}$$

and thus

$$z_{r_{j+1}} = z_1^{3^{j+1}} = \left(z_1^{3^j}\right)^3 = (z_{r_j})^3.$$

Hence, in the series (17.16) of roots, each root is the cube of the preceding root. The terms from decompositions (17.11) and (17.12) appear in even and odd positions of series (17.16), respectively.

The integer 3 in the above text is chosen because 3 is the smallest primitive root modulo 17, and thus is a generator of all the nonzero residues modulo 17 (compare with Remark 2.15).

Remark 17.5. For a given segment whose length is a it is not possible, in general, to find a Euclidean construction of a segment whose length is $\sqrt[k]{a}$, unless k is a power of two. Let us point out that the ancient Greeks tried unsuccessfully to construct $\sqrt[3]{2}$ (the so-called problem of the duplicated cube; see Figure 17.12).

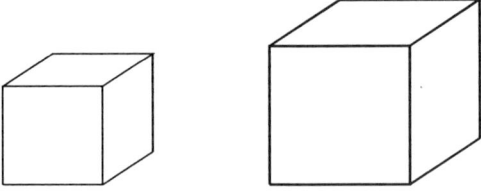

Figure 17.12. For a given cube it is impossible to construct the edge of a cube with duplicated volume by ruler and compass.

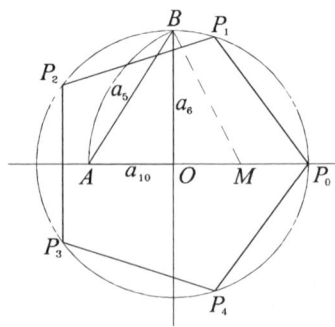

Figure 17.13. Euclidean construction of the regular pentagon.

Remark 17.6. Construction of the regular pentagon (see, e.g., [Šofr]):

Denote by $P_0, P_1, \ldots, P_4$ its unknown vertices on a circle with center O. Let P_0 be given. Denote by M the midpoint of the straight line segment OP_0. On the circle choose a point B such that $BO \perp OP_0$. Finally, construct the right triangle ABO such that $|BM| = |AM|$, as sketched in Figure 17.13. Then $|AB|$ is the length of the side of the regular pentagon.

Remark 17.7. By [Šofr], it is easy to find that (see Figure 17.13)

$$a_5^2 = a_6^2 + a_{10}^2,$$

where a_n is the length of the side of the regular n-gon. In the case $a_6 = 1$, we have

$$a = a_5 = \sqrt{\frac{5 - \sqrt{5}}{2}}, \quad a_{10} = \frac{\sqrt{5} - 1}{2}.$$

Moreover, it is well known that a_5 is the larger part of the diagonal of the regular pentagon when the diagonal is divided into two line segments the ratio of whose lengths is the *golden section* (see Figure 17.14), i.e,

$$\frac{a}{d} = \frac{d - a}{a} = \frac{\sqrt{5} - 1}{2},$$

where d is the length of the diagonal.

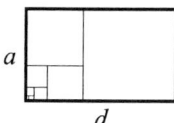

Figure 17.14. Aesthetic beauty is commonly attributed to the rectangle the ratio of whose sides is the golden section, i.e., $a : d = d : (a + d)$.

Remark 17.8. Construction of the regular heptadecagon ($=$ heptakaidecagon):

We shall describe one of the simplest known constructions of the regular 17-gon, originally given by Richmond (see [Richmond, 1893, 1909], [Beiler], [Stewart, I., 1973, p. 188]). Denote by $P_0, P_1, \ldots, P_{16}$ its unknown vertices on the circle with center O. Let P_0 be given (see Figure 17.15). We will construct only P_3 and P_5, corresponding to the angles $6\pi/17$ and $10\pi/17$, respectively, since the other vertices can be easily completed (i.e., the proposed relation (17.3) for $\alpha = 2\pi/17$ is not directly employed here, since we do not construct the point P_1). Let B be the end-point of the radius, which is perpendicular to OP_0, and let us construct the point I of the segment OB for which $|OI| = \frac{1}{4}|OB|$. Let us find the point E on the segment OP_0 such that $|\sphericalangle OIE| = \frac{1}{4}|\sphericalangle OIP_0|$. On the half-line opposite to OP_0 find the point F for which $|\sphericalangle FIE| = \frac{1}{4}\pi$. Now denote by K the intersection of the segment OB and the circle with the diameter FP_0. Another circle with center E and radius $|EK|$ cuts the line OP_0 at the points N_3 (between the points O and P_0) and N_5. These are orthogonal projections of the vertices P_3 and P_5 onto the line OP_0.

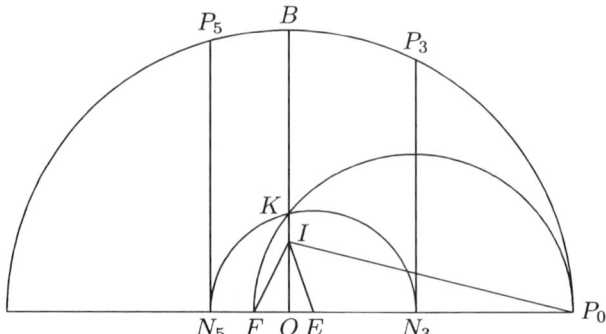

Figure 17.15. Euclidean construction of the regular 17-gon.

Remark 17.9. In Figure 17.16 the regular 257-gon is illustrated. An original description of its Euclidean construction is given in [Richelot]. It contains more than 80 pages. A much shorter analysis of this problem is in [Gottlieb].

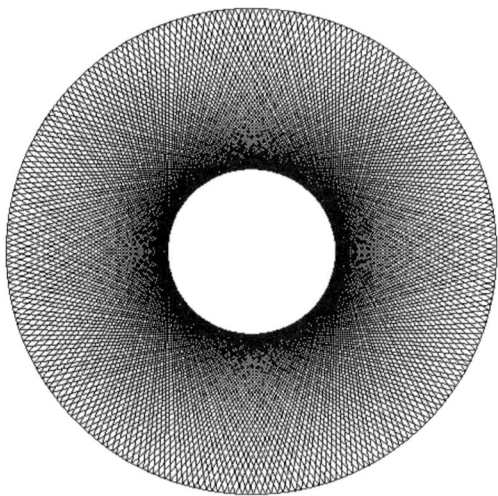

Figure 17.16. The regular 257-gon and some of its diagonals.

Remark 17.10. A manuscript on the construction of the regular 65537-gon can be found in [Hermes]. For a short note see also [Gottlieb]. According to [Schroeder, p. 242], there is a "suitcase" at the Mathematics Institute of the University of Göttingen that is jam-packed with the details of constructing the regular 65537-gon.

The construction of the other regular n-gons is based on the following two assertions.

Lemma 17.11. *Let p and q be coprime positive integers. Then there exist positive integers x and y such that*

$$(17.17) \qquad\qquad px - qy = 1.$$

P r o o f . According to Proposition 2.4 (for $d = 1$, $m = p$, and $n = -q$) there exist integers x' and y' such that $px' - qy' = 1$ and $p(x' + jq) - q(y' + jp) = 1$ for any integer j. Hence, there exist j such that $y = y' + jp$ is positive. Setting $x = x' + jq$, we see that (17.17) holds and x is also positive.   □

Proposition 17.12. *Let $n = pq$, where $p \geq 2$ and $q \geq 2$ are coprime integers. Then there exists a Euclidean construction of the roots of the equation $z^n - 1 = 0$ if and only if there exists a Euclidean construction of the roots of the equations $z^p - 1 = 0$ and $z^q - 1 = 0$.*

P r o o f . Let $p \geq 2$ and $q \geq 2$ be coprime. Then by Lemma 17.11 there exist positive integers x, y satisfying (17.17), and we have

$$\frac{x}{q} - \frac{y}{p} = \frac{1}{pq};$$

i.e., if we are able to construct $1/q$ and $1/p$ of 2π, then we also can construct $1/(pq)$ of 2π.

The converse implication is trivial.   □

Example 17.13. Let $p = 5$ and $q = 3$. Then from (17.17) we find that $x = 2$, $y = 3$, and $2/3 - 3/5 = 1/15$. Therefore, we can construct one-fifteenth of 2π and, hence also a regular 15-gon.

Remark 17.14. Since we know how to bisect any circular arc by ruler and compass, we are able to divide it into 2, 4, 8, ..., 2^i equal parts. Therefore, if we can construct the regular n-gon, we can also construct the regular $2^i n$-gon. (This also follows from Proposition 17.12.)

Remark 17.15. It is not possible to trisect the angle $2\pi/9$ by ruler and compass, because $\phi(9)$ is not a power of 2. Therefore, there is no Euclidean construction of the regular 9-gon. It is also not possible to construct any regular p^k-gon for any integer $k > 1$ and a prime $p > 2$, as follows from the equivalence stated in Gauss's Theorem 4.3 (see also Theorem 4.5). C. F. Gauss in Chapter 7 of his *Disquisitiones arithmeticae* [Gauss] proved only the sufficiency of his condition. The proof of the necessity was given later in [Wantzel] and also in [Pierpont].

Figure 17.17. Constructible regular polygons for $n \leq 10$.

Remark 17.16. The old geometric problem concerning constructions of regular polygons is thus transformed into an algebraic problem. The regular n-gon for

$n \leq 100$ can be constructed by ruler and compass if and only if

$$n \in \{3, \ 4, \ 5, \ 6, \ 8, \ 10, \ 12, \ 15, \ 16, \ 17, \ 20, \ 24, \ 30,$$
$$32, \ 34, \ 40, \ 48, \ 51, 60, \ 64, \ 68, \ 80, \ 85, \ 96\}$$

(compare with Figure 17.17). According to [Josephy], the number of positive integers $n \leq M$ for which the regular n-gon is constructible by ruler and compass would be close to $\frac{1}{2}(\log_2 M)^2$, assuming it were true that all Fermat numbers are prime.

Remark 17.17. Although hundreds of factors of the Fermat numbers and many necessary and sufficient conditions for the primality of F_m are known (see, e.g., Chapters 4, 5, 10), no one has been able to discover a general principle that would lead to a definitive answer to the question whether F_4 is the largest Fermat prime. Therefore, up to now we still do not know whether the current list of constructible regular polygons is complete.

Appendix A
Tables of Fermat Numbers and Their Prime Factors

The problem of distinguishing prime numbers from composite numbers and of resolving the latter into their prime factors is known to be one of the most important and useful in arithmetic.

Carl Friedrich Gauss
Disquisitiones arithmeticae, Sec. 329

Fermat Numbers

$F_0 = 3,$

$F_1 = 5,$

$F_2 = 17,$

$F_3 = 257,$

$F_4 = 65537,$

$F_5 = 4294967297,$

$F_6 = 18446744073709551617,$

$F_7 = 340282366920938463463374607431768211457,$

$F_8 = 11579208923731619542357098500868790785326998466656405640394575840079131296399937,$

$F_9 = 1340780792994259709957402499820584612747936582059239337772356144372176403007354697680187429816690342769003185818648605085375388281194656994643364900608409,$

$F_{10} = 179769313486231590772930519078902473361797697894230657273430081157732675805500963132708477322407536021120113879871393357658789768814416622492847430639474124377767893424865485276302219601246094119453082952085005768838150682342462881473913110540827237163350510684586298239947245938479716304835356329624224137217.$

The only known Fermat primes are $F_0, \ldots, F_4$.

Completely Factored Composite Fermat Numbers

m	prime factor	year	discoverer
5	641	1732	Euler
5	6700417	1732	Euler
6	274177	1855	Clausen
6	67280421310721*	1855	Clausen
7	59649589127497217	1970	Morrison, Brillhart
7	5704689200685129054721	1970	Morrison, Brillhart
8	1238926361552897	1980	Brent, Pollard
8	p_{62}^{**}	1980	Brent, Pollard
9	2424833	1903	Western
9	p_{49}	1990	Lenstra, Lenstra, Jr., Manasse, Pollard
9	p_{99}^{***}	1990	Lenstra, Lenstra, Jr., Manasse, Pollard
10	45592577	1953	Selfridge
10	6487031809	1962	Brillhart
10	p_{40}	1995	Brent
10	p_{252}	1995	Brent
11	319489	1899	Cunningham
11	974849	1899	Cunningham
11	167988556341760475137	1988	Brent
11	3560841906445833920513	1988	Brent
11	p_{564}^{****}	1988	Brent

Table A.1. The only known completely factored composite Fermat numbers F_m, $5 \leq m \leq 11$. The primality was proved by * Landry, Le Lasseur, and Gérardin; ** Williams; *** Odlyzko; and **** Morain. The numbers p_j stand for primes with j digits, which are given below:

$p_{62} = 93461639715357977769163558199606896658405123754163818858 0280321,$

$p_{49} = 7455602825647884208337395736200454918783366342657,$

$p_{99} = 741640062627530801524787141901937474059940781097519 023905821316144415759504705008092818711693940737,$

$p_{40} =$ 4659775785220018543264560743076778192897,

$p_{252} =$ 13043987440548818972748476879650990394660853084161189218689529577683241625147186357414022797757310489589878392884292384483114903291379872908860161794609411944901059590671013053190617101835449160961919391248853811608071229967232280621782075312701442457 7,

$p_{564} =$ 173462447179147555430258970864309778377421844723664084649347019061363579192879108857591038330408837177983810868451546421940712978306134189864280826014542758708589243873685563973118948869399158545506611147420216132557017260564139394366945793220968665108959685482705388072645828554151936401912464931182546092879815733057795573358504982279280090942872567591518912118622751714319229788100979251036035496917279912663527358783236647193154777091427745377038294584918917590325110939381322486044298573971650711059244462177542540706913047034664643603491382441723306598834177.

Composite Fermat Numbers Without Any Known Prime Factor

m	status	year	discoverer
14	composite	1961	Selfridge, Hurwitz
20	composite	1987	Young, Buell
22	composite	1993	Crandall, Doenias, Norrie, Young
24	composite	1999	Crandall, Mayer, Papadopoulos

Table A.2. Fermat numbers that are known to be composite.

Factors of Fermat Numbers

m	prime factor	year	discoverer
12	114689	1877	Lucas, Pervouchine
12	26017793	1903	Western
12	63766529	1903	Western
12	190274191361	1974	Hallyburton, Brillhart
12	1256132134125569	1986	Baillie
13	2710954639361	1974	Hallyburton, Brillhart
13	2663848877152141313	1991	Crandall
13	3603109844542291969	1991	Crandall
13	319546020820551643220672513	1995	Brent
15	1214251009	1925	Kraïtchik
15	2327042503868417	1987	Gostin
15	168768817029516972383024127016961	1997	Crandall, Van Halewyn
16	825753601	1953	Selfridge
16	188981757975021318420037633	1996	Crandall, Dilcher
17	31065037602817	1978	Gostin
18	13631489	1903	Western
18	81274690703860512587777	1999	McIntosh, Tardif
19	70525124609	1962	Riesel
19	646730219521	1963	Wrathall
21	4485296422913	1963	Wrathall
23	167772161	1878	Pervouchine
25	25991531462657	1963	Wrathall
25	204393464266227713	1985	Gostin
25	2170072644496392193	1987	McLaughlin
26	76861124116481	1963	Wrathall
27	151413703311361	1963	Wrathall
27	231292694251438081	1985	Gostin
28	1766730974551267606529	1997	Taura
29	2405286912458753	1980	Gostin, McLaughlin
30	640126220763137	1963	Wrathall
30	1095981164658689	1963	Wrathall

Table A.3. Known prime factors of the Fermat numbers F_m, $12 \leq m \leq 30$.

Prime Factors $p = k2^n + 1$ of Fermat Numbers F_m

m	p	k	n
5	641	5	7
5	6700417	52347	7
6	274177	1071	8
6	67280421310721	262814145745	8
7	59649589127497217	116503103764643	9
7	5704689200685129054721	11141971095088142685	9
8	1238926361552897	604944512477	11
8	p_{62}	[59 digits]	11
9	2424833	37	16
9	p_{49}	[46 digits]	11
9	p_{99}	[96 digits] .	11
10	45592577	11131	12
10	6487031809	395937	14
10	p_{40}	[37 digits]	12
10	p_{252}	[248 digits]	13
11	319489	39	13
11	974849	119	13
11	167988556341760475137	10253207784531279	14
11	3560841906445833920513	434673084282938711	13
11	p_{564}	[560 digits]	13
12	114689	7	14
12	26017793	397	16
12	63766529	973	16
12	190274191361	11613415	14
12	1256132134125569	76668221077	14
13	2710954639361	41365885	16
13	2663848877152141313	20323554055421	17
13	3603109844542291969	6872386635861	19
13	319546020820551643220672513	6094856665932753836099	19
15	1214251009	579	21
15	2327042503868417	17753925353	17
15	168768 ... 016961	12876038896905286589281015555	17

Table A.4. The form $p = k2^n + 1$ of prime factors of the Fermat numbers F_m, $5 \le m \le 15$. The primes p_j are listed after Table A.1.

m	p	k	n
16	825753601	1575	19
16	18898175797502131842003763	180227048850079840107	20
17	31065037602817	59251857	19
18	13631489	13	20
18	81274690703860512587777	9688698137266697	23
19	70525124609	33629	21
19	646730219521	308385	21
21	4485296422913	534689	23
23	167772161	5	25
25	25991531462657	48413	29
25	204393464266227713	1522849979	27
25	2170072644496392193	16168301139	27
26	76861124116481	143165	29
27	151413703311361	141015	30
27	231292694251438081	430816215	29
28	1766730974551267606529	25709319373	36
29	2405286912458753	1120049	31
30	640126220763137	149041	32
30	1095981164658689	127589	33

Table A.5. The form $p = k2^n + 1$ of prime factors of the Fermat numbers F_m, $16 \leq m \leq 30$.

Further factors of F_m can be found in [Brillhart, Lehmer, Selfridge, Tuckerman, Wagstaff] and [www1].

Appendix B
Mersenne Numbers

> *The numbers $2^n - 1$ are prime for*
> $n = 2, 3, 5, 7, 13, 17, 19, 31, 67, 127, 257$
> *and composite for all other*
> *positive integers $n < 257$.*
>
> Incorrect statement by
> Father Marin Mersenne,
> from Preface to his *Cogitata*
> *Physica-Mathematica* (1644)

The number $M_p = 2^p - 1$, where p is prime, is called a *Mersenne number*. If $2^p - 1$ itself is prime, then it is called a *Mersenne prime*. Primes of this type were investigated by the French mathematician Marin Mersenne (1588–1648).

Notice that the number $2^{ij} - 1$ for integers $i > 1$ and $j > 1$ can be written as a product of two nontrivial factors:

$$(B.1) \qquad 2^{ij} - 1 = (2^i - 1)(2^{i(j-1)} + 2^{i(j-2)} + \cdots + 2^i + 1).$$

This is why we require that the exponent p of the Mersenne number $2^p - 1$ be prime. By a contradiction argument, factorization (B.1) immediately leads to the following theorem, which was already known by Pierre de Fermat (see [Dickson, p. 12], [Mahoney, p. 294]).

Theorem B.1. *If $2^p - 1$ is prime, then so is p.*

We see that the first four prime exponents $p = 2$, 3, 5, 7 yield the primes 3, 7, 31, 127. However, for $p = 11$ the number $2^{11} - 1 = 2047$ is divisible by 23. Hence, the converse of Theorem B.1 does not hold. The foregoing example with $p = 11$ is generalized in Theorem B.4 below.

At the present time, almost 40 Mersenne primes have been discovered, but very little is know about their distribution (some empirical formulae are surveyed, e.g., in [Schroeder]). The number M_p is prime if

$p =$2, 3, 5, 7, 13, 17, 19, 31, 61, 89,

    107, 127, 521, 607, 1279, 2203, 2281, 3217, 4253, 4423,

    9689, 9941, 11213, 19937, 21701, 23209, 44497, 86243, 110503, 132049,

    216091, 756839, 859433, 1257787, 1398269, 2976221, 3021377, ?, 6972593, ...

Here the symbol ? indicates that as of 2000 not all lower exponents have been checked, i.e., it was not known whether 6972593 is the next Mersenne prime exponent after 3021377. According to [Kraïtchik, 1952], Fermat himself factored $2^p - 1$ for $p = 11, 23, 37$. His results led him to the discovery of Fermat's little theorem.

Let us denote by $M(n)$ the nth Mersenne prime, i.e., $M(1) = 2^2 - 1 = 3$, $M(2) = 2^3 - 1 = 7$, $M(3) = 2^5 - 1 = 31$, $M(4) = 2^7 - 1 = 127, \ldots$. In Figure B.1 we observe an interesting pattern in the distribution of the Mersenne primes $M(n)$.

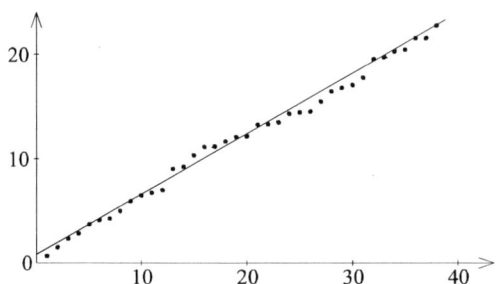

Figure B.1. The values of $\log_2(\log_2 M(n))$ versus n.

The 37th Mersenne prime number $M(37) = 2^{3021377} - 1$ was discovered in 1998. It has more than nine hundred thousand digits. The largest known Mersenne prime, $2^{6972593} - 1$, was discovered in 1999 and has more than two million digits. This is the first known prime with more than one million digits. A table of discoverers of Mersenne primes including the year of the discovery is, e.g., in [Ribenboim, 1996, p. 94] and in [www2].

Theorem B.2 (Lucas–Lehmer Test). *Let* $S_1 = 4$ *and* $S_{k+1} = S_k^2 - 2$ *for* $k = 1, 2, \ldots$. *Then, for* $p > 2$, *the Mersenne number* $M_p = 2^p - 1$ *is prime if and only if* M_p *divides* S_{p-1}.

For a proof see [Lehmer, 1930, Theorem 5.4] or [Riesel, 1985, p. 126].

A table of factors of the Mersenne numbers M_p, $p \leq 257$ prime, is contained in [Riesel, 1985]. For more extensive tables see [Brillhart, Lehmer, Selfridge, Tuckerman, Wagstaff]. A general form of possible divisors of Mersenne numbers is given by the following theorem, which was also known by Fermat (see [Dickson, p. 12], [Mahoney, p. 294]).

Theorem B.3. *Let* $p > 2$ *be a prime. Then all prime divisors of* $2^p - 1$ *have the form* $2kp + 1$.

P r o o f . Let q be a prime divisor of $2^p - 1$. Then $2^p \equiv 1 \pmod{q}$. Since p is prime and $2^1 \not\equiv 1 \pmod{q}$, we derive that $\mathrm{ord}_q 2 = p$. By Fermat's little theorem (i.e., $2^{q-1} \equiv 1 \pmod{q}$) we get $p \mid q - 1$. Thus there exists j such that $jp = q - 1$. Since p is odd and $q - 1$ is even, $j = 2k$ for some integer k.   □

The following theorem was also suggested by Fermat and later proved by Euler and independently also by Lagrange.

Theorem B.4. *Let* p *be a prime such that* $p \equiv 3 \pmod{4}$. *Then* $2p + 1 \mid M_p$ *if and only if* $2p + 1$ *is prime.*

For a proof see, e.g., [Ribenboim, 1996, pp. 90–91], [Robbins, p. 149]. Thus if $p = 11, 23, 83, \ldots$, then M_p has a factor $23, 47, 167, \ldots$ (compare with Remark 5.32 on the Sophie Germain primes).

Theorem B.5. *If $n \mid M_p$ and $p > 2$, then $n \equiv \pm 1 \pmod 8$.*

For the proof see [Ribenboim, 1991, p. 66].

There is an interesting connection between Mersenne primes and the perfect numbers. Recall that a natural number n is said to be *perfect* if it is equal to the sum of all its divisors less than n. For example, the numbers 6 and 28 are perfect, since $6 = 1 + 2 + 3$ and $28 = 1 + 2 + 4 + 7 + 14$. Let n be an arbitrary natural number. Denote by $\sigma(n)$ the sum of all its positive divisors. Then we have an equivalent definition, namely, n is perfect if and only if $\sigma(n) = 2n$. A necessary and sufficient condition for an even number n to be perfect is that it be of the form $n = 2^{p-1}(2^p - 1)$, where $p > 1$ is a natural number and $2^p - 1$ is a prime (i.e., p is also prime). Euclid (4th–3rd century B.C.) already knew that this condition is sufficient, but did not know whether it is also necessary. This question was answered two millennia later by Leonhard Euler (1707–1783), who proved its necessity.

Theorem B.6 (Euclid). *If $2^p - 1$ is prime, then the number $n = 2^{p-1}(2^p - 1)$ is perfect.*

P r o o f . We have

$$\sigma(n) = \sigma(2^{p-1})\sigma(2^p - 1) = \frac{2^p - 1}{2 - 1}(1 + 2^p - 1) = (2^p - 1)2^p = 2n,$$

and thus n is perfect.   □

Theorem B.7 (Euler). *All even perfect numbers are of the form*

$$n = 2^{p-1}(2^p - 1),$$

where $p > 1$ and $2^p - 1$ is a prime.

P r o o f . If n is even, then we can write $n = 2^{p-1}u$, where $p > 1$ and u is odd. Since 2^{p-1} and u are coprime, the sum of the divisors of n is equal to

$$\sigma(n) = \sigma(2^{p-1})\sigma(u) = (2^p - 1)\sigma(u).$$

If n is perfect, we have

$$\sigma(n) = 2n = 2^p u,$$

and thus

$$(2^p - 1)\sigma(u) = 2^p u.$$

Since $2^p - 1$ and 2^p are coprime, we see that $\sigma(u) = 2^p t$ and $u = (2^p - 1)t$, where t is a natural number. However, since u has at least the divisors $1, t, 2^p - 1$, and $t(2^p - 1)$ for $t > 1$, the sum of the divisors of u satisfies the inequality

$$\sigma(u) \geq 1 + t + 2^p - 1 + t(2^p - 1) = 2^p(1 + t),$$

which contradicts $\sigma(u) = 2^p t$. Therefore, $t = 1$. But then $\sigma(u) \geq 1 + 2^p - 1 = 2^p$, and the required equality becomes true only if $2^p - 1$ is a prime. □

According to Theorems B.6 and B.7, there is another interesting relation between the even perfect numbers n and the Mersenne primes M_p, namely,

$$n = 2^{p-1}(2^p - 1) = \frac{2^p}{2}(2^p - 1) = 1 + 2 + \cdots + (2^p - 1) = \sum_{i=1}^{M_p} i.$$

Theorem B.8. *If you sum the digits of any even perfect number greater than 6, then sum the digits of the resulting number, and repeat this process until you get a single digit, then that digit will be one.*

For the proof see [www2].

The following theorem can be found in [Kraïtchik, 1952].

Theorem B.9 (Heath). *Every even perfect number* $2^{p-1}(2^p - 1)$ *for* $p > 2$ *is the sum of cubes of* $2^{(p-1)/2}$ *odd numbers.*

P r o o f . First note that by Theorems B.1 and B.7, p is prime. Let $p > 2$. Setting $k = (p-1)/2$ and $m = 2^k$, we get

$$s = 1^3 + 3^3 + 5^3 + \cdots + (2m - 1)^3 = \sum_{k=1}^{m}(2k - 1)^3 = \sum_{k=1}^{m}(8k^3 - 12k^2 + 6k - 1)$$

$$= 8\frac{m^2(m+1)^2}{4} - 12\frac{m(m+1)(2m+1)}{6} + 6\frac{m(m+1)}{2} - m$$

$$= m^2(2m^2 - 1).$$

Now we see that $s = 2^{2k}(2^{2k+1} - 1) = 2^{p-1}(2^p - 1)$. □

Recall (see Theorem 5.11) that all Mersenne numbers are primes or pseudoprimes, that is,

$$2^{M_p} \equiv 2 \pmod{M_p}.$$

There are many open problems concerning Mersenne numbers. It is conjectured that there are infinitely many Mersenne primes, and thus infinitely many perfect numbers. However, up to now we do not know whether there is an odd perfect number. There are only necessary conditions for such a number to exist. For instance, it has been proved that each odd perfect number is larger than 10^{300} and has at least 8 different prime divisors. We also know that each odd perfect number has the form $12j + 1$ or $36j + 9$ for a suitable integer j.

It has also been conjectured that the prime M_p yields another prime $M_{M_p} = 2^{M_p} - 1$. However, a counterexample was found for $p = 13$, since $M_{13} = 8191$ is prime, whereas $2^{8191} - 1$ is composite (see [Ribenboim, 1987]). Anyway, there is still another unsolved conjecture: whether the sequence $m_{k+1} = 2^{m_k} - 1$ starting from $m_1 = 2$ contains only primes. Indeed, the first five terms $m_1 = 2$, $m_2 = 2^2 - 1 = 3$, $m_3 = 2^3 - 1 = 7$, $m_4 = 2^7 - 1 = 127$, and

$$m_5 = 2^{127} - 1 = 170141183460469231731687303715884105727$$

are the Mersenne primes M_1, M_2, M_3, M_7, and M_{127}. The character of m_6 is unknown at the present time. However, if m_k were to be composite for some k, then m_{k+1} would also be composite due to (B.1).

Other well-known conjectures include the following: Are there infinitely many composite Mersenne numbers? Is every Mersenne number square-free? (Cf. [Rotkiewicz, 1965] and also later [Warren, Bray].) We know only that if a prime p divides a Mersenne number M_q then

$$p^2 \mid M_q \quad \Longleftrightarrow \quad 2^{p-1} \equiv 1 \pmod{p^2} \quad \text{(Wieferich's congruence)}.$$

For more information about the Mersenne numbers see, e.g., [Dickson], [Ribenboim, 1996], or [www3]. The Mersenne number transform, which is defined similarly to the Fermat number transform (15.1), is examined, e.g., in [Crandall, Fagin], [Dimitrov, Cooklev, Donevsky], [Elliott, Rao, p. 425], [Gorshkov, 1994b], [Kučera].

Figure B.2. Memorial plaque of Marin Mersenne at his birthplace in Oizé (dépt. Sarthe, formerly dépt. Maine, France).

Appendix C
Remembrance of Pierre de Fermat

*Fermat, l'un des plus beaux génies
qui aient illustré la France.*
1839
CAUCHY

Inscription on the base of Fermat's
statue in Beaumont-de-Lomagne.

Figure C.1. Birthplace of Pierre de Fermat in Beaumont-de-Lomagne.

Figure C.2. Memorial plaque of Pierre de Fermat at his birthplace.

Figure C.3. Statue of Pierre de Fermat in his native Beaumont-de-Lomagne.

Figure C.4. Fermat and a muse in the "Salle des Illustres" in Capitole of Toulouse (see [Hiriart-Urruty, p. 53] for details).

Figure C.5. Bust of Pierre de Fermat in the "Salle Henri-Martin" in Capitole of Toulouse.

Figure C.6. Lycée Pierre de Fermat and Collège Pierre de Fermat in Toulouse.

Figure C.7. Statue of Pierre de Fermat in "Musée des Augustins" in Toulouse.

Figure C.8. Portrait of Pierre de Fermat by Roland Lefévre in the Narbonne City Museums, France.

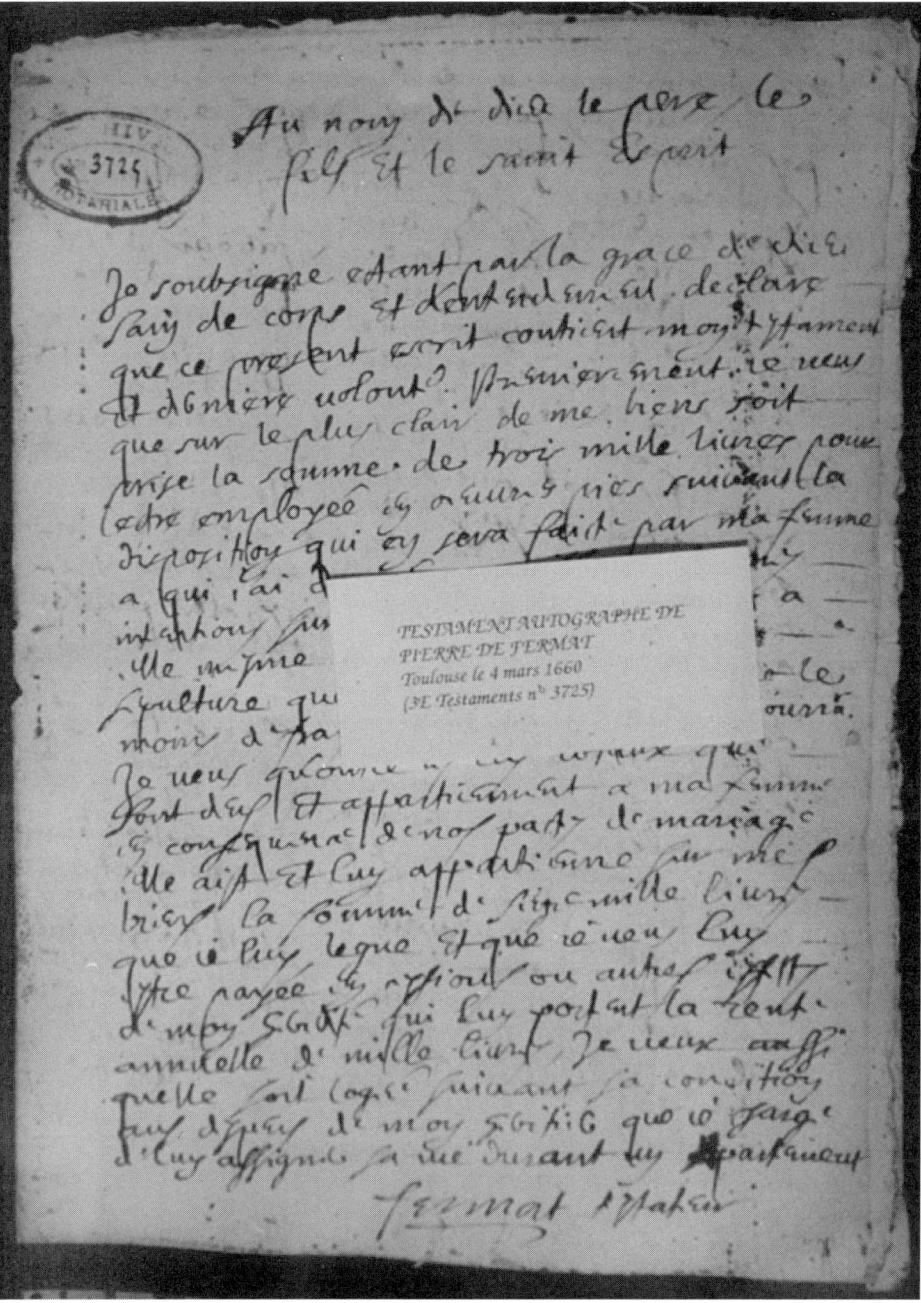

Figure C.9. Fermat's autographical testament (with his signature) saved in the Museum of Pierre de Fermat in Beaumont-de-Lomagne.

References

Adleman, L. M., Pomerance, C., Rumely, R. S., *On distinguishing prime numbers from composite numbers*, Ann. of Math. (2) **117** (1983), 173–206. MR 84e:10008.

Agarwal, R. C., Burrus, C. S., *Fast digital convolution using Fermat transforms*, Southwest IEEE Conf. Rec., Houston, Texas, 1973, 538–543.

Agarwal, R. C., Burrus, C. S., *Fast convolution using Fermat number transforms with applications to digital filtering*, IEEE Trans. Acoust. Speech Signal Processing **22** (1974), 87–97. MR 53 #2501.

Aigner, A., *On prime numbers for which (almost) all Fermat numbers are quadratic nonresidues (German)*, Monatsh. Math. **101** (1986), 85–93. MR 87g:11010.

Akushskiĭ, I. Ya., Burtsev, V. M., *Realization of primality tests for Mersenne and Fermat numbers (Russian)*, Vestnik Akad. Nauk Kazakh. SSR (1986), 52–59. MR 87f:11106.

Alford, W. R., Granville, A., Pomerance, C., *There are infinitely many Carmichael numbers*, Ann. of Math. **140** (1994), 703–722. MR 95k:11114.

André-Jeannin, R., *Irrationalité de la somme des inverses de certaines suites récurrentes*, C. R. Acad. Sci. Paris Sér. I Math. **308** (1989), 539–541. MR 90b:11012.

Antonyuk, P. N., Stanyukovich, K. P., *Periodic solutions of the logistic difference equation (Russian)*, Dokl. Akad. Nauk SSSR **313** (1990a), 1033–1036, English translation in Soviet Math. Dokl. **42** (1991), 116–119. MR 92f:39003.

Antonyuk, P. N., Stanyukovich, K. P., *The logistic difference equation. Period doublings and Fermat numbers (Russian)*, Dokl. Akad. Nauk SSSR **313** (1990b), 1289–1292, English translation in Soviet Math. Dokl. **42** (1991), 138–141. MR 92d:11019.

Artjuhov, M. M., *Certain criteria for primality of numbers connected with the little Fermat theorem (Russian)*, Acta Arith. **12** (1966/67), 355–364. MR 35 #4153.

Arya, S. P., *Fermat numbers*, Math. Ed. **6** (1989), 5–6.

Arya, S. P., *More about Fermat numbers*, Math. Ed. **7** (1990), 139–141.

Asadulla, S., *A note on Fermat numbers*, J. Natur. Sci. Math. **17** (1977), 113–118. MR 56 #11886.

Atkin, A. O. L., Morain, F., *Elliptic curves and primality proving*, Math. Comp. **61** (1993), 29–68. MR 93m:11136.

Atkin, A. O. L., Rickert, N. W., *Some factors of Fermat numbers*, Abstracts Amer. Math. Soc. **1** (1980), 211.

Badea, C., *The irrationality of certain infinite series*, Glasgow Math. J. **29** (1987), 221–228. MR 88i:11044.

Baillie, R., *New primes of the form $k2^n + 1$*, Math. Comp. **33** (1979), 1333–1336. MR 80h:10009.

Baillie, R., Cormack, G., Williams, H. C., *The problem of Sierpiński concerning $k \cdot 2^n + 1$*, Math. Comp. **37** (1981), 229–231. MR 83a:10006a; Corrigenda ibid. **39** (1982), 308. MR 83a:10006b.

Baker, A., *Linear forms in the logarithms of algebraic numbers*, Mathematika **13** (1966), 204–216. MR 36 #3732.

Baker, A., *The theory of linear forms in logarithms*, Transcendence Theory: Advances and Applications, Academic Press, London, 1977, 1–27. MR 58 #16543.

Balasubramanian, R., *Number theory and primality testing.*, Workshop on Mathematics of Computer Algorithms (Madras, 1986), Inst. Math. Sci., Madras, 1986, 1–29. MR 89i:11140.

Beeger, N. G. W. H., *On even numbers m dividing $2^m - 2$*, Amer. Math. Monthly **58** (1951), 553–555. MR 13,320d.

Beiler, A. H., *Recreations in the theory of numbers*, Dover Publications, New York, 1964, 1966. Zbl 154.04001.

Bellon, M. P., Maillard, J.-M., Rollet, G., Viallet, C.-M., *Deformations of dynamics associated to the chiral Potts model*, Internat. J. Modern Phys. **B 6** (1992), 3575–3584. MR 93m:82028.

Biermann, K.-R., *Thomas Clausen, Mathematiker und Astronom*, J. Reine Angew. Math. **216** (1964), 159–198. MR 29 #2153.

Birkhoff, G. D., Vandiver, H. S., *On the integral divisors of $a^n - b^n$*, Ann. of Math. (2) **5** (1904), 173–180.

Björn, A., Riesel, H., *Factors of generalized Fermat numbers*, Math. Comp. **67** (1998), 441–446. MR 98e:11008.

Borwein, P., *On the irrationality of $\sum 1/(q^n + r)$*, J. Number Theory **37** (1991), 253–259. MR 92b:11046.

Bosma, W., *Explicit primality criteria for $h2^k + 1$*, Math. Comp. **61** (1993), 97–109, S7–S9. MR 94c:11005.

Brent, R. P., *Succinct proofs of primality for the factors of some Fermat numbers*, Math. Comp. **38** (1982), 253–255. MR 82k:10002.

Brent, R. P., *Factorization of the eleventh Fermat number*, Abstracts Amer. Math. Soc. **10** (1989), 176–177.

Brent, R. P., *Parallel algorithms for integer factorisation*, Number theory and cryptography (Sydney, 1989), London Math. Soc. Lecture Note Ser., 154, Cambridge Univ. Press, Cambridge, 1990, 26–37. MR 91h:11148.

Brent, R. P., *Factorization of the tenth Fermat number*, Math. Comp. **68** (1999), 429–451. MR 99e:11154.

Brent, R. P., Crandall, R. E., Dilcher, K., Van Halewyn, C., *Three new factors of Fermat numbers*, Math. Comp. **69** (2000), 1297–1304. MR 2000j:11194.

Brent, R. P., Pollard, J. M., *Factorization of the eighth Fermat number*, Math. Comp. **36** (1981), 627–630. MR 83h:10014.

Bressoud, D. M., *Factorization and primality testing*, Springer, New York, 1989. MR 91e:11150.

Brillhart, J., Lehmer, D. H., Selfridge, J. L., *New primality criteria and factorizations of $2^m \pm 1$*, Math. Comp. **29** (1975), 620–647. MR 52#5546.

Brillhart, J., Lehmer, D. H., Selfridge, J. L., Tuckerman, B., Wagstaff, S. S., *Factorization of $b^n \pm 1$, $b = 2, 3, 5, 6, 7, 10, 11, 12$ up to high powers*, Contemporary Math. vol. 22, second edition, Amer. Math. Soc., Providence, 1988. MR 90d:11009.

Brillhart, J., Selfridge, J. L., *Some factorizations of $2^m \pm 1$ and related results*, Math. Comp. **21** (1967), 87–96, 751. MR 37 #131.

Brun, V., *Ein Satz über Irrationalität*, Arch. for Math. og Naturvideskab (Kristiania) **31** (1910), 3.

Buchmann, J., Düllmann, S., *A probabilistic class group and regulator algorithm and its implementation*, Computational Number Theory (Debrecen, 1989), de Gruyter, Berlin, 1991, 53–72. MR 92m:11150.

Buell, D. A., Young, J., *Some large primes and the Sierpiński problem*, SRL Technical Report 88-004, Supercomputing Research Center, Lanham, Maryland, May, 1988.

Bugeaud, Y., Mignotte, M., *Sur l'equation diophantienne $\frac{x^n-1}{x-1} = y^q$, II*, C. R. Acad. Sci. Paris Sér. I Math. **328** (1999), 741–744.

Burton, D. M., *Elementary number theory*, fourth edition, McGraw-Hill, New York, 1989, 1998. MR 90e:11001.

Canals, I., *Fermat numbers and the limitation of computers (Spanish)*, Acta Mexicana Ci. Tecn. **7** (1973), 29–30. MR 51 #8009.

Carlip, W., Jacobson, E., Somer, L., *Pseudoprimes, perfect numbers, and a problem of Lehmer*, Fibonacci Quart. **36** (1998), 361–371. MR 99g:11013.

Carmichael, R. D., *Note on a new number theory function*, Bull. Amer. Math. Soc. **16** (1910), 232–238.

Carmichael, R. D., *On composite numbers P which satisfy the Fermat congruence $a^{p-1} \equiv 1 \pmod{P}$*, Amer. Math. Monthly **19** (1912), 22–27.

Carmichael, R. D., *On the numerical factors of arithmetic forms $\alpha^n \pm \beta^n$*, Ann. of Math. **15** (1913), 30–70.

Chang, C. C., *An ordered minimal perfect hashing scheme based upon Euler's theorem*, Inform. Sci. **32** (1984), 165–172. MR 85f:68012.

Cipolla, M., *Sui numeri composti P, che verificano la congruenza di Fermat $a^{P-1} \equiv 1 \pmod{P}$*, Annali di Matematica (3) **9** (1904), 139–160.

Cohen, H., Lenstra, H. W., *Primality testing and Jacobi sums*, Math. Comp. **42** (1984), 297–330. MR 86g:11078.

Conway, J. H., Guy, R. K., *The book of numbers*, Springer-Verlag, New York, 1996. MR 98g:00004.

Conway, J. H., Guy, R. K., Schneeberger, W. A., Sloane, N. J. A., *The primary pretenders*, Acta Arith. **LXXVIII** (1997), 307–313. Zbl 863.11005.

Cooley, J. W., Tukey, J. W., *An algorithm for the machine calculation of complex Fourier series*, Math. Comp. **19** (1965), 297–301. MR 31#2843.

Cormack, G. V., Williams, H. C., *Some very large primes of the form $k2^m + 1$*, Math. Comp. **35** (1980), 1419–1421. MR 81i:10011.

Coxeter, H. S. M., *Introduction to geometry*, second edition, John Wiley & Sons, New York, 1969. MR 49#11369, MR 90a:51001.

Crandall, R. E., *Topics in advanced scientific computation*, Springer, Berlin, 1996. MR 97g:65005.

Crandall, R., Dilcher, K., Pomerance, C., *A search for Wieferich and Wilson primes*, Math. Comp. **66** (1997), 433–449. MR 97c:11004.

Crandall, R., Doenias, J., Norrie, C., Young, J., *The twenty-second Fermat number is composite*, Math. Comp. **64** (1995), 863–868. MR 95f:11104.

Crandall, R., Fagin, B., *Discrete weighted transforms and large-integer arithmetic*, Math. Comp. **62** (1994), 305–324. MR 94c:11123.

Crandall, R. E., Mayer, E., Papadopoulos, J., *The twenty-fourth Fermat number is composite*, Math. Comp., submitted, 1999, 1–21.

Crandall, R. E., Pomerance, C., *Prime numbers. A computational perspective*, Springer, New York, 2001.

Creutzburg, R., *Application of Fermat-number transform to fast digital correlation*, Proc. of the 4th Internat. Meeting of Young Comput. Scientists, IMYCS '86 (Smolenice Castle, 1986). Tanulmányok—MTA Számitástech. Automat. Kutató Int. Budapest No. 185 (1986), 121–126.

Creutzburg, R., Grundmann, H.-J., *Schnelle digitale Korrelation von Matrizen mittels Fermattransformation*, Beiträge zur Optik und Quantenphysik **8** (1983a), 126–127.

Creutzburg, R., Grundmann, H.-J., *The Fermat transform and its application in the fast computation of digital convolutions (German)*, Rostock. Math. Kolloq. No. 24 (1983b), 77–98. MR 85k:94008.

Creutzburg, R., Grundmann, H.-J., *Fast digital convolution via Fermat number transform (German)*, Elektron. Informationsverarb. Kybernet. **21** (1985), 35–46. MR 87d:94010.

Creutzburg, R., Tasche, M., *Number-theoretic transformations and primitive roots of unity in a residue class ring modulo m, Parts I, II (German)*, Rostock. Math. Kolloq. No. 25 (1984), 4–22, No. 26 (1984), 103–109. MR 87f:11003a,b.

Cullen, J., *Question 15897*, Math. Quest. Educ. Times **9** (1905), 534.

Cunningham, A. J., *Solution of question 15897*, Math. Quest. Educ. Times **10** (1906), 44–47.

Cunningham, A. J., Western, A. E., *On Fermat's numbers*, Proc. London Math. Soc. **2(1)** (1904), 175.

Dickson, L. E., *History of the theory of numbers, vol. I, Divisibility and primality*, Carnegie Inst., Washington, 1919.

Diffie, W., Hellman, M. E., *New directions in cryptography*, IEEE Trans. Inform. Theory **22** (1976), 644–654. MR 55 #10141.

Dilcher, K., *Fermat numbers, Wieferich and Wilson primes: Computations and generalizations*, Proc. Conf. on Computational Number Theory and Public Key Cryptography (Warsaw, Sept. 2000), 1–22.

Dimitrov, V. S., Cooklev, T. V., Donevsky, B. D., *Generalized Fermat–Mersenne number theoretic transform*, IEEE Trans. Circuits and Systems II, Analog Digit. Signal Process. **41** (1994), 133–139. Zbl 808.65146.

Dirichlet, P. G. L., *Beweis des Satzes dass jede unbegrenzte arithmetische Progression, deren erstes Glied und Differenz ganze Zahlen ohne gemeinschaftlichen Factor sind, unendlich viele Primzahlen enthält*, Abh. d. Königl. Akad. d. Wiss. (1837), 45–81; reprinted in Werke, vol. 1, 315–350, G. Reimer, Berlin, 1889.

Dubner, H., *Generalized Fermat primes*, J. Recreational Math. **18** (1985/86), 279–280.

Dubner, H., Keller, W., *Factors of generalized Fermat numbers*, Math. Comp. **64** (1995), 397–405. MR 95c:11010.

Dudek, J., *On bisemilattices. III.*, Math. Sem. Notes Kobe Univ. **10** (1982), 275–279. MR 84h:06005.

Duparc, H. J. A., *On Carmichael numbers, Poulet numbers, Mersenne numbers and Fermat numbers*, Rapport ZW 1953-004, Math. Centrum Amsterdam, 1953, 1–7. MR 15,933j.

Dyson, F., *The sixth Fermat number and palindromic continued fractions*, Enseign. Math. (2) **46** (2000), 385–389.

Elliott, D. F., Rao, K. R., *Fast transforms. Algorithms, analyses, applications*, Academic Press, London, 1982. MR 85e:94001.

Erdős, P., *On arithmetical properties of Lambert series*, J. Indian Math. Soc. (N. S.) **12** (1948), 63–66. MR 10,594c.

Erdős, P., *On the converse of Fermat's theorem*, Amer. Math. Monthly **56** (1949), 623–624. MR 11,131g.

Erdős, P., *On almost primes*, Amer. Math. Monthly **57** (1950), 404–407. MR 12,80i.

Erdős, P., *Some problems and results on the irrationality of the sum of infinite series*, J. Math. Sci. **10** (1975), 1–7. MR 80k:10029.

Erdős, P., Graham, R. L., *Old and new problems and results in combinatorial number theory*, Université de Genève, L'Enseignement Mathématique, Imprimerie Kunding, 1980. MR 82j:10001.

Erdős, P., Odlyzko, A. M., *On the density of odd integers of the form $(p-1)2^{-n}$ and related questions*, J. Number Theory **11** (1979), 257–263. MR 80i:10077.

Erdős, P., Straus, E. G., *On the irrationality of certain Ahmes series*, J. Indian Math. Soc. (N. S.) **27** (1963), 129–133. MR 31#124.

Euler, L., *Observationes de theoremate quodam Fermatiano aliisque ad numeros primos spectantibus*, Comment. Acad. Sci. Petropol. **6,** ad annos 1732-33 (1738), 103–107.

Euler, L., *Theoremata circa divisores numerorum*, Novi Comment. Acad. Sci. Petropol. **1,** ad annos 1747–48 (1750), 20–48.

Fehér, J., Kiss, P., *Note on super pseudoprime numbers*, Ann. Univ. Sci. Budapest. Eötvös Sect. Math. **26** (1983), 157–159. MR 85c:11008.

Feigenbaum, M. J., *Quantitative universality for a class of nonlinear transformations*, J. Stat. Phys. **19** (1978), 25–52. MR 58#18601.

Ferentinou-Nicolacopoulou, J., *Une propriété des diviseurs du nombre $r^{r^m}+1$. Applications au dernier théorème de Fermat*, Bull. Greek Math. Soc. **4** (1963), 121–126. MR 29#68.

Flammenkamp, A., Luca, F., *Binomial coefficients and Lucas sequences*, J. Number Theory, accepted in 2001, 1–30.

Gardner, M., *Mathematical carnival: A new round-up of tantalizers and puzzles from Scientific American*, Vintage Books, New York, 1977, 1989. MR 90d:00006.

Gauss, C. F., *Disquisitiones arithmeticae*, Springer, Berlin, 1986. MR 87f:01105.

Golomb, S. W., *Sets of primes with intermediate density*, Math. Scand. **3** (1955), 264–274. MR 17,828d.

Golomb, S. W., *On the sum of the reciprocals of the Fermat numbers and related irrationalities*, Canad. J. Math. **15** (1963), 475–478. MR 27 #105.

Golomb, S. W., *Properties of the sequence $3 \cdot 2^n + 1$*, Math. Comp. **30** (1976), 657–663. MR 53 #7933, MR 82m:10025.

Good, I. J., *A reciprocal series of Fibonacci numbers*, Fibonacci Quart. **12** (1974), 346. MR 50#4465.

Gorshkov, A. S., *Method of fast multiplication modulo Fermat primes (Russian)*, Dokl. Akad. Nauk SSSR **336** (1994a), 175–178, English translation in Soviet Phys. Dokl. **39** (1994a), 314–317. Zbl 939.68963.

Gorshkov, A. S., *On the method of the number-theoretic Mersenne transform (Russian)*, Dokl. Akad. Nauk **336** (1994b), 33–34, English translation in Phys. Dokl. **39** (1994b), 312–313. MR 95i:11003.

Gorshkov, A. S., Kravchenko, V. F., *Fermat numbers in digital signal processing (Russian)*, Dokl. Akad. Nauk SSSR **320** (1991), 835–838, English translation in Soviet Phys. Dokl. **36** (1991), 669–671. Zbl 753.94004.

Gorshkov, A. S., Kravchenko, V. F., Rvachev, V. A., Rvachev, V. L., *On a number-theoretic method for the fast Fourier transform in the Fermat ring (Russian)*, Dokl. Akad. Nauk SSSR **320** (1991), 303–306, English translation in Soviet Phys. Dokl. **36** (1991), 616–618 MR 93g:65171.

Gostin, G. B., *A factor of F_{17}*, Math. Comp. **35** (1980), 975–976. MR 81f:10010.

Gostin, G. B., *New factors of Fermat numbers*, Math. Comp. **64** (1995), 393–395. MR 95c:11151.

Gostin, G. B., McLaughlin, P. B., *Six new factors of Fermat numbers*, Math. Comp. **38** (1982), 645–649. MR 83c:10003.

Gottlieb, C., *The simple and straightforward construction of the regular 257-gon*, Math. Intelligencer **21** (1999), 31–37. MR 2000c:12006.

Granville, A., *Primality testing and Carmichael numbers*, Notices Amer. Math. Soc. **39** (1992), 696–700.

Grytczuk, A., *Some remarks on Fermat numbers*, Discuss. Math. **13** (1993), 69–73. MR 94k:11028.

Grytczuk, A., Grytczuk, J., *A primality test for Fermat numbers*, Acta Acad. Paedagog. Agriensis, Sect. Mat. **23** (1995), 33–35. Zbl 881.11012.

Grytczuk, A., Luca, F., Wójtowicz, M., *Another note on the greatest prime factors of Fermat numbers*, Southeast Asian Bull. Math. **25** (2001), 111–115.

Gulliver, T. A., *Self-reciprocal polynomials and generalized Fermat numbers*, IEEE Trans. Inform. Theory **38** (1992), 1149–1154. MR 93h:11135.

Gutfreund, H., Little, W. A., *Physicist's proof of Fermat's theorem of primes*, Amer. J. Phys. **50** (1982), 219–220.

Guy, R. K., *The primes 1093 and 3511*, Math. Student **35** (1967), 205–206. MR 42 #4473.

Guy, R. K., *The strong law of small numbers*, Amer. Math. Monthly **95** (1988), 697–712. MR 90c:11002.

Guy, R. K., *The second strong law of small numbers*, Math. Mag. **63** (1990), 3–20. MR 91a:11001.

Guy, R. K., *Unsolved problems in number theory*, second edition, Springer, Berlin, 1994. MR 96e:11002.

Hallyburton, J. C., Brillhart, J., *Two new factors of Fermat numbers*, Math. Comp. **29** (1975), 109–112. MR 51 #5460. Corrigenda ibid. **30** (1976), 198. MR 52 #13599.

Harborth, H., *Ein Primzahlkriterium nach Mann und Shanks*, Arch. Math. (Basel) **27** (1976), 290–294. MR 54 #5099.

Harborth, H., *Prime number criteria in Pascal's triangle*, J. London Math. Soc. (2) **16** (1977), 184–190. MR 57 #16182.

Hardy, G. H., Wright, E. M., *An introduction to the theory of numbers*, Clarendon Press, Oxford, 1945, 1954, 1960, 1979. MR 16,673c, MR 81i:10002.

Hermes, J., *Über die Teilung des Kreises in 65537 gleiche Teile*, Nachr. Königl. Gesellsch. Wissensch. Göttingen, Math.-Phys. Klasse, 1894, 170–186.

Hewgill, D., *A relationship between Pascal's triangle and Fermat's numbers*, Fibonacci Quart. **15** (1977), 183–184. MR 55 #10275.

Hilton, P., Pedersen, J., *On folding instructions for products of Fermat numbers*, Southeast Asian Bull. Math. **18** (1994), 19–27. MR 96e:11005.

Hiriart-Urruty, J.-B., *Historical associations of Fermat in Beaumont and Toulouse, France*, Math. Intelligencer **12** (2) (1990), 52–53. MR 90m:01068.

Hoggatt, E. V., Bicknell, M., *A reciprocal series of Fibonacci numbers with subscripts $2^n k$*, Fibonacci Quart. **14** (1976), 453–455. MR 54#216.

Hooley, C., *Applications of sieve methods to the theory of numbers*, Cambridge Tracts in Mathematics, No. 70, Cambridge Univ. Press, Cambridge, 1976. MR 53 #7976.

Huard, J. G., Spearman, B., Williams, K., *Pascal's triangle modulo 8*, European J. Combin. **19** (1998), 45–62. MR 99b:11012.

Hungerford, T. W., *Algebra. Graduate texts in mathematics*, Vol. 73, Springer-Verlag, 1980. MR 82a:00006.

Hurwitz, A., Selfridge, J. L., *Fermat numbers and perfect numbers*, Notices Amer. Math. Soc. **8** (1961), 601.

Inkeri, K., *Tests for primality*, Ann. Acad. Sci. Fenn. Ser. A I No. 279 (1960), 1–19. MR 22 #7984.

Ireland, K., Rosen, M., *A classical introduction to modern number theory*, second edition, Springer, New York, 1990. MR 92e:11001.

Jaeschke, G., *Reciprocal hashing: A method for generating perfect hashing functions*, Comm. ACM **24** (1981), 829–833. MR 83f:68013.

Jaeschke, G., *On the smallest k such that all $k \cdot 2^N + 1$ are composite*, Math. Comp. **40** (1983), 381–384. MR 84k:10006; Corrigendum ibid. **45** (1985), 637. MR 87b:11009.

Jarden, D., *Existence of an infinitude of composite n for which $2^{n-1} \equiv 1 \pmod{n}$ (Hebrew, Engl. Summary)*, Riveon Lematematika **4** (1950), 65–67. MR 12,481e.

Jeans, J. H., *The converse of Fermat's theorem*, Messenger of Mathematics **27** (1897/98), 174.

Jiang, Z. R., Yu, P. N., *A mixed algorithm for fast polynomial transforms and Fermat number transforms of hyperlarge-scale two-dimensional cyclic convolutions (Chinese)*, Gaoxiao Yingyong Shuxue Xuebao vol 6 (1991), 530–537. MR 92m:65177.

Jiménez Calvo, I., *A note on factors of generalized Fermat numbers*, Appl. Math. Lett. **13** (2000), 1–5. MR 2001b:11007.

Jones, R., Pearce, J., *A postmodern view of fractions and the reciprocals of Fermat primes*, Math. Mag. **73** (2000), 83–97.

Josephy, M., *An afterthought of Gauss on cyclotomy*, Proc. of the 2nd Gauss Symposium. Conference A: Mathematics and Theoretical Physics (Munich, 1993), de Gruyter, Berlin, 1995, 147–150. MR 96e:11003.

Keller, W., *Factors of Fermat numbers and large primes of the form $k2^n + 1$*, Math. Comp. **41** (1983), 661–673. MR 85b:11117.

Keller, W., *Whence come the largest presently known primes? (German)*, Mitt. Math. Ges. Hamburg **12** (1991), 211–229. MR 92j:11006.

Keller, W., *Factors of Fermat numbers and large primes of the form $k.2^n + 1$, II*, Preprint Univ. of Hamburg, 1992, 1–40.

Keller, W., *New Cullen primes*, Math. Comp. **64** (1995), 1733–1741. MR 95m:11015.

Kiss, E., *Notes on János Bolyai's researches in number theory*, Historia Math. **26** (1999), 68–76. MR 2000a:01017.

Klein, F., *Famous problems of elementary geometry*, Chelsea Publ. Company, New York, 1955. MR 17,445b.

Knuth, D. E., *The art of computer programming, vol. 2: Seminumerical algorithms*, Addison-Wesley, Reading, Mass., 1969. MR 44 #3531; 1981, MR 83i:68003.

Koblitz, N., *A course in number theory and cryptography*, second edition, Springer, New York, 1994. MR 88i:94001, MR 95h:94023.

Koch, M., *Einige Primzahlkriterien im Pascaldreieck*, Ph. D. dissertation, Braunschweig, 1979.

Korselt, A., *Problème chinois*, L'Interm. des Math. **6** (1899), 143.

Kraïtchik, M., *Théorie des nombres, vol. 2*, Gauthier-Villars, Paris, 1926.

Kraïtchik, M., *On the factorization of $2^n \pm 1$*, Scripta Math. **18** (1952), 39–52. MR 14,121e.

Krishna, H. V., *On Mersenne and Fermat numbers*, Math. Student **39** (1971), 51–52. MR 48 #5989.

Křížek, M., *On Fermat numbers (Czech)*, Pokroky Mat. Fyz. Astronom. **40** (1995), 243–253. MR 97b:11005.

Křížek, M., Chleboun, J., *A note on factorization of the Fermat numbers and their factors of the form $3h2^n + 1$*, Math. Bohem. **119** (1994), 437–445. MR 95k:11006.

Křížek, M., Chleboun, J., *Is any composite Fermat number divisible by the factor $5h2^n + 1$?*, Tatra Mt. Math. Publ. **11** (1997), 17–21. MR 98j:11003.

Křížek, M., Křížek, P., *Magic dodecahedron (Czech)*, Rozhledy mat.-fyz. **74** (1997), 234–238.

Křížek, M., Luca, F., Somer, L., *On the convergence of series of reciprocals of primes related to the Fermat numbers*, J. Number Theory, accepted in 2001.

Křížek, M., Somer, L., *A necessary and sufficient condition for the primality of Fermat numbers*, Math. Bohem. **126** (2001), 541–549.

Kučera, R., *Computation of the discrete convolution by means of number theoretic transforms (Czech)*, Elektrotechn. časopis **38** (1987), 50–60.

Kummer, E. E., *Über die Ergänzungssätze zu den allgemeinen Reciprocitätsgesetzen*, J. Reine Angew. Math. **44** (1852), 93–146.

Labunets, V. G., *Algebraic theory of signals and systems. Digital processing of signals (Russian)*, Krasnoyarsk. Gos. Univ., Krasnoyarsk, 1984, MR 87c:94003.

Landry, F., *Sur la décomposition du nombre $2^{64} + 1$*, C. R. Acad. Sci. Paris **91** (1880a), 138.

Landry, F., *Méthode de décomposition des nombres en facteurs premiers*, Assoc. Française Avance. Sci. Comptes Rendus **9** (1880b), 185–189.

Larras, J., *Sur la primarité des nombres de Fermat*, C. R. Acad. Sci. Paris Sér. I Math. **242** (1956), 2203–2204. MR 17,1055f.

Laššák, M., Porubský, Š., *Fermat–Euler theorem in algebraic number fields*, J. Number Theory **60** (1996), 254–290. MR 97f:11086.

Le, M., *A note on the greatest prime factors of Fermat numbers*, Southeast Asian Bull. Math. **22** (1998), 41–44. MR 2000a:11015.

Lebesgue, V. A., *Sur l'impossibilité en nombres entiers de l'equation $x^m = y^2 + 1$*, Nouv. Annal. des Math. **9** (1850), 178–181.

Lee, Y. C., Min, B. K., Suk, M., *Realization of adaptive digital filtering using the Fermat number transform*, IEEE Trans. Acoust. Speech Signal Processing **33** (1985), 1036–1039.

Lehmer, D. H., *Tests for primality by the converse of Fermat's theorem*, Bull. Amer. Math. Soc. **33** (1927), 327–340.

Lehmer, D. H., *An extended theory of Lucas' functions*, Ann. of Math. **31** (1930), 419–448.

Lehmer, D. H., *On the converse of Fermat's theorem*, Amer. Math. Monthly **43** (1936), 346–354.

Leibowitz, L. M., *A simplified binary arithmetic for the Fermat number transform*, IEEE Trans. Acoust. Speech Signal Processing **24** (1976), 356–359.

Lenstra, A. K., Lenstra, H. W. (eds.), *The development of the number field sieve*, Lecture Notes in Math. 1554, Springer, Berlin, 1993. MR 96m:11116.

Lenstra, A. K., Lenstra, H. W., Jr., Manasse, M. S., Pollard, J. M., *The factorization of the ninth Fermat number*, Math. Comp. **61** (1993), 319–349. MR 93k:11116. Addendum ibid. **64** (1995), 1357.

Lenstra, H. W., *Factoring integers with elliptic curves*, Ann. of Math. **126** (1987), 649–673. MR 89g:11125.

Lenstra, H. W., Pomerance, C., *A rigorous time bound for factoring integers*, J. Amer. Math. Soc. **5** (1992), 483–516. MR 92m:11145.

Lepka, K., *History of the Fermat quotients (Czech)*, Prometheus, Prague, 2000.

LeVeque, W. J., *Fundamentals of number theory*, Dover, Mineola, N.Y., 1996 (reprint of the 1977 original, MR 58 #465). MR 97a:11002.

Leyendekkers, J. V., Shannon, A. G., *Fermat and Mersenne numbers and some related factors*, Internat. J. Math. Ed. Sci. Tech. **30** (1999), 627–629. MR 2000f:11006.

Li, L., *A survey of research on prime factorizations and fast primality testing algorithms (Chinese)*, Math. Practice Theory (1989), 83–87. MR 91b:11145.

Li, W., Peterson, A. M., *FIR filtering by the modified Fermat number transform*, IEEE Trans. Acoust. Speech Signal Processing **38** (1990), 1641–1645. Zbl 707.65108.

Lidl, R., Niederreiter, H., *Finite fields. Encyclopedia of mathematics and its applications*, Vol. 20, second edition, Cambridge Univ. Press, Cambridge, 1997.

MR 97i:11115.

Ligh, S., Jones, P., *Generalized Fermat and Mersenne numbers*, Fibonacci Quart. **20** (1982), 12–16. MR 83f:10015.

Ligh, S., Neal, L., *A note on Mersenne numbers*, Math. Mag. **47** (1974), 231–233. MR 50 #230.

Liu, P., *An application of Fermat numbers to group theory (Chinese)*, Xinan Shifan Daxue Xuebao Ziran Kexue Ban **23** (1998), 273–277. MR 2000h:11011.

Luca, F., *The anti-social Fermat number*, Amer. Math. Monthly **107** (2000a), 171–173. MR 2000k:11015.

Luca, F., *Equations involving arithmetic functions of Fibonacci and Lucas numbers*, Fibonacci Quart. **38** (2000b), 49–55. MR 2000i:11009.

Luca, F., *On the equation $\phi(|x^m - y^m|) = 2^n$*, Math. Bohem. **125** (2000c), 465–479.

Luca, F., *Pascal's triangle and constructible polygons*, Util. Math. **58** (2000d), 209–214.

Luca, F., *Fermat numbers in the Pascal triangle*, submitted, 2000e.

Luca, F., *Fermat numbers and Heron triangles with prime power sides*, Amer. Math. Monthly, accepted, 2000f.

Luca, F., *Multiply perfect numbers in Lucas sequences with odd parameters*, Publ. Math. Debrecen **58** (2001), 121–155.

Luca, F., Křížek, M., *On the solutions of the congruence $n^2 \equiv 1 \pmod{\phi^2(n)}$*, Proc. Amer. Math. Soc. **129** (2001), 2191–2196.

Luca, F., Somer, L., *A remark on a question of Rotkiewicz*, Colloq. Math., submitted, 2000, 1–5.

Lucas, E., *Sur la division de la circonférence en parties égales*, C. R. Acad. Sci. Paris **85** (1877), 136–139.

Lucas, E., *Sur les congruences des nombres eulériens et des coefficients différentiels des fonctions trigonométriques, suivant un module premier*, Bull. Soc. Math. France **6** (1877–78), 49–54.

Lucas, E., *Théorie des fonctions numériques simplement périodiques*, Amer. J. Math. **1** (1878a), 184–240, 289–321.

Lucas, E., *Théorèmes d'arithmétique*, Atti della Reale Accademia delle Scienze di Torino **13** (1878b), 271–284.

Lucas, E., *Question 453*, Nouv. Corresp. Math. **5** (1879), 137.

Lucas, E., *Théorie des nombres*, Gauthier-Villars, Paris, 1891; Reprinted by A. Blanchard, Paris, 1961.

Lucká, M., Creutzburg, R., Grundmann, H.-J., Vajteršic, M., *Parallel SIMD convolution using the Fermat number transform*, ZKI Inf., Akad. Wiss. DDR **2** (1988), 67–86. Zbl 699.10007.

Lucká, M., Vajteršic, M., Creutzburg, R., Grundmann, H.-J., *Parallel associative fast Fermat number transform implementation*, Comput. Artificial Intelligence **8** (1989), 267–280. Zbl 734.68042.

Mahler, K., *On the transcendency of the solutions of a special class of functional equations*, Bull. Australian Math. Soc. **13** (1975), 389–410. MR 53#2850.

Mahnke, D., *Leibniz auf der Suche nach einer allgemeinen Primzahlgleichung*, Bibliotheca Math. **13** (1913), 29–61.

Mahoney, M. S., *The mathematical career of Pierre de Fermat (1601–1665)*, Princeton Univ. Press, 1973, 1994. MR 58 # 10055, MR 95g:01015.

Mąkowski, A., *On a problem of Rotkiewicz on pseudoprime numbers*, Elem. Math. **29** (1974), 13. MR 49 #206.

Malm, D. E. G., *On Monte-Carlo primality tests*, Notices Amer. Math. Soc. **24** (1977), A–529, abstract 77T–A22.

Malo, E., *Nombres qui sans être premiers, verifient exceptionnellement une congruence de Fermat*, L'Interm. des Math. **10** (1903), 88.

Mandelbrot, B., *The fractal geometry of nature*, Freeman, New York, 1977.

Mann, H. B., Shanks, D., *A necessary and sufficient condition for primality and its source*, J. Combin. Theory Ser. A **13** (1972), 131–134. MR 46 #5225.

Martzloff, J.-C., *The history of Chinese mathematics*, Springer, Berlin, 1997. MR 98a:01005.

Maruyama, S., Kawatani, T., *On the Fermat numbers (Japanese)*, Res. Rep., Kitakyushu Coll. Technol. **20** (1987), 119–127. Zbl 627.10005.

McClellan, J. H., *Hardware realization of a Fermat number transform*, IEEE Trans. Acoust. Speech Signal Processing **24** (1976), 216–225.

McDaniel, W. L., *The gcd in Lucas and Lehmer number sequences*, Fibonacci Quart. **29** (1991), 24–29. MR 91m:11008.

McIntosh, R., *A necessary and sufficient condition for the primality of Fermat numbers*, Amer. Math. Monthly **90** (1983), 98–99. MR 85c:11022.

Mignotte, M., *Quelques problèmes d'effectivité en théorie des nombres*, Thesis, Univ. Paris XIII, Paris, 1974.

Miller, G. L., *Riemann's hypothesis and tests for primality*, J. Comput. System Sci. **13** (1976), 300–317. MR 58 #470a.

Monier, L., *Evaluation and comparison of two efficient probabilistic primality testing algorithms*, Theor. Comput. Sci. **12** (1980), 97–108. MR 82a:68078.

Montgomery, P. L., *Speeding the Pollard and elliptic curve methods of factorization*, Math. Comp. **48** (1987), 243–264. MR 88e:11130.

Montgomery, P. L., *New solutions of $a^{p-1} \equiv 1 \pmod{p^2}$*, Math. Comp. **61** (1993), 361–363. MR 94d:11003.

Montgomery, P. L., *A survey of modern integer factorization algorithms*, CWI Quarterly **7** (1994), 337–366. MR 96b:11161.

Morehead, J. C., *Note on Fermat's numbers*, Bull. Amer. Math. Soc. **11** (1905), 543–545.

Morehead, J. C., *Note on the factors of Fermat's numbers*, Bull. Amer. Math. Soc. **12** (1906), 449–451.

Morehead, J. C., Western, A. E., *Note on Fermat's numbers*, Bull. Amer. Math. Soc. **16** (1910), 1–6.

Morháč, M., *Precise deconvolution using the Fermat number transform*, Comput. Math. Appl. Part A **12** (1986), 319–329. MR 87g:65171.

Morháč, M., *k-dimensional error-free deconvolution using the Fermat number transform*, Comput. Math. Appl. **18** (1989), 1023–1032. MR 90i:65256.

Morikawa, Y., Hamada, H., Yamane, N., *Fast Fourier transform algorithm using Fermat number transform*, Systems-Comput.-Controls **13** (1982), 12–21. MR 86b:94005.

Morimoto, M., *On primes of Fermat type (Japanese)*, Sûgaku **38** (1986), 350–354. MR 88h:11007.

Morimoto, M., Kida, Y., *Factorization of cyclotomic numbers (Japanese)*, Sophia Kokyuroku in Math. **26** (1987), 1–240. Zbl 632.10001.

Morrison, M. A., Brillhart, J., *The factorization of F_7*, Bull. Amer. Math. Soc. **77** (1971), 264. MR 42 #3012.

Morrison, M. A., Brillhart, J., *A method of factoring and the factorization of F_7*, Math. Comp. **29** (1975), 183–205. MR 51 #8017.

Narkiewicz, W., *The development of prime number theory. From Euclid to Hardy and Littlewood*, Springer, Berlin, 2000. MR 2001c:11098.

Naur, T., *New integer factorizations*, Math. Comp. **41** (1983), 687–695. MR 85c:11123.

Niven, I., Zuckerman, H. S., Montgomery, H. L., *An introduction to the theory of numbers*, fifth edition, John Wiley & Sons, New York, 1991. MR 91i:11001.

Nussbaumer, H. J., *Complex convolutions via Fermat number transforms*, IBM J. Res. Develop. **20** (1976), 282–284. MR 54 #12394.

Nussbaumer, H. J., *Digital filtering using pseudo Fermat number transforms*, IEEE Trans. Acoust. Speech Signal Process. **25** (1977), 79–83. Zbl 374.94003.

Nussbaumer, H. J., *Fast Fourier transform and convolution algorithms*, Springer Series in Information Sci. 2, Springer, Berlin, 1981, 1982. MR 83e:65219.

Papademetrios, I., *Concerning Fermat's numbers and Euclid's perfect numbers (Greek)*, Bull. Soc. Math. Grèce **24** (1949), 103–110. MR 12,243a.

Paxson, G. A., *The compositeness of the thirteenth Fermat number*, Math. Comp. **15** (1961), 420. MR 23 #A1578.

Pepin, P., *Sur la formule $2^{2^n} + 1$*, C. R. Acad. Sci. **85** (1877), 329–331.

Pethe, S., Horadam, A. F., *Generalized Gaussian Lucas primordial functions*, Fibonacci Quart. **26** (1988), 20–30. MR 89m:11018.

Petr, K., *Geometrical proof of Wilson's theorem (Czech)*, Časopis Pěst. Mat. Fyz. **34** (1905), 164–166.

Pierpont, J., *On an undemostrated theorem of the Disquisitiones Arithmeticæ*, Bull. Amer. Math. Soc. **2** (1895/96), 77–83.

Pocklington, H. C., *The determination of the prime or composite nature of large numbers by Fermat's theorem*, Proc. Cambridge Philos. Soc. **18** (1914–1916), 29–30.

Pollard, J. M., *Theorems on factorization and primality testing*, Math. Proc. Cambridge Philos. Soc. **76** (1974), 521–528. MR 50 #6992.

Pomerance, C., *On the distribution of pseudoprimes*, Math. Comp. **37** (1981), 587–593. MR 83k:10009.

Pomerance, C., *A new lower bound for the pseudoprime counting function*, Illinois J. Math. **26** (1982), 4–9. MR 83h:10012.

Pomerance, C., *Factoring*, Cryptology and computational number theory (Boulder, CO, 1989), Proc. Sympos. Appl. Math., 42, Amer. Math. Soc., Providence, RI, 1990, 27–47. MR 92b:11089.

Pomerance, C., *A tale of two sieves*, Notices Amer. Math. Soc. **43** (1996), 1473–1485. MR 97f:11100.

Pomerance, C., Selfridge, J. L., Wagstaff, S. S., *The pseudoprimes to $25 \cdot 10^9$*, Math. Comp. **35** (1980), 1003–1026. MR 82g:10030.

Poulet, P., *Table des nombres composés vérifiant le théorème de Fermat pour le module 2 jusqu'à 100.000.000*, Sphinx **8** (1938), 42–52. Errata in Math. Comp.

25 (1971), 944-945, Math. Comp. **26** (1972), 814. MR 58 #31707.

Proth, F., *Correspondance*, Nouv. Corresp. Math. **4** (1878a), 210–211.

Proth, F., *Mémoires presentés*, C. R. Acad. Sci. Paris **87** (1878b), 374.

Proth, F., *Théorèmes sur les nombres premiers*, C. R. Acad. Sci. Paris **87** (1878c), 926.

Rabin, M. O., *Probabilistic algorithms for testing primality*, J. Number. Theory **12** (1980), 128–138. MR 81f:10003.

Racliş, N., *Théorème pour les nombres de Fermat*, Bull. École Polytech. Bucarest **14** (1943), 3–9. MR 7,47g.

Rademacher, H., *Lectures on elementary number theory*, Robert E. Krieger Publ. Company, New York, 1977. MR 58 #10677.

Radovici-Mărculescu, P., *Diophantine equations without solutions, (Romanian)*, Gaz. Mat. Mat. Inform. **1** (1980), 115–117. MR 83m:10007.

Reed, I. S., Scholtz, R. A., Truong, T. K., Welch, L. R., *The fast decoding of Reed–Solomon codes using Fermat theoretic transforms and continued fractions*, IEEE Trans. Inform. Theory **24** (1978), 100–106. MR 58#20794.

Reed, I. S., Truong, T. K., Welch, L. R., *The fast decoding of Reed-Solomon codes using Fermat transforms*, IEEE Trans. Inform. Theory **24** (1978), 497–499. MR 58#20795.

Reid, C., *From zero to infinity. What makes numbers interesting*, MAA Spectrum. Math. Association of America, Washington, DC, 1992. MR 93g:00006.

Ribenboim, P., *On the square factors of the numbers of Fermat and Ferentinou-Nicolacopoulou*, Bull. Greek Math. Soc. **20** (1979a), 81–92. MR 83f:10016.

Ribenboim, P., 13 *lectures on Fermat's last theorem*, Springer, New York, 1979b. MR 81f:10023.

Ribenboim, P., *Prime number records (a new chapter for the Guinness book of records) (Russian)*, Uspekhi Mat. Nauk **42** (1987), 119–176. MR 89c:11181.

Ribenboim, P., *The book of prime number records*, Springer, New York, 1988, 1989. MR 89e:11052, MR 90g:11127.

Ribenboim, P., *The little book of big primes*, Springer, Berlin, 1991. MR 92i:11008.

Ribenboim, P., *Catalan's conjecture. Are 8 and 9 the only consecutive powers?*, Academic Press, London, 1994. MR 95a:11029.

Ribenboim, P., *The new book of prime number records*, Springer, New York, 1996. MR 96k:11112.

Richelot, F. J., *De resolutione algebraica aequationis $x^{257} = 1$, sive de divisione circuli per bisectionam anguli septies repetitam in partes 257 inter se aequales commentatio coronata*, Crelle's Journal **IX** (1832), 1–26, 146–161, 209–230, 337-356.

Richmond, H. W., *A construction for a regular polygon of seventeen sides*, Quart. J. Pure Appl. Math. **26** (1893), 206–207.

Richmond, H. W., *To construct a regular polygon of 17 sides*, Math. Ann. **67** (1909), 459–461.

Riesel, H., *A factor of the Fermat number F_{19}*, Math. Comp. **17** (1963), 458. Zbl 115.26204.

Riesel, H., *Common prime factors of the numbers $A_n = a^{2^n} + 1$*, Nordisk Tidskr. Informationsbehandling (BIT) **9** (1969), 264–269. MR 41 #3381.

Riesel, H., *Prime numbers and computer methods for factorization*, Birkhäuser, Boston-Basel-Stuttgart, 1985, 1994. MR 88k:11002, MR 95h:11142.

Riesel, H., Björn, A., *Generalized Fermat numbers*, Mathematics of Computation 1943–1993: a half-century of computational mathematics (Vancouver, BC, 1993), 583–587, Proc. Sympos. Appl. Math., 48 (ed. W. Gautschi), Amer. Math. Soc., Providence, RI, 1994, 583–587. MR 95j:11006.

Ripley, B. D., *Stochastic simulations*, John Wiley & Sons, New York, 1987. MR 88b:68181.

Rivest, R. L., Shamir, A., Adleman, L. A., *A method for obtaining digital signatures and public key cryptosystems*, Comm. ACM **21** (1978), 120–126. MR 83m:94003.

Robbins, N., *Beginning number theory*, Dubuque, IA: Wm. C. Brown Publishers, 1993. Zbl 824.11001.

Robinson, R. M., *Mersenne and Fermat numbers*, Proc. Amer. Math. Soc. **5** (1954), 842–846. MR 16,335b.

Robinson, R. M., *Factors of Fermat numbers*, Math. Tables Aids Comput. **11** (1957a), 21–22. MR 19,14d.

Robinson, R. M., *The converse of Fermat's theorem*, Amer. Math. Monthly **64** (1957b), 703–710. MR 20 #4520.

Robinson, R. M., *A report on primes of the form $k2^n + 1$ and on factors of Fermat numbers*, Proc. Amer. Math. Soc. **9** (1958), 673–681. MR 20 #3097.

Rosen, M., *Abel's theorem on the lemniscate*, Amer. Math. Monthly **88** (1981), 387–395. MR 82g:14041.

Rosser, J. B., Schoenfeld, L., *Approximate formulas for some functions of prime numbers*, Illinois J. Math. **6** (1962), 64–94. MR 25 #1139.

Rotkiewicz, A., *Sur les nombres pseudopremiers de la forme $ax + b$*, C. R. Acad. Sci. Paris Sér. I Math. **257** (1963), 2601–2604. MR 29 #61.

Rotkiewicz, A., *Sur les formules donnant des nombres pseudopremiers*, Colloq. Math. **12** (1964a), 69–72. MR 29 #3416.

Rotkiewicz, A., *Remarque sur un théorème de F. Proth*, Mat. Vesnik **1 (16)** (1964b), 244–245. MR 32 #7483b.

Rotkiewicz, A., *Sur les nombres de Mersenne dépourvus de facteurs carres et sur les nombres naturels n tells que $n^2 \mid 2^n - 2$*, Mat. Vesnik **2 (17)** (1965), 78–80. MR 33 #2596.

Rotkiewicz, A., *On the pseudoprimes of the form $ax + b$*, Proc. Cambridge Philos. Soc. **63** (1967), 389–392. MR 35 #122.

Rotkiewicz, A., *On the prime factors of the numbers $2^{p-1} - 1$*, Glasgow Math. J. **9** (1968), 83–86. MR 38 #2078.

Rotkiewicz, A., *Pseudoprime numbers and their generalizations*, Stud. Assoc. Fac. Sci. Univ. Novi Sad, 1972. MR 48 #8373.

Rotkiewicz, A., *Solved and unsolved problems on pseudoprime numbers*, in: Applications of Fibonacci Numbers, vol. 8 (ed. F. T. Howard), Kluwer Academic Publishers, Dordrecht, 1999, 293–306. MR 2000j:11010.

Sándor, J., *Some classes of irrational numbers*, Studia Univ. Babeş-Bolyai Math. **29** (1984), 3–12. MR 86i:11035.

Satyanarayana, M., *A note on Fermat and Mersenne's numbers*, Math. Student **26** (1958), 177–178. MR 22 #4660.

Scharlau, W., Opolka, H., *From Fermat to Minkowski. A course on number theory and its development (German)*, Springer, Berlin, 1980. MR 82g:10001.

Schinzel, A., *On primitive prime factors of $a^n - b^n$*, Proc. Cambridge Philos. Soc. **58** (1962), 555–562. MR 26 #1280.

Schönhage, A., Strassen, V., *Fast multiplication of large numbers (German)*, Computing **7** (1971), 281–292. Zbl 223.68007.

Schram, J. M., *A recurrence relation for Fermat numbers*, J. Recreational Math. **16** (1984), 195–197. Zbl 579.10005.

Schroeder, M. R., *Number theory in science and communication*, Springer Series in Information Sci. 7, second edition, Springer, Berlin, 1986. MR 85j:11003.

Selfridge, J. L., *Factors of Fermat numbers*, Math. Tables Aids Comput. **7** (1953), 274–275.

Selfridge, J. L., Hurwitz, A., *Fermat numbers and Mersenne numbers*, Math. Comp. **18** (1964), 146–148. MR 28 #2991.

Shanks, D., *Solved and unsolved problems in number theory*, Chelsea, New York, 1962, 1978, 1985. MR 86j:11001.

Shippee, D. E., *Four new factors of Fermat numbers*, Math. Comp. **32** (1978), 941. MR 57 #12359.

Shorey, T. N., Stewart, C. L., *On divisors of Fermat, Fibonacci, Lucas and Lehmer numbers. II*, J. London Math. Soc., II. **23** (1981), 17–23. MR 82m:10025.

Sierpiński, W., *Remarque sur une hypothèse des Chinois concernant les nombres $(2^n - 2)/n$*, Colloq. Math. 1 (1948), 9. MR 9,331a.

Sierpiński, W., *Theory of numbers (Polish)*, Warszawa, 1950. MR 13,821e.

Sierpiński, W., *Les nombres de Mersenne et de Fermat*, Matematiche (Catania) **10** (1955), 80–91. MR 17,711c.

Sierpiński, W., *Sur les nombres premiers de la forme $n^n + 1$*, Enseign. Math. (2) **4** (1958), 211–212. MR 21#29.

Sierpiński, W., *Sur un problème concernant les nombres $k \cdot 2^n + 1$*, Elem. Math. **15** (1960), 73–74. MR 22 #7983.

Sierpiński, W., *Sur un théorème de F. Proth*, Mat. Vesnik **1 (16)** (1964a), 243–244. MR 32 #7483a.

Sierpiński, W., *Elementary theory of numbers*, Państwowe Wydaw. Naukowe, Warszawa, 1964b. MR 89f:11003.

Sierpiński, W., *250 problems in elementary number theory*, American Elsevier, New York, 1970. MR 42 #4475.

Sierpiński, W., *Elementary theory of numbers, 2nd Engl. ed. revised and enlarged by A. Schinzel*, Państwowe Wydaw. Naukowe, Warszawa, 1988. MR 89f:11003.

Skula, L., *Inclusion among special Stickelberger subideals*, Tatra Mt. Math. Publ. **11** (1997), 147–158. MR 98m:11126.

Šofr, B., *Euclidean geometric constructions (Slovak)*, ALFA, Bratislava, 1976.

Solovay, R., Strassen, V. A., *A fast Monte-Carlo test for primality*, SIAM J. Comput. **6** (1977), 84–85. MR 55 #2732.

Somer, L., *On Fermat d-pseudoprimes*, In: Théorie des nombres (éd. J.-M. De Koninck & C. Levesque), Walter de Gruyter, Berlin, New York, 1989, 841–860. MR 90j:11006.

Somer, L., *On Lucas d-pseudoprimes*, In: Applications of Fibonacci numbers, vol. 7 (eds. G. E. Bergum, A. N. Philippou, A. F. Horadam), Kluwer Academic Publishers, Dordrecht, 1998, 369–375. MR 2000a:11027.

Somer, L., *On super-pseudoprimes*, Preprint, 2001, 1–8.

Steuerwald, R., *Über die Kongruenz $2^{n-1} \equiv 1 \pmod{n}$*, S.- B. Math.-Nat. Kl., Bayer. Akad. Wiss., 1947, 177. MR 11,11e.

Stewart, C. L., *On divisors of Fermat, Fibonacci, Lucas, and Lehmer numbers*, Proc. London Math. Soc., III. **35** (1977), 425–447. MR 58 #10694.

Stewart, C. L., *On divisors of Fermat, Fibonacci, Lucas and Lehmer numbers. III*, J. London Math. Soc., II. **28** (1983), 211–217. MR 85g:11021.

Stewart, I., *Galois theory*, Chapman and Hall, London, 1973, 1989. MR 48 #8460, MR 90j:12001.

Stewart, I., *Geometry finds factors faster*, Nature **325** (1987), 199.

Suyama, H., *Searching for prime factors of Fermat numbers with a microcomputer (Japanese)*, BIT (Tokyo) **13** (1981), 240–245. MR 82c:10012.

Suyama, H., *A note on the factors of Fermat numbers, II*, Abstracts Amer. Math. Soc. **5** (1984a), 132.

Suyama, H., *The cofactor of F_{15} is composite*, Abstracts Amer. Math. Soc. **5** (1984b), 271–272.

Szalay, L., *A discrete iteration in number theory (Hungarian)*, BDTF Tud. Közl. VIII. Természettudományok 3., Szombathely, 1992, 71–91.

Szymiczek, K., *On prime numbers p, q, and r such that pq, pr, and qr are pseudoprimes*, Colloq. Math. **13** (1965), 259–263. MR 31 #4757.

Szymiczek, K., *Note on Fermat numbers*, Elem. Math. **21** (1966a), 59. MR 33 #1278.

Szymiczek, K., *Several theorems on pseudoprime numbers (Polish)*, Zeszyty Nauk. Wyż. Szkol. Ped. w Katowicach Sekc. Mat. Nr. 5 (1966b), 39–46. MR 51 #336.

Szymiczek, K., *On pseudoprimes which are products of distinct primes*, Amer. Math. Monthly **74** (1967), 35–37. MR 34 #5746.

Taylor, R., Wiles, A., *Ring-theoretic properties of certain Hecke algebras*, Ann. of Math. **141** (1995), 553–572. MR 96d:11072.

Trevisan, V., Carvalho, J. B., *The composite character of the twenty-second Fermat number*, J. Supercomput. **9** (1995), 179–182. MR 87j:11146.

Truong, T. K., Chang, J. J., Hsu, I. S., Pei, D. Y., Reed, I. S., *Techniques for computing the discrete Fourier transform using the quadratic residue Fermat number systems*, IEEE Trans. Comput. **35** (1986), 1008–1012. MR 87j:11146.

Vaidya, A. M., *On Mersenne's, Fermat's and triangular numbers*, Math. Student **37** (1969), 101–103. MR 42 #185.

van Maanen, J., *Euler and Goldbach on Fermat's numbers: $F_n = 2^{2^n} + 1$ (Dutch)*, Euclides (Groningen) **57** (1981/82), 347–356. MR 85i:01014.

Varshney, A. K., *An extension of Fermat's numbers*, Proc. Math. Soc. **7** (1991), 163–164. MR 94c:11007.

Vasilenko, O. N., *On some properties of Fermat numbers (Russian)*, Vestnik Moskov. Univ. Ser. I Mat. Mekh., no. 5 (1998), 56–58. MR 2000g:11006.

Vassilev-Missana, M., *The numbers which cannot be values of Euler's function ϕ*, Notes Number Theory Discrete Math. **2** (1996), 41–48. MR 97m:11012.

Voorhees, B., *Geometry and arithmetic of a simple cellular automaton*, Complex Systems **5** (1991), 169–182. MR 93g:68099.

Wantzel, P. L., *Recherches sur les moyens de reconnaître si un Problème de Géométrie peut se résoudre avec la règle et le compas*, J. Math. **2** (1837), 366–372.

Warren, L. R. J., Bray, H. G., *On the square-freeness of Fermat and Mersenne numbers*, Pacific J. Math. **22** (1967), 563–564. MR 36 #3718.

Watabe, M., *On class numbers of some cyclotomic fields*, J. Reine Angew. Math. **301** (1978), 212–215. MR 80h:12005.

Weil, A., *Number theory. An approach through history. From Hammurapi to Legendre*, Birkhäuser Boston, Inc., Boston, Mass., 1984. MR 85c:01004.

Western, A. E., *Notes and corrections*, Proc. London Math. Soc. **3(2)** (1905), xxi–xxii.

Wiedemann, D., *An iterated quadratic extension of GF(2)*, Fibonacci Quart. **26** (1988), 290–295. MR 89m:11122.

Wieferich, A., *Beweis des Satzes, dass sich eine jede ganze Zahl als Summe von höchstens neun positiven Kuben darstellen lässt*, Math. Ann. **66** (1909), 95–101.

Wiles, A., *Modular elliptic curves and Fermat's last theorem*, Ann. of Math. **141** (1995), 443-551. MR 96d:11071.

Williams, H. C., *Primality testing on a computer*, Ars Combin. **5** (1978), 127–185. MR 80d:10002.

Williams, H. C., *A note on the primality of $6^{2^n} + 1$ and $10^{2^n} + 1$*, Fibonacci Quart. **26** (1988), 296–305. MR 89i:11013.

Williams, H. C., *How was F_6 factored?*, Math. Comp. **61** (1993), 463–474. MR 93k:01046.

Williams, H. C., *Édouard Lucas and primality testing*, Canad. Math. Soc. series of monographs and advanced texts, vol. 22, John Wiley & Sons, New York, 1998. MR 2000b:11139.

Williams, H. C., Judd, J. S., *Some algorithms for prime testing using generalized Lehmer functions*, Math. Comp. **30** (1976), 867–886. MR 54 #2574.

Williams, H. C., Zarnke, C. R., *A report on prime numbers of the forms $M = (6a + 1)2^{2m-1} - 1$ and $M' = (6a - 1)2^{2m} - 1$*, Math. Comp. **22** (1968), 420–422. MR 37#2680.

Wrathall, C. P., *New factors of Fermat numbers*, Math. Comp. **18** (1964), 324–325. MR 29 #1167.

Yang, W. Q., *A new algorithm for the rapid computation of the equal-size multidimensional Fermat number transform (Chinese)*, Sichuan Daxue Xuebao **25** (1988), 62–69. MR 90b:65257.

Young, J., *Large primes and Fermat factors*, Math. Comp. **67** (1998), 1735–1738. MR 99a:11010.

Young, J., Buell, D. A., *The twentieth Fermat number is composite*, Math. Comp. **50** (1988), 261–263. MR 89b:11012.

Zsigmondy, K., *Zur Theorie der Potenzreste*, Monatsh. Math. **3** (1892), 265–284.

Web Site Sources
Valid as of May 17, 2001

[www1] http://www.prothsearch.net/fermat.html

[www2] http://www.utm.edu/research/primes/glossary/Mersennes.html

[www3] http://www.mersenne.org

[www4] www.utm.edu/research/primes/glossary/SierpinskiNumber.html

[www5] http://www.prothsearch.net/sierp.html

[www6] www-groups.dcs.st-and.ac.uk/~history/Mathematicians/Fermat.html

[www7] http://www.ams.org

[www8] http://www.emis.de/ZMATH

[www9] http://forum.swarthmore.edu/dr.math/faq/formulas/faq.regpoly.html

[www10] http://www.research.att.com/~njas/sequences/

[www11] http://www.math.uga.edu/~ntheory/web.html

[www12] http://www.seanet.com/~ksbrown/inumber.htm

[www13] http://www.dmf.mathematics.dk/clausen_en.html

[www14] http://www.utm.edu/research/primes/glossary/Cullens.html

[www15] http://www.prothsearch.net/cullen.html

[www16] www.utm.edu/cgi-bin/caldwell/primes.cgi/Generalized%20Fermat

[www17] http://perso.wanadoo.fr/yves.gallot/primes/gfn.html

Name Index

Subject Index

Rue Fermat in Paris.